CAMERA GEOLOGICA

Siobhan Angus

CAMERA GEOLOGICA

An Elemental History of Photography

DUKE UNIVERSITY PRESS
Durham and London
2024

Project Editor: Ihsan Taylor
Designed by A. Mattson Gallagher
Typeset in Portrait Text, Degular, and Georama by
Westchester Publishing Services

Library of Congress Cataloging-in-Publication Data
Names: Angus, Siobhan, author.
Title: Camera geologica : an elemental history of photography /
Siobhan Angus.
Description: Durham : Duke University Press, 2024. | Includes
bibliographical references and index.
Identifiers: LCCN 2023028158 (print)
LCCN 2023028159 (ebook)
ISBN 9781478030188 (paperback)
ISBN 9781478025931 (hardcover)
ISBN 9781478059172 (ebook)
Subjects: LCSH: Photography—History. | Photography—
Environmental aspects. | Photography—Equipment and supplies.
Classification: LCC TR147 .A548 2024 (print) | LCC TR147 (ebook) |
DDC 771/.5—dc23/eng/20231212
LC record available at https://lccn.loc.gov/2023028158
LC ebook record available at https://lccn.loc.gov/2023028159

Cover art: Tsēmā Igharas, *Black Gold Infinity*, 2018. Digital
photo collage of bitumen core sample. Courtesy of the artist.

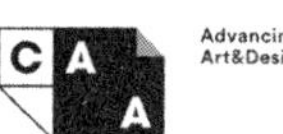

PUBLICATION OF THIS BOOK HAS BEEN
AIDED BY A GRANT FROM THE
WYETH FOUNDATION FOR AMERICAN
ART PUBLICATION FUND OF CAA.

CONTENTS

Plates

ACKNOWLEDGMENTS

This project was shaped by the contributions of so many activists, artists, and scholars, and I am deeply grateful for the depth of engagement. The foundation of this project emerged from conversations with Warren Cariou, whose art and scholarship transformed my thinking about extraction and ethics, and who, in the process, taught me how to see differently. I am grateful for his generosity and for the model of engaged scholarship he embodies. This manuscript was written during a postdoctoral fellowship at Yale, and it is deeply indebted to Jennifer Raab who has been an invaluable interlocutor in thinking through ways of seeing and extraction. For her perceptive feedback, keen insight, and friendship, I will forever be grateful. Sarah Parsons played a critical role in shaping the project: archival work at George Eastman Museum and making daguerreotypes are some of my fonder memories of the research process. It is unlikely that this project would have happened without Kevin Coleman's intellectual generosity and commitment to studying the visual cultures of labor. The nucleus of this project emerged from formative conversations with Ian Radforth around labor, mining, and images. My interest in mining photography was sparked two decades ago by Louie Palu, who first showed me that photography's raw materials come from the mine.

My deepest thanks go to Ken Wissoker for his faith in this project. His incisive vision and generous engagement immeasurably shaped the manuscript. This project wouldn't have been possible without the innovative and politically engaged scholarship that Ken has nurtured and championed. Ryan Kendall, Ihsan Taylor, James Moore, Chad Royal, Laura Sell, A. Mattson Gallagher, Erika Jackson, Lalitree Darnielle, and the rest of Duke University Press's staff were integral to bringing the book through production. Thanks, as well, to Jenn Bennett-Genthner, Eric and Doreen Anderson, and Robert and Cynthia Swanson. I am particularly grateful to the three anonymous reviewers of the manuscript for their insightful and thoughtful comments, which improved the book immensely. I have tried to address their helpful suggestions, and any faults that remain are only mine.

I have been blessed with a remarkable network of mentors and friends in the fields of art history, labor studies, and environmental justice organizing. Two of them deserve particular mention: Patrick DeDauw's ruthless rigor and incredibly careful engagement with the manuscript brought the core arguments into view and strengthened the focus of the project. Elizabeth Keto nuanced and expanded my thinking, sharpened my writing, and a number of the key insights in the book emerged in dialogue with Elizabeth.

The manuscript was refined in dialogue with colleagues in the History of Art and the Environmental Humanities at Yale, with particular thanks to Carol Armstrong, Jill Jarvis, Joanna Fiduccia, Milette Gaifman, Morgan Ng, Paul Sabin, Royce Young Wolf, and Tim Barringer as well as Aaron Levin, Abigail Fields, Angela Chen, Blair Betik, Caterina Franciosi, Caitlin Kossmann, Kevin Hong, Manon Gaudet, Michelle Donnelly, Savannah Sather Marquardt, Vu Horwitz, and Yechen Zhao. Thanks to Paul Messier and Katherine Mintie at Yale's Lens Media Lab whose enthusiasm for and insight into material histories infuses this book. Thanks as well to George Miles at the Beinecke for his generosity and engagement. A remarkable group of students nuanced my thinking on material histories, with thanks to Ada Griffin, Adriana Ballinger, Angela Higuera, Cassidy Arrington, Elizabeth Levie, Kapp Singer, Nithy Baskaran, Seyma Kaya, and Sofia Kouri, while Elaina Foley, Ariana Habibi, Catherine Webb, and Jisoo Choi enriched my thinking on the possibilities of art in the context of ecological crisis.

My thinking on post-extractive futures was clarified during a research residency at the Center for Creative Ecologies at the University of California at Santa Cruz, with thanks to T. J. Demos, A. Laurie Palmer, and the members of the ecosocialist working group. Roxana Marcoci is an ongoing inspiration and a model of feminist mentorship. Joan Greer and Jesse Thomas have been energizing collaborators. My colleagues at Carleton University provided a fresh infusion of ideas as I wrapped up the manuscript, with thanks to Armond Towns, Emily Hiltz, Ira Wagman, Irena Knezevic, Joshua Greenberg, Merlyna Lim, Liam Cole Young, Rena Bivens, Tracey Lauriault, Sandra Robinson, Sheryl Hamilton, and Vincent Andrisani. I am particularly grateful for the friendship of Susana Vargas Cervantes who has been an essential interlocutor in thinking about photography. Thanks to my collaborators at the *Goose*, Alec Follett, Anita Girvan, David Huebert, Julien Defraeye, Melanie Dennis Unrau, and Rina Garcia Chua.

My research has been supported by fellowships awarded by the American Philosophical Society, the Library Company of Philadelphia, the Social Sciences and Research Council of Canada, the Science History Institute, the Paul Mellon Centre, the Yale Center for British Art, and York University. My postdoctoral work was funded by a Banting Postdoctoral Fellowship through the Canadian Institutes of Health Research. This book's publication has been made possible in part with support from the College Art Association's Wyeth Foundation for American Art Publication Grant, Yale's Environmental Humanities program, and the Faculty of Public Affairs at Carleton University.

Crucial research support was provided by a number of archivists and curators. Mike Robinson offered a wealth of insights into the daguerreotype process. Research for chapter 2 was aided by George Miles at the Beinecke Library, Miranda Mims and Melinda Wallington at the Rush Rhees Library, and Stephanie Hofner at the George Eastman Museum. Chapter 4 was written during a fellowship at the Library Company of Philadelphia, and research support by Erika Piola, Cornelia King, and Sarah Weatherwax was invaluable. Paul Messier, Katherine Mintie, and the team at the Lens Media Lab provided critical insights into paper substrates. Chapter 6 was written during a fellowship at the Science History Institute, and my work was informed by conversations with Isabelle Held, Megan Piorko, Charlotte Abney Salomon, and Gustave Lester.

For valuable feedback that enriched the manuscript, I thank the organizers and participants of lectures and symposia at College Art Association, Comité international d'histoire de l'art World Congress, the Courtauld Institute of Art, Harvard University, the Institut National d'Histoire de l'Art, the University of Kansas, the Library Company of Philadelphia, Newnham College at the University of Cambridge, the University of Oregon, the Paul Mellon Center for British Art, the Rachel Carson Center, the Science History Institute, the University of Toronto's Jackman Humanities Institute, the Performance Studies Working Group at Yale University, the Photography Network, the History and Philosophy of Science Working Group at the University of Cambridge, and the seminar series on El Dorado hosted by the Americas Society, Museo Amparo, and Fundación PROA.

I am continually inspired by the artistic, activist, and intellectual work of my friends. Jason Cyrus has brought so much joy to my life: I am infinitely blessed by his radiant energy and passion. Andrew Gayed continually inspires me with his warmth, enthusiasm, and drive. Ivana Dizdar's excellent humor and sharp wit continually provide perspective. Martabel Wasserman lives her commitment to building a better and more enchanted world. Vanessa Nicholas brings rigor, wisdom, and style to everything she does. Samantha Spady has consistently enriched my thinking about things both subterranean and celestial. Marina Dumont's generosity of spirit provides a consistent reminder of the simple pleasures of life and research. Mary Soroka's creativity and compassion are a guiding light in my life. To Anastazja Krynska and Johannes Krause: I am so thankful for everything. To Aaron Katzeman, Caroline Duffy, Erica Toffoli, Kalie Richardson, Isabelle Held, Marnie Bjornson, Simon Cheesman, Mathieu Belanger, Megan Davies, Mike O'Brien, Nathan Isberg, Stephanie Coffey, Ilya Klymkiw, Stephanie Nakitsas, Jeremy Withers, Steve McClellan, Vanessa Lakewood, and Ryan Le: I am grateful to call you friends. Many thanks go to Julia Wawrzyniak-Beyer and Asher Hartman for their guidance.

My family reminds me every day that it is through community, organizing, and love that we will build better worlds—and just how close that world is. To Brit Griffin, Charlie Angus, Lola Angus, Mariah Griffin-Angus, Anne Angus, and Alex Bird: I love you all more than you could know.

Introduction

IN 1871, the North Bloomfield Gravel Mining Company hired Carleton Watkins to document a hydraulic mining operation at Malakoff Diggins in the Sierra Nevada mountains of California. A photograph of the scene holds industry and nature in a delicate balance (plate 1). The arc of water from the four hydraulic hoses resembles a set of rainbows, gently cascading in front of the rock cliffs. The curvilinear ridge frames the scene, while the hazy vapor softens the hardness of the exposed rock. In the foreground, smooth water streams through rugged stone. A bridge intrudes into the landscape, while the rock face dwarfs five men who pose casually, looking toward the camera as pressurized water reshapes the ecosystem. The miners are diminutive compared to the water and the rock, but by harnessing water as a productive force, they dominate the landscape. The water cannons blasted sixteen thousand gallons of water per minute, removing four thousand cubic yards of earth every day. The dramatic, large-format "mammoth" print has a remarkable precision that brings the scene into sharp focus and creates a sense of tactility, emphasizing the spray of water and the striation in the rock cliff. The detail enhances the tenuous equilibrium of the photograph, but a closer look reveals a set of forces in conflict within the image.

The striking formal beauty of the image stands in contrast to contemporary descriptions of Malakoff Diggins, which was described as desolate and forbidding, a "battlefield" where "nature here reminds one of a princess fallen into the hands of robbers who cut off her fingers for the jewels she wears."[1] The image sustains a fantasy of nature and industry held in balance because the more damaging impacts of blasting are illegible within the photographic frame.[2] Hydraulicking was pioneered in the 1850s during the California Gold Rush to facilitate the search for new ore bodies by blasting "overburden"—everything above the ore (the silver or gold)—with pressurized water. As a concept, overburden reveals a particular way of seeing nature that prioritizes extraction. At Malakoff Diggins, it is estimated that twelve billion tons of debris were washed downstream, damaging forty thousand acres of land in the vicinity of the mining operation.[3] The demands of capitalist extraction recast trees, rocks, flowers, plants, insects, and animals as obstacles, a burden to be blasted away to facilitate the removal and processing of parts of an ecosystem for profit. The image renders the violence of blasting quietly aesthetic.

The mining company commissioned Watkins to document the site in order to raise capital investment for the company; encouraging further extraction is the intended function of the image.[4] Within the scene, the nonmining histories of this landscape are unevenly represented, if at all. The foundational act of violence that made hydraulic blasting possible was the expropriation of Indigenous land, as the Sierra Nevada region is the traditional territory of the Nisenan people. The Gold Rush brought a flood of migration and changes in land use that resulted in dispossession. This history of the violent seizure of Indigenous land is not visible in the image, though the fruits of the labor of Chinese migrant workers is. Malakoff Diggins's infrastructure was partly built by three hundred Chinese laborers who dug reservoirs out of the mountains to provide the water for the blasting, and built 5,276 miles of flumes, canals, and ditches to supply water to the mines. This infrastructure harnessed water as a productive force, putting it to work. The spectacular, immediate destructiveness of the hydraulic blast, killing plants, insects, animals, and people, led to flooding and mercury contamination as mud, gravel, and debris clogged waterways. There are no visible traces of animal, insect, or plant life in the frame. Nor can the photograph show us what was to come, a landscape shattered by the pressure of blasting. The effects of hydraulic mining still mark the eroded landscape of Malakoff Diggins known as an "industrial Grand Canyon."[5] The decimated landscape led to the first environmental legal decision issued in the United States, as California banned hydraulic blasting in 1884.[6]

Watkins's photograph stages a confrontation between deep time and industry that makes visible what Karl Marx called a "metabolic rift": a fracture between the natural world and the human society that grew out of it, an internal rift driven by the perpetually accelerating growth imperatives of capitalism. For Marx, humans were part of nature: "Man lives on nature. . . . That man's physical and spiritual life is linked to nature means simply that nature is linked to itself, for man is a part of nature."[7] The extraction of more and more energy (both as natural resources and labor) to create endless accumulations of capital disrupts, for Marx, the temporal complementarity—or metabolism—of human activity in its relation to the other ecological processes making up life and landscape on Earth. Marx links ecological thought with economic theory, highlighting the incompatibility of capitalism's imperative for growth and speed with natural processes requiring limits to use and periods of recovery in order to successfully renew. The rapid speed of industrial growth jolted human society out of sync, out of time. Three temporalities thus coexist within Watkins's image: the gradual layering of geological time visible in the stratified rock, the hyperacceleration of industrial mining, and the flash of light that set the photo in silver.

Watkins's visually striking photograph also gestures to the materiality of the print itself. Metals used in photography—silver, gold, and mercury—were mined in Malakoff Diggins. Silver forms light-sensitive silver halides, gold is a toner, and mercury is a developing agent. Ecosystems like Malakoff Diggins provided the raw materials for photography, establishing an essential link between the arts of photography and the multisited work of mining. The material and social connection between mining and photography lies at the heart of this book. I analyze the extraction of materials by focusing on images whose subject matter and photographic materiality tell us something about these processes and their extension into social and ecological worlds of work and despoliation that make them possible.

Different materials have been extracted from the earth at various points in photography's history to facilitate image-making. In this book, I focus on six: bitumen, silver, platinum, iron, uranium, and rare earth elements. Exploring the role played by materials in photographic processes allows us, in turn, to consider how these images help make sense of the social relations that sustain our world as it is. The history of photography cannot be reduced to the history of the mine, but new histories emerge when we reorient our vantage point: How does photography look from the perspective of the mine?

A focus on mining enacts a reorientation of vision that directs our attention underground. Light is often emphasized in photographic discourse, reflected in its name, "written with light." But light is only part of the story. It is the interaction of light with metals that makes photography possible. *Camera Geologica* shifts the focus from light to metals to consider the histories of labor and environment that underpin the photographic object. An emphasis on mined materials draws into view the connective tissue between geology, raw materials, labor, empire, colonization, and art. Thinking materially alongside representation—considering photographs as both objects and images—yields new insights into the history of photography and environmental change. Rather than take the ubiquity of mined materials for granted, this book focuses on their materiality, the work of their extraction, and the social, cultural, and political imaginaries that accompany them. Narratively, each material is laden with histories, both symbolic and concrete. Materials bring these histories into the everyday. Once set in the photograph, they communicate meaning. As such, photographs are a powerful means to illuminate ecological catastrophe in the present and to conceptualize how to redress such harm for the future.[8]

At the root of this book is a simple premise: that the mine is a necessary precondition for photography as a medium. Since its inception, photography, both analog and digital, has relied on both small- and large-scale extraction. In 1863, Oliver Wendell Holmes—a poet and physician who wrote extensively on photography—vividly described the Messrs. E. & H. T. Anthony factory in New York, one of the largest photographic supply companies in the nineteenth century. Of the chemical substances used in photography, he summarized that "to give an idea of the scale on which these are required, we may state that the estimate of the annual consumption of the precious metals for photographic purposes, in this country, is set down at ten tons for silver and half a ton for gold."[9] Holmes noted photography's industrial nature and the expansive work processes that enabled production: the factory was powered by steam and the "labor [was] greatly subdivided, [the workers became] wonderfully adroit in doing a fraction of something." That "fraction of something" was embodied, for instance, by the rows of young women processing eggs, which acted as a chemical binding agent in albumen prints.[10] Here, we see how gender and class divides structured the labor within the factory. As photography became a mass medium by the end of the nineteenth century, this demand for materials escalated.

What materials, besides metals and eggs, were required for photography? As Kodak summarized, film is "animal, vegetable, and mineral," borrowing the slogan of alchemy.[11] Paper was made from plant fibers like cotton and linen, while cotton was essential to celluloid for film stock.[12] By 1929 Kodak used more than five million pounds of cotton annually.[13] The gelatin in film stock was made from the hide, bones, cartilage, ligaments, and connective tissue of calves (considered the very best), sheep (less desirable), and other animals who passed through the slaughterhouse.[14] Six kilograms of bone went into a single kilogram of gelatin. Eventually, the demands of photographic industries generated so much need for animal byproducts that slaughterhouses became integrated into the photographic production chain.[15] Controlling the supply chain became key to Kodak's success. In 1882, as Kodak began to grow as a company, widespread complaints of fogged and darkened plates stopped production. The crisis almost ruined Kodak financially and resulted in the company tightly monitoring the animal by-products used in gelatin. Decades later, a Kodak emulsion scientist discovered that cattle who consumed mustard seed metabolized a sulfuric substance, enhancing the light sensitivity of silver halides and enabling better film speeds. The poor-quality gelatin in 1882 was due to the *lack* of mustard seeds in the cows' diet. The head of research at Kodak, Dr. C. E. Kenneth Mees, concluded, "If cows didn't like mustard there wouldn't be any movies at all."[16] By controlling the diet of cows who were used to make gelatin, Kodak ensured the quality of its film stock. As literary scholar Nicole Shukin reflects, there is a "transfer of life from animal body to technological media."[17] The image comes alive through animal death, carried along by the work of ranchers, meatpackers, and Kodak production workers.

In addition to the extensive use of organic materials, photography is also synthetic, tracking the rise of the chemical industry in the late nineteenth century.[18] By the mid-twentieth century, the Kodak Park plant produced hundreds of different chemicals for use in photography and thousands of research chemicals.[19] Photographic film was one of the earliest applications of plastics, as cellulose nitrate, the first semisynthetic polymer, coated glass plates and transparent roll film.[20] Cellulose nitrate is perhaps best known as gun cotton—a mild explosive that was also quickly applied in mining.

The material realities of photographic production undermine many of the stories photography likes to tell about itself—whether about its ease of use, its lack of mediation, or its nonorganic technological sophistication. Photography has long been invested in appearing immaterial. For instance, Oliver Wendell Holmes, who invented a streamlined stereoscope that illusionistically rendered three-dimensional views of photographs, wrote, "Form is henceforth divorced from matter. In fact, matter as a visible object is of no great use any longer, except as the mold on which form is shaped. Give us a few negatives of a thing worth seeing, taken from different points of view, and that is all we want of it."[21] As we saw from his description of the factory, Holmes was very aware of the material foundations of photography when he laid forth this immaterial fantasy; it was not ignorance of material realities that drove his narrative but a desire to transcend them.

This pursuit runs throughout photographic discourse, as if the medium itself aspires to transparency.[22] Photographers are not unique in the desire to transcend banal materiality: artists often try to overcome paint and canvas to produce something more meaningful, creating a rift between the object and the masterpiece, which is located somewhere in the immaterial idea.[23] Photography, however, so often promises the possibility of an unmediated lens onto the world. A lightness, pure vision, unchained from the earthly work of production and reproduction. Of all mediums, it is the most effective at concealing its materiality.

Photography's mechanical reproduction promises to divorce the possibilities of representation from the expense and limitations of matter and space. This is significant, for as Holmes went on to clarify, "Matter in large masses must always be fixed and dear; form is cheap and transportable."[24] In this framing, photography allows the reproducibility of form across distances and thus annihilated the friction and costs of matter. This conceptual separation of form from matter—which works to make certain industrial processes less visible—has tangible impacts.[25] Holmes concludes that photography is a new system of value, which could create a "universal currency of these bank-notes . . . which the sun has engraved for the great Bank of Nature" and invites readers to "fill out a blank check on the future as they like."[26] Throughout the nineteenth and twentieth centuries, this extractive way of understanding nature—a blank check on the future—subtly implied that the mining that made the technology possible was both necessary and natural. Holmes invokes the sun but also currency, which historically was made of metals, but had begun to shift to paper. Reproducibility—of the image and of currency— thus seems to promise the possibility of moving past the messy material processes of mining.

Photography's desire to transcend its material origins parallels oft-hidden extraction processes. Despite the centrality of mining to economies, extractive capitalism functions by making industrial production largely invisible: it shifts attention to the commodity, not the labor or materials that make it. Consumer capitalism encourages us to forget how commodities are made. Photography's narratives of effortless ease invite comparisons to oil, which likewise promised the ability to overcome the limits to growth bounded by the productivity of land and human labor, transformatively reducing the cost of bringing any particular thing to where it needed to be.

The networked or so-called dematerialized world of the present is profoundly reliant on resource extraction. In the context of climate crisis, extraction is a pressing material problem. According to a recent United Nations report, resource extraction is a primary driver of global climate change, responsible for half of the world's carbon emissions and more than 80 percent of its biodiversity loss. Despite the increasing awareness of the impact of extractive activities in contributing to climate change, the annual global extraction of materials by ton is increasing by 3.2 percent per year.[27] While the world's population has doubled since the 1970s, resource extraction has tripled. Once extracted from the earth, raw natural resources are eventually transformed into consumer goods, which bear little evidence of the complex networks of human and nonhuman labor that brought them into being. In the process, extraction's histories of labor, displacement, and ecosystem destruction are cast out of view. Scientist Stefanie Hellweg describes how consumer products hide the cost of extraction, observing that "resources are hiding behind products."[28] Mined materials are omnipresent in our daily lives to the degree that we rarely notice them, or at least stop asking where they came from and what the consequences of this extraction are to humans and ecosystems.

The seductive fantasy that the world has become altogether dematerialized is only possible because corporations deliberately aim to obscure the labor of extraction and its environmental costs. These consequences are felt most acutely in what cultural theorist Macarena Gómez-Barris calls "extractive zones": resource-rich regions that are often far from cities and centers of power.[29] Corporations and governments turn these landscapes into sacrifice zones where their original inhabitants are displaced, exposed to toxicity, or both. Working-class people mine and refine raw materials, often experiencing industrial disease and the precarious livelihoods that mark mining's boom-and-bust economies. Industrial production's toxic refuse is typically processed or dumped in racialized, low-income neighborhoods. Environmental racism has global dimensions: corporations have largely outsourced extraction and waste dumping to the Global South. This neocolonial process alleviates the immediate violence of extraction in the countries that bear the most responsibility for climate breakdown. Environmental injustice tracks the fault lines of race, class, and geography.

Taking the integral relationship between form and matter as my starting point, I propose a reorientation of perspective that restores the photograph to histories of materials, land, property, and labor.[30] A structural analysis that considers the people, places, and beings that bear the costs of hyperextraction is necessary to address the urgent challenges of the present. The cumulative and successive crises of climate change have such vast temporal and spatial dimensions that they transcend comprehension and pose challenges for understanding how we got here and where we go from here. Shifting our relationship to resource extraction, by staring back at it through the representation it makes possible, is one place to start.

Vast amounts of earthly materials have to be dredged up to make photography seem weightless. While metals and fossil fuels are used in many, if not all, artistic mediums, *Camera Geologica* makes a case for medium specificity. In doing so, I follow historians Kevin Coleman and Daniel James, who argue that photography and capitalism are premised on the "fiction of endless accumulation in a finite world" while being characterized by a "nervous vibration between the concrete and the abstract."[31] Photography's material and symbolic links to extraction—and its emergence as a technological form coincident with the rise of large-scale industry and the spread of global capitalism—make it a particularly productive angle from which to consider the complex imbrication of extraction in daily life.

Photography emerged within a rapidly industrializing world. In 1784, the Scottish inventor James Watt patented the steam engine, which connected coal fire to the continuous motion of the wheel, transforming heat into energy. The eighteenth and nineteenth centuries became known as the Era of Steam, though it is more accurately the beginning of the era of "Fossil Capital."[32] Quickly, the United Kingdom transitioned to a mineral-based economy, in which burning coal produced a power source so potent that it allowed burgeoning sites of capitalist production to break free, relatively, of the limits to growth imposed by the productivity of land and human labor in particular places.[33] The breakneck increase in industrial development coupled its rapidly accelerating production of wealth with devastating accumulations of environmental and social damage. And the system fed on itself: more factories in more places demanded other mined materials in addition to coal, as the needs of industry expanded the traditional use of metals in agriculture, the military, and currency.[34] These transformations required, technically and culturally, new forms of representation and meaning-making, and photography emerged, in part, in response to these socioeconomic shifts.

The coal-fueled socioecological transformations of the nineteenth century are part of a longer continuum of extraction. Still, the rise of fossil fuels as primary energy sources marked a rapid escalation in the human impact on the environment. In 1848, Marx and Friedrich Engels described how "modern bourgeois society, with its relations of production, of exchange and of property, a society that has conjured up such gigantic means of production and of exchange, is like the sorcerer who is no longer able to control the powers of the nether world whom he has called up by his spells."[35] The reference to the netherworld is not only metaphorical, importantly: coal is quite literally extracted from deposits deep in the earth.[36] Marx and Engels's framing gestures to an implicit perversion, for coal is carbon based, which is the basic element of life—fossil fuels are the remains of ancient life—so to "conjure" coal is to disturb the dead. We witness here a very particular twist on the revivification of dead labor in production.

The rise of an intensively mineral-based economy also complicated emergent photographic processes in material ways, as atmospheric pollution caused the degradation of photographic prints. Silver is the most common material used in analog photography. The high light-sensitivity of silver halides allows for short exposure times, which in turn provides for instantaneous image capture. The relative chemical stability of silver was another asset, as the metal doesn't react to air or water. However, silver does tarnish when exposed to sulfur compounds. In the coal-fueled Victorian period, this was a significant problem. Reactions with atmospheric sulfur pollution damaged silver prints, and many nineteenth-century photographs printed in silver degraded into a faded, brownish tinge.[37] In 1880, the *Photographic News* described the polluted, sulfur-filled atmosphere of industrial cities as one of the primary challenges facing photographers, reminding readers that "the photograph is, after all, but a thin film of metallic silver, and silver is of all metals one of those most prone to suffer from the action of sulphurous acid."[38] This highlights that the photograph's materiality—and its meaning—changes as it moves through the industrial world. Photographers sought alternatives to silver-based processes because of the challenges of fixing a silver print in the polluted atmosphere.

Photography as and against Extraction

This book turns to the questions of expropriation at the heart of photography. Extraction is both a material process and a worldview.[39] Materially, extraction provides the raw materials that give our world form. Fundamentally, mining is a problem of *material* production: the raw materials, machinery, facilities, and labor used to produce goods. Extraction describes the physical processes of taking raw natural materials from the earth, which, under extractive capitalism, often results in the violent dispossession of Indigenous peoples and the destruction of lands, waters, and nonhuman species. Extraction is the first step in an accumulative process through which materials are transformed into wealth.

Some of these raw materials become the material foundations for art-making. This link is made explicit in the Prussian metallurgist Georgius Agricola's groundbreaking sixteenth-century pedagogical guide to mining and metallurgy, *De Re Metallica* (*On the Nature of Metals*), which argues that the mine is a precondition for art. Agricola observed mining techniques during the Central European mining boom (1451–1540), and his text marked a transition within mining from artisanal practice to codified engineering knowledge. *De Re Metallica* was lavishly illustrated with woodcuts and created deliberate visual motifs intended to shape ways of relating to land and labor under industrial mining. In one woodcut, the engraver presents nature as a resource: water is diverted, trees are cut down, and ore is removed. This image shows an apparatus that captures wind to provide air to miners underground. The wind is anthropomorphized with a face, borrowing from cartographers' conventions. Even wind, the woodcut suggests, participates in mining. In the background, the city signifies the necessity of metals for economies and, of course, art. Photo theorist and photographer Allan Sekula observes that mining was one of the first industries to be pictured visually, while the geologist Martin J. S. Rudwick argued that developing a visual language was central to producing geological knowledge.[40] Specific modes of visual representation perform crucial work in the context of developing the means and social relations required to enable mining on an expanded scale.

The first section of *De Re Metallica* restates and responds to the criticisms of mining by ancient and early modern writers who censured the practice for promoting avarice, war, and the destruction of the earth. In the *Metamorphoses*, for instance, the first-century Roman poet Ovid linked mining to conflict: "Men descended into the entrails of the earth, and they dug up the riches, those incentives to vice, which the earth had hidden and had removed to the Stygian shades. Then destructive iron came forth, and gold, more destructive than iron; then war came forth."[41]

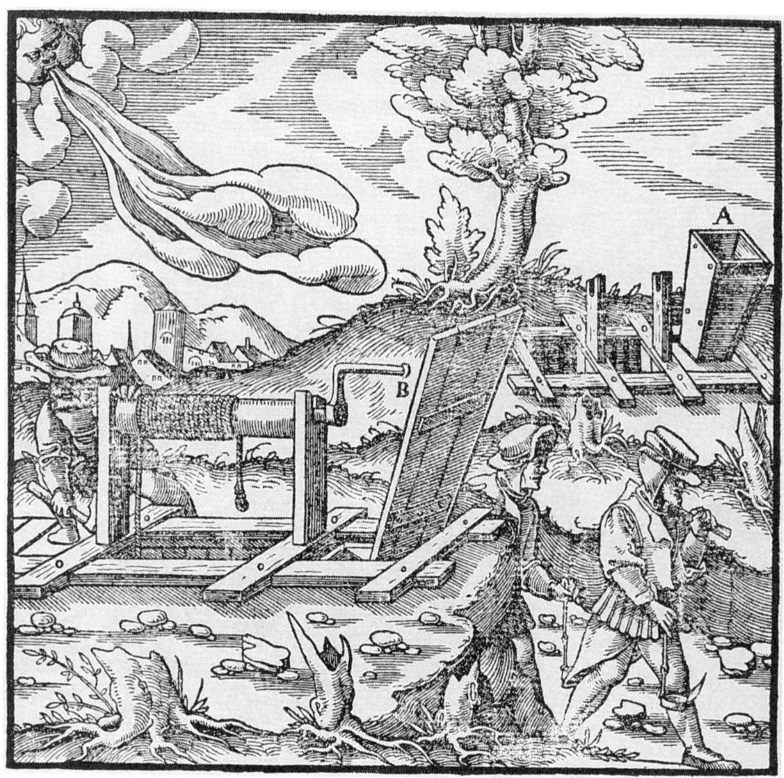

Ecofeminist scholar Carolyn Merchant documents how sixteenth-century descriptions of nature as a nurturing mother in literature and philosophy operated as an ethical constraint on mining.[42] In this context, Merchant suggests that Agricola's treatise functions as an attempt to free mining from the moral restrictions imposed by such an understanding of the natural world, presenting a more instrumental conception of nature that would allow for forms of development required by mercantilist policy. In his defense of mining, Agricola pointed to the role that mining played in art production, noting that extraction enabled the production of metal-based pigments and tools while artists used mined materials decoratively to make "elegant, embellished, elaborate, useful" works of art.[43] He concludes with the reflection: "How few artists could make anything that is

I.1   Georgius Agricola, illustration from *De Re Metallica*, book 6.

beautiful and perfect without using metals?"[44] In an early example of art-washing, Agricola invokes art to make climate-damaging practices seem more culturally acceptable. It subtly implies that art justifies mining and that aesthetics justify the environmental and human costs.

Agricola's text responds to shifting cultural values around mining and nature, highlighting how materials are enmeshed with cultural and economic systems that change over time.[45] Nature is not a static condition that lies outside of culture: the imaginaries of materials take on different cultural valences in different contexts and times. In the Romantic period, the fascination with the natural world and the proliferation of developments in geology lent a profound cultural significance to mining, particularly in Germany and England. However, variable forms of mining resulted in very different cultural connotations. Germany was the primary source of precious metals in Europe. Still, it was a century behind England in industrializing mining, as Germany did not develop the coal fields of the Ruhr Valley until the mid-nineteenth century. The smaller scale of production gave greater autonomy to the individual worker and was less destructive to the environment. As a result, within German Romantic literature and art, the mine was imagined as a place of "mysterious caves, wise miners, hidden secrets and ancient knowledge," linking the mine to the hero's journey.[46] This conceptual link between underground descent and the journey of the soul dates back to ancient literature: Odysseus, Aeneas, and Dante all travel to the underworld on heroic quests. The symbolic power of the mine did not displace its material realities: many German romantic artists trained as mining engineers.[47] In contrast, the rapid development of the coal and iron mines in England was tied to the emergence of steam power and was marred by pollution and social dislocations.[48] The tensions that accompanied industrial growth enabled by mining were perhaps most famously evoked by William Blake in 1804, who decried the "dark, Satanic Mills" of industrial England and the resultant destruction of nature and human relationships.[49] The mine as an imaginary emerges from the material realities of labor and environment. As such, this book does not posit one singular relationship to extraction: the forms of labor and technology required to extract, process, and transform these materials are different, resulting in disparate cultural imaginaries and political possibilities.

Agricola's text links extraction to culture, highlighting that extraction is not simply a material reality but a cultural problem: it is a way of seeing and understanding the world. As a worldview, extraction views nature—and the people understood as part of nature—as a resource to be expropriated. This way of seeing promotes economic growth as an all-consuming priority. Nature becomes a backdrop to human activity and a storehouse of resources. Turning land into property that can be expropriated is a prerequisite to extraction, and this way of seeing nature is thus foundational to empire, settler colonialism, and neocolonialism. European settler colonialism in the Americas, for instance, evidences the world-building power of the extractive gaze. European colonial powers established colonial-capitalist systems by transforming natural resources into commodities traded on global markets.[50] Throughout the Americas, settler states invoked the terra nullius (nobody's land) doctrine to argue that the land being conquered was legally empty because its inhabitants were not using it according to a particular standard of productivity and, therefore, could be expropriated regardless of the existing settlement. In this context, productive land use meant logging, agriculture, and mining. As Glen Coulthard (Yellowknives Dene) has shown, in the settler-colonial states of the Americas, the state's attempt to secure access to land cannot be reduced to a historical event but instead forms an ongoing and constantly renewed structural relationship.[51]

Notions of the "right to property" that underpinned European settlement in the Americas are entangled with the histories of extraction in the other geographies that this book traces. Still, the transnational and ever-changing geographies of extraction make determining a singular explanation of the motive forces that underpin extractive capitalist colonialism difficult. While property is a core thread, the specific histories of the extraction of labor and land look different in each context.[52] Still, we can broadly conclude that to frame land as "empty" and to turn life into "commodities" is a deliberate refusal to see a world teeming with life (and value that exists outside of economic calculation) that lies at the heart of extraction.[53]

What are the visual imperatives of extraction, and how do these imperatives shape how we see and know? Media studies scholar Nicholas Mirzoeff describes this way of seeing nature as "Anthropocene visuality," a way of seeing that obscures rather than reveals environmental and social injustices, rooted in an understanding of the human relationship with nature as a conquest. Mirzoeff suggests that "Anthropocene visuality allows us to move on, to see nothing and keep circulating commodities, despite the destruction of the biosphere."[54] One example of Anthropocene visuality, or the extractive gaze, is a 1918 advertisement for a mining explosives company, the Hercules Powder Co., that ran in the *Saturday Evening Post*. Coal is emblazoned over a cartoonish drawing of a stegosaurus in front of rock cliffs (see figure I.2). The ad exults:

> Millions of years before the advent of man, Nature was preparing for his comfort. In the gray dawn of the world—when gigantic saurians dragged their ungainly bodies through thickets of giant ferns, when mighty tempests beat to earth trees as tall as cathedral spires, when flying reptiles bigger than aeroplanes rushed screaming through the air—She was laying the foundations of our coal beds.

The ad goes on to describe the importance of dynamite to coal miners, who are, in turn, central to the United States war effort in World War I. The anthropocentrism of the ad is hyperbolic: the ad looks back hundreds of millions of years and concludes that the sole destiny of the dinosaur was to transform itself into fossil fuel for the future comfort of human beings. The metaphors deployed reframe natural phenomena in human terms: trees become cathedral spires; animals become airplanes. The ad narrates a guiding hand prefiguring the eventual dominion of humans over nature. In this framework, coal mining is not just necessary—or even a necessary evil—but right, preordained. The stark phrasing of *"to get it out"* crystalizes the crude realities of extraction. Although the cartoonish stegosaurus is charming—charismatic megafauna if there ever was any—most fossil fuels are formed by plants, trees, and tiny marine organisms. The ad taps into "dino-fascination" by dressing coal in the charisma and grandeur of the prehistoric great beasts.[55] In the process, it diverts attention away from the messier realities of coal mining to frame the use of fossil fuels as benign, even preordained. This way of seeing the world recasts environmental degradation as progress under the guise of technological innovation, economic development, and increasing quality of life. At the same time, it appeals to ideas of nature and the natural to legitimize the exploitation of humans and the natural world. A key area of focus in this book is how abstract conceptions of nature come to justify extraction. As the ad shows us, extractive capitalism did not just change materials into commodities

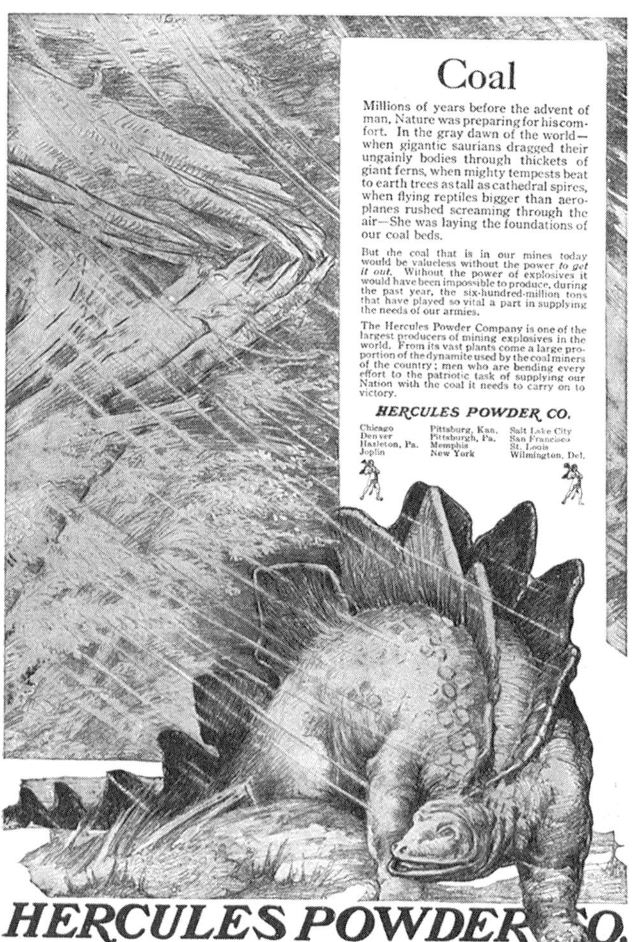

but also assembled a constellation of images that made large-scale extraction seem not only necessary but natural. Embedded within the image are more ambivalent messages, however. In popular culture, the story we tell about the mass extinction of dinosaurs centers on the dependence of life on its planetary environs, and thus on the fragility of that reliance. The ad therefore links fossil fuels to a precursory moment of mass extinction. Within the valorization of the extraction and burning of fossil fuels, then,

I.2   Hercules Powder Company, "Coal," *Saturday Evening Post*, October 12, 1918.

are the seeds of its undoing, a—perhaps unconscious—recognition of its overreach. We can also locate a link to photography: Hercules Powder Co. supplied papers to photographic industries, highlighting the complex imbrication of the chemical, extractive, and photographic industries.

Extraction is thus not only an economic process but a way of engaging with the world around us. Fundamentally, *Camera Geologica* is concerned with how extraction shapes how we see and know. While considering the cultural forms of extraction, land and labor remain central to my focus. Photography scholars have observed the violent language of photography: shoot, take, aim, capture, trigger.[56] This language of the hunt is often confirmed in extractive image-making processes. The culturally extractive nature of photography as a medium is well established: for instance, a photographer with institutional power extracts meaning, beauty, or pain from the subject, transforming these intangible things into the art world's marketable commodities. However, Imre Szeman and Jennifer Wenzel call attention to the loss of meaning when extraction is deployed as a metaphor (a cultural and ideological problem) rather than a material process: it is attention to material processes that makes extractivism a useful analytic.[57] As such, I examine how cultural forms of extraction shape ways of seeing nature, but I keep the material realities of mining in view to consider how cultural forms function to naturalize and thus facilitate extraction—resituating photography within histories of mining and industry means not thinking about visual regimes alone but situates them within the violent human-nature interactions, with the viciously and unevenly distributed burdens and benefits, that they make possible.

While the extractive gaze teaches us to see nature as a resource, there are alternative viewpoints within this system of visuality that enable us to see nature and human relations in different ways. Extraction is never an all-encompassing process that displaces all other forms of living and relating.[58] In part, seeing extractively is rooted in deliberate forgetting, in a refusal to reckon seriously with the inheritances of the past. By naming the processes and structures that have produced the interlocking crises of the current historical moment, artists can and have intervened in an extractive visuality, redirecting the same materials toward the undoing of the ecocidal and genocidal projects they have underpinned.

There is no guarantee, however, that environmentally activist photographs result in empathy or action. Indeed, in many cases, they may aestheticize or anesthetize. In the context of climate change, images play a complex role. Paradoxically, the same chemicals that cause harm can document extractive practices and processes, making visible what extractive capitalism renders invisible. Despite its historic and material complicities, photography can challenge extraction as a worldview.[59] As with any liberatory action taken within a broader set of exploitative relations, materials and practices from those relations must unavoidably be repurposed toward other ends—as there is no world-changing action from nowhere within that same world.[60] It is photography's implication in damaging systems that makes it a productive site to initiate critique.

The Ecology of Photography

Ecology is another thread that runs throughout the book. The use of photography to explore environmental issues has a long history. Perhaps the most famous points of intersection between photography and the environment center on wilderness landscape photography. Carleton Watkins, whose work opened this book, rendered the vastness and grandeur of the glacial valleys, cascading waterfalls, and ancient rock faces of the western United States in exquisite detail. Watkins arrived in California in 1851 and documented the birth of industrial mining alongside his famed wilderness landscape photos of nearby Yosemite, which were instrumental in establishing the American visual vocabulary of wilderness.[61] His photographs of Yosemite would establish him as the preeminent landscape photographer in the United States and bring him national fame. It was partly due to the rugged sublimity of these photographs that President Lincoln set aside Yosemite for conservation and public use in 1864, creating the blueprint for the National Park System.

In the early twentieth century, Ansel Adams's explicitly environmentalist work for the Sierra Club, a nonprofit organization dedicated to environmental conservation, used vast vistas in sharp focus to promote a particular vision of the wilderness there was to conserve. Similarly, Eliot Porter's closely framed photographs of flora and fauna—also taken for the Sierra Club—celebrated the Northeastern United States on an intimate scale. As art historian Robin Kelsey has written, the widespread use of photographs to promote conservation confirms the assumption "that the value of those places was primarily visual."[62] It would seem that the motivation behind many strands of environmentalism was at least partially aesthetic—the desire to maintain a more attractive place to live and play.

The ties between conservationist landscape aesthetics and extraction are stronger than they initially might appear. Wilderness landscape photography played a dual role that points to the tensions and interconnections between preservationist and instrumental approaches to nature. These contradictory myths—the spiritual call of virgin wilderness that needed to be protected and the promise of an extractive frontier that would fuel development—both find their natural expression in landscape photography. As Watkins's work and biography reveal, wilderness played a more complicated role than simply celebrating ancient and unchanging nature. These pristine sites often had direct links to extraction. The Gold Rush brought settlers like Watkins to California. Watkins began his commercial career by photographing California's rapidly developing mining industry, taking photographs to promote extraction as well as to serve an evidentiary function in mining land-claim disputes. Watkins first visited Yosemite in 1861 with his patron, the mine owner Trenor Park. Later, Watkins became a photographer for the California Geological Survey and Clarence King's US Geological Exploration of the Fortieth Parallel. The photographs produced for the geological survey played a pivotal role in transforming land into settler property, in both private and public forms. While Yosemite became a protected wilderness site, whose status as a "conserved" landscape was predicated on the displacement of Indigenous peoples and nations, Malakoff Diggins became an environmental sacrifice zone, an ecosystem destroyed for profit. These two ways of seeing are integrally connected.[63] In Watkins's oeuvre, we witness the visual emergence of a binary yet mutually constitutive understanding of ecosystems marked for protection or exploitation.

With the rise of environmental art history as a field of study, there has been considerable analysis of how the visual shapes our understanding of the natural world. I turn to photographs of extraction, which make visible the violent interactions of humans with nature, to consider how the products of extraction ultimately come to form the image itself. Here, my approach is informed by art historian T. J. Demos, who proposes we read spectacular, aestheticizing images of climate breakdown against the grain by resituating these images in the relational realm of ecology.[64] Rather than thinking of such images as "pictures of ecology," we might place them within an "ecology of pictures." A more capacious understanding of the image brings into view the structural causes and gross inequalities at the heart of ecological crisis, which converge in images—if not representationally, then relationally. Such convergences may not always be visible *in* the image as representation but reading the photograph as an object *and* an image reveals these histories are *in* the physical materiality of the image.

In ecology, *ecotone* describes a junction between two distinct ecosystems. In a process called the *edge effect*, the two ecosystems meet and integrate. These are locations of high biodiversity but they contain characteristics of two different zones, likewise considered zones of tension.[65] As a medium, photography is analogous to an ecotone. It is at once an art, science, and technology; evidentiary and aesthetic; a material object and representation; fixed and contingent. In keeping with this, I approach photography as a zone of tension rather than trying to locate photography in any one category. Photography's overlapping spheres of influence and unstable boundaries make it a fruitful site for ecological thought.

Materials and Materialism

Mined materials are the organizing structure of the book: each chapter centers on the extractive, material, and visual history of one metal or fossil fuel used in photographic processes, ranging from their discovery to the present. As we follow materials and the social relations required for their production and use, unexpected connections between images emerge across genres, geographies, and temporalities. As such, the structure of this book employs an iterative approach to consider how nineteenth-century image-objects reveal something about our present condition. In turn, I explore how artists in the present are critically reactivating these analog methods. These images emphasize their corporeality, drawing explicit links between materiality and meaning. With the exception of the final chapter, I focus on materials that were used as light-sensitive materials or, in the case of uranium, light sources. Here, we are reminded that even a seemingly immaterial element of photography—light—is intrinsically bound up with its material characteristics.

To think materially about photography, I use Marx's historical-materialist framework, which draws attention to the impossibility of transcending the material world of nature, infrastructure, and the power relations that structure the social reproduction of human life. Ideology and culture are the products of material realities, and politics and culture emerge from and in material circumstances, even as they take on relatively autonomous lives of "their own." Marx employs the analogy of the camera obscura to show that idealist philosophy approaches the world backward, writing, "If in all ideology men and their circumstances appear upside-down as in a *camera obscura*, this phenomenon arises just as much from their historical life-process as the inversion of objects on the retina does from their physical life-process."[66] In a materialist framework, the world of concepts and images emerges from the tangible world of changing forms of human interaction with nature and, in turn, responds to and shapes material conditions. Following Marx, then, in this book, "we ascend from earth to heaven": we move from materials to representation, from the mine to meaning.[67] The interplay between material and representation is a complex dialectic, however. In "Theses on the Philosophy of History," Walter Benjamin, building from Marx and referencing class struggle, reflects on "the fight for the crude and material things, without which no fine and spiritual things could exist."[68] Throughout this book, I think about the "crude and material" alongside the "fine and spiritual." Such an approach centers labor, process, and the contingency of making, as well as the aesthetic power of the image.

The image-objects explored in this book have very different material foundations, technological histories, and visual forms, drawing attention to the base fact that, *pace* Holmes, form can never be divorced from matter. Thinking these objects alongside each other destabilizes a clear notion of what constitutes photography. What these different visual forms I study share is an interest, whether intentional or unintentional, in exploring the consequences of extraction. My focus on extraction does not suggest that photography can be reduced to its materials, its production, or its commodity status. It is precisely by bringing together these inextricable but often ignored aspects of photography's material and social origins that we can take seriously the affective and artistic value that transcends their production as commodities. The photographer Simon Starling (whose work is discussed in chapter 3) reflects:

A photograph is invariably a symptom of the forces that brought it into being—the institutions that surround it, the economics that fuelled its making, etc.—but also that photography's particularities—its conflation of chemistry and optics, its phantasmagorical relationship to history—can in some way transcend those institutional boundaries, to be "itself" in one way or another.[69]

The particularities and paradoxes that make photography so compelling come into the foreground when we put materiality into dialogue with representation, and each chapter makes both a material and visual argument—both of which, as we shall see, are profoundly social. As Starling's practice shows us, tracing the chains of photographic production reveals unexpected interconnections between the natural and cultural world and between the miner and the artist.

My emphasis on materials may sound aligned to the goals of new materialism, but the history of materials themselves is not the project's aim. The chemical properties of materials are only part of the story. In studying materials in their chemical and physical specificity, I am interested in how they mediate relations between people through sociospatial relations of production and cultural forms, as well as how they enable certain processes of production that result in variable, geographically uneven forms of labor exploitation and environmental pollutions. As such, I do not make arguments about the agency of materials but rather employ a historical-materialist approach that explores how processes and relations of labor transform and refine the innate chemical possibilities of materials, adapting them to forms of utility dictated by and powerful within the social structures of power and relation that make these processes possible. Materials can be an entry point to explore capitalism, labor, and land as they relate to photography: extraction emphasizes structural questions that exist in the political sphere of human organization, in all of its conflictual differentiation, across the planet's surface.

An emphasis on materials also enables a different engagement with the past. The interplay of image and material forges a tangible connection between the past and the present. Walter Benjamin used photography's relationship to time as an analogy for a historical project that sought to challenge the understanding of time as linear and progressive. Benjamin argued that moments from the past could "blast forward" into the present, introducing the possibility of future action.[70] Making the past active in the present is a method to open up the rewriting of history from below. This conception of photography and history suggests that the past, particularly the often overlooked history of labor, forms an integral part of the present in material and symbolic ways. Benjamin argues that history, like the photograph, is contingent. He reflected:

> No matter how artful the photograph, no matter how carefully posed his subject, the beholder feels an irresistible urge to search such a picture for the tiny spark of contingency, of the here and now, with which reality has (so to speak) seared the subject, to find the inconspicuous spot where in the immediacy of that long-forgotten moment the future nests so eloquently that we, looking back, may rediscover it.[71]

Drawing from the contingency that lies at the root of Benjamin's conception of history and the photograph, I approach the photograph as something that contains multiple and unruly meanings. Even the most staged photograph contains information that evades the control of the photographer, and this "tiny spark of contingency" makes the photograph a powerful resource for critical inquiry. Within each photograph is a confusion of intersecting histories concretized through the image's materiality. The photograph's meaning is never fixed; meaning is always negotiated and changing. In light of this, I argue that something is recoverable from the past, even from its more complicated legacies. Often these important movements are found in the "overburden"—to borrow a mining term—that has been stripped away and obscured by discourses of transparency, immediacy, and immateriality.

For photography to play an activist role in the context of environmental justice, however, its intimate links to extraction must be understood. Photographs often fail in the context of environmental justice: images of nature tend to make the scene seem timeless and outside of human history. Scenes of climate catastrophe become abstracted into spectacular, tragic forms that blame everyone and no one, shifting attention away from the specific set of choices that were and are being made to prioritize wealth-production over sustainable worlds. In both cases, a politically informed, historically rooted indictment of the consequences of a particular set of economic relations becomes transformed into evidence of suffering that speaks to human nature and the human condition, a belief reflecting in the very naming of the Anthropocene.[72] This critique of photography's transhistoricizing tendency has long roots. Bertolt Brecht, for instance, argued that photography is often superficial, functioning to aestheticize and abstract social relations.[73] For Brecht, the social and economic relationships structured by capitalism are not easily made visible within photography: a photograph of a factory struggles to show how the factory functions within capitalist social relations. As he summarizes, photography's abstract, aesthetic universalism results in significant problems for political action: "It seems impossible to alter what has long not been altered. We are always coming on things that are too obvious for us to bother to understand them."[74] The factory is one such site; so is the mine. In the process, historically contingent social relations transfigure into something eternal: it becomes easier to imagine the end of the world than the end of capitalism.[75]

Brecht, however, also provides a model for mobilizing art and spectatorship for political change. His most famous work outlined a model of historical materialist theater that could provoke politically engaged consciousness among spectators. By emphasizing, through a variety of techniques, how the social world depicted in the play is continuous with the world that made the play possible—the theater, the actors, the wood used for the stage—*and* how the dilemmas within the social world depicted on the stage are the product of changeable human social relations, this type of art functions through alienation: it allows us to recognize its subject but makes it at once unfamiliar and broader than what we see before us. By alienating the familiar, it is possible to develop a critical eye that sees how society could change, and indeed already is changing through human activity. By showing the construction of the scene, its historical and material specificity comes into view: the social relations depicted transform into something socially constructed and historically contingent. Following Brecht, a dialectical materialist art must "put some artistry into the act of showing"; the processes, inconsistencies, and seams, all of which are seeds for change that is yet to unfold, should be visible.[76] A form of Brecht's method here is applied by many of the critical contemporary artists I examine in this book, but it also guides my own method as a spectator and critic. Examining any piece of art for the historical-social relations that made it possible—both in its aesthetic form and in its material constitution—is precisely the kind of critical dialectical work that art historical criticism can enable, situating our practice within a broader struggle over the shaping of these same earth-choking social relations of industrialized human sacrifice.[77]

The majority of photographs in this book do not register as photographic in the sense of a neutral window onto the world. Rather, they make their objectness—and by extension their historical and material specificity—clear. The materials themselves are not the agent of meaning, but artists, by making the materials visible, render the obvious scene strange. Situating images of extraction within the historical and material contexts of production reveals dissimilarity and contradiction. *Camera Geologica* assembles an archive of images that denaturalize photography, making the production and illusions of the image visible. For instance, LaToya Ruby Frazier's cyanotypes of the steel industry are obviously artificial, rendered unnaturally in shades of blue, denying the absorption of the image into a canon of liberal humanist images of labor. In doing so, Frazier's cyanotypes shift attention from a human tragedy to a structural critique of organized abandonment, as is shown in chapter 4. More broadly, *Camera Geologica* works toward a methodology to name and analyze the complex networks of materials and labor that make images possible. In doing so, it proposes a mode of critical spectatorship that generates questions about how extraction makes our world and how these processes are historically contingent choices based in what society has chosen to value. Recognizing contingency means that these things can change: we can shift our relationship to capitalist extraction. By generating a critical attitude toward the scene, art can encourage the spectator toward transformation by developing a consciousness of the contradictions of life under capitalism. Art of this kind can be one entry point into the multisited and durational process of large-scale social change.

Chapter 1 begins with bitumen, the light-sensitive material in the first photograph taken by Nicéphore Niépce in 1826. Taking as a case study Warren Cariou's petrographs of the Athabasca tar sands in Western Canada, the chapter proposes a shift in focus from light to minerals, considering the complex interplay between time, fossils, solarity, and labor that bitumen introduces. I situate Cariou's very material photographs within the hidden-in-plain-sight visual culture of oil, reading Cariou alongside work by Tsēmā, Edward Burtynsky, and Allan Sekula. Crucially, Cariou's petrographs move toward a land-based photography, bringing into view the complex networks of settler colonialism, petrocapitalism, and consumption that make the image possible while proposing other ways of seeing human relations with territory. In doing so, Cariou makes a case for photography as a critical site of anti-extractive world-making.

Chapter 2 turns to silver, the most important material used in analog photography. Silver's remarkable light sensitivity, relatively low cost, and ubiquity enabled the rise of photography as an industry. Focusing on scale, this chapter traces a long historical arc, moving from the fifteenth-century discovery of silver in Potosí (now Bolivia) to Timothy O'Sullivan's photographs of silver at Comstock Lode, Nevada, in the 1860s, concluding with Eastman Kodak Company and the rise of photography as a mass medium. In the process, we see how socially contested changes in currency standards, industrial uses, and recycling impacted the supply of silver that could then be conscripted into the scaled-up production required for Kodak to become a household name.

Chapter 3 turns to platinum and the theme of atmosphere. The pictorialists championed the atmospheric aesthetics of platinum prints, but platinum and atmosphere also have a material dimension: platinum prints were a chemically stable alternative to silver prints, which were vulnerable to growing industrial air pollution. Tracing platinum's supply chains to South Africa, I do an atmospheric reading of platinum prints by David Goldblatt and Simon Starling to show how the metal's promise of stable boundaries is undermined by the dust and particles that atmosphere carries between bodies and landscapes. I conclude with Larry McNeil's exploration of coal mining and atmosphere in the western United States to contrast the futurity promised by the stability of the platinum print with the reality that polluted atmosphere is foreclosing collective futures on this planet.

The theme of unstable boundaries is developed in chapter 4, which centers on iron and cyanotypes, or blueprint photography, which I argue materially register industrial growth. Reading Anna Atkins's cyanotypes of algae and ferns through Walter Benjamin's writing on the links between iron, metabolism, and industry reveals the connections between the print, the plant, plantation slavery, and industrial growth. I then turn to railroad photography in Pennsylvania's Steel Belt during the Second Industrial Revolution to consider the rise of blueprint photography, contrasting blueprints with William Rau's albumen prints. The chapter concludes with LaToya Ruby Frazier's cyanotypes, which explore embodied histories of deindustrialization in the Rust Belt. Throughout, I show how iron as a material moves between registers—the plant, the body, and infrastructure—enabling both biological and industrial growth alongside differentially distributed costs to sacrificed life.

Chapter 5 explores how uranium pushes photography beyond that which is visible to the human eye. Centering on the problem of slow violence, the argument in this chapter is twofold.[78] First, experiments by Niépce de Saint-Victor, Wilhelm Röntgen, and Henri Becquerel show that photography is central to the development of atomic culture—just as many of the qualities of radiation were first perceived on photographic paper. Photography was deliberately used to direct attention from the violence caused by the atomic bomb to the spectacular imaginary of the bomb itself. At the same time, photographs made with uranium can make visible forms of attritional violence that otherwise can't be seen. Materially, uranium highlights the limits and possibilities of seeing and visibility in the context of violence, both slow and spectacular.

Chapter 6 turns to the digital world to consider the extractive and visual image economies of the present. I focus on rare earth elements, seventeen chemically similar minerals that make technologies brighter, faster, and lighter. Rare earths serve as our guide to digital images because they are used in lenses and screens and to build color. Following rare earths from their extraction in Baotou, Mongolia, through the very material infrastructure of the Cloud, which runs through cables deep under the ocean, to Agbogbloshie, an e-waste dump in Accra, Ghana, highlights the environmental and labor costs of seemingly immaterial images and points to the open global contests over what new forms of image-making will do, and who, where, will pay the price.

Throughout, materials and photographs form an entry point into the broader system of extractive capitalism. The structure of this book is kaleidoscopic: each chapter makes a stand-alone argument, but when the fragments are placed alongside the others and filtered through a lens (photography and extraction), a more complex picture of photography's implication within, and potential to resist, extraction emerges. Works of art are both enmeshed within and enable geopolitical and economic systems, a position of complicity that is relational *and* material. More productively, the ability of art to produce and form worlds reveals that art's role is not passive—pointing to possible, if so often unrealized, political potential.[79]

Extraction reveals complicated networks of implication: the universities that fund research to develop extractive industries, the nation-states that subsidize mining, and the museums that greenwash corporations through exhibition funding and boards of trustees, to name a few sites of intersection. Those who are insulated from the *immediate* violences of extraction often prefer not to think about their implication in this component of capitalist industry more generally.[80] For some of us, it is a choice not to look, not to do the reconnecting work. I am a white settler in the settler-colonial state of Canada, which is one of the primary drivers of extraction worldwide. Over 75 percent of mining corporations worldwide are headquartered in Canada, and the Athabasca tar sands are one of the largest single contributors to global climate change.

On a more personal level, mining shaped my family's history. My great-grandfather Joseph MacNeil worked in the coal fields of the British Empire Steel Company's Dominion #6 Colliery in Glace Bay, Nova Scotia, and, later, deep underground in the McIntyre Mines in Schumacher, Ontario, one of the richest gold finds in history. He broke his back during a cave-in underground and occupational disease slowly poisoned him. My other great-grandfather, Charlie Angus, a miner and socialist organizer, died in an accident underground at the Hollinger Mine in Timmins. At the same time, mining labor was a critical organizing site for working-class people, which, coupled with the investment in the welfare state in the postwar period, enabled my grandparents to move into the middle class.[81] The benefits that have shaped my life, like public education and national health care, are in part funded by Canada's extractive history and present—which is predicated on the theft and ongoing occupation of Indigenous land. The legacies of these violences are responsibilities I have inherited as a settler.[82] Generational distance from mining labor likewise did not prevent exposure to toxins like lead and arsenic, industrial histories indexed in my body. The ubiquity of toxicity shows that no one is insulated from the slow violence that accompanies the extraction and refining of materials, processes that release toxins that mutate bodies and transform ecosystems globally. I have benefited from extraction and have been damaged by it. I write this book from a position of entanglement and implication underpinned by a commitment to ecosocialism and environmental justice. Consciousness of complicated histories and the complex path forward to a more ethical relationship with the natural world are the first steps to transformation.

While my focus on materials might suggest a suspicion of the visual, I am deeply invested in representation. Activists and artists—most often people working at the intersection of both—have shaped my thinking on extraction. In the context of catastrophe, slowing down and looking closely is important to diagnose the historical roots of our current crisis so that, collectively, we can chart more just paths forward. Given that the science on climate change is clear and yet political change has been lethally slow, it is evident that we need new narratives, new stories, and new ways of seeing. This is what artists do: they can transform how we see our world, helping all of us take the many actions we need to remake it into a world worth living in.

1

Bitumen and a Reorientation of Vision

This is just to say

we've burned up all the oil

and poisoned the air

you were probably hoping to breathe.

Forgive us.

It was delicious

the way it burned

so bright and

so fast.

Warren Cariou, "Tarhands: A Messy Manifesto" (2012)

A GOLDEN-HUED PHOTOGRAPH SHIMMERS as it catches the light. The photograph, *Syncrude Plant and Tailings Pond Reflection* (2015) (plate 2), documents the infrastructure of a bitumen mine in the Athabasca tar sands of western Canada. Oil extraction is the subject of the image, but bitumen, an unrefined form of petroleum, is also the material used to produce it. Warren Cariou, an artist and academic of Métis and European descent, replicates a labor-intensive early photographic process, heliography, to create photographs made with oil from the tar sands. The bitumen is fixed on the surface of a polished metal plate, where it registers a material trace of extraction. The opacity and texture of the hardened bitumen form a striking contrast to the smooth surface of polished aluminum. Cariou deploys the bitumen to capture the extraction of bitumen, bringing forward questions of process, labor, and material.

Within the image, the infrastructure of the tar sands comes into view. The visual signifier of the smokestack marks the scene as industrial while the title reveals that the water in the foreground is a tailings pond. In the monochrome photograph, the toxic slurry—a mixture of salts, residual bitumen, and chemical compounds, including acids, benzene, hydrocarbons, silt, and clay—is neutralized into a general impression of gleaming liquid. The Syncrude Plant is mirrored in the reflective surface of the tailings pond, an apt linking of cause and effect. But the stable formal symmetry between the top and bottom half of the image is constantly disrupted, as the engraving is set on the highly reflective surface of a polished aluminum plate. From some angles, the evanescent plate functions as a mirror, superimposing the human body into the landscape and destabilizing the boundary between the inside and the outside of the frame. The viewer is immersed within a petroleum-coated landscape, upon which they have a visible effect for the duration of looking.

The tar sands are located on Indigenous territory subject to Treaty 8, signed in 1899 and 1900 between the Canadian Crown and the Cree and Dene nations of the Athabasca region.[1] The largest industrial project in the world, the tar sands are the planet's third-largest proven oil reserve, with an estimated 1.84 trillion barrels of crude bitumen.[2] The spatial scale of extraction is immense, spanning 140,200 square kilometers—of which 220 is tailings ponds or manufactured lakes managed by dams and dykes that hold the toxic byproducts of extraction.[3] The tar-like bitumen mixed with sand is difficult and energy-intensive to extract, making the tar sands one of the world's most ecologically destructive industrial sites. If Canada is to reach the relatively low bar of its commitments to the Paris Agreement, a legally binding international treaty on climate change, scientists have shown that the Athabasca tar sands bitumen must stay in the ground. Resistance to extraction by Indigenous land and water protectors and environmentalists has made the tar sands a space of contestation between Indigenous communities, petrochemical companies, and the government. The discourse about this oil sits at the intersection of climate change, settler colonialism, and extractive capitalism.

While Cariou's photographs use bitumen to make a clear political statement about the contemporary environmental devastation caused by resource extraction, the interrelation between image-making technologies and bitumen, specifically, can be traced back to the emergence of photography. Bitumen was the light-sensitive material in the earliest known camera-generated photograph. In 1826, the French inventor Nicéphore Niépce captured an image using bitumen of Judea on a pewter plate (plate 3). Niépce described his experiment, the earliest known to successfully fix an image from light, as "the first uncertain step in a completely new direction," which "consists in the automatic reproduction, by the action of light, with their gradations of tones from black to white, of the images obtained in the camera obscura."[4] The bitumen image, *View from the Window at Le Gras*, was shot from a window of Niépce's country estate near Chalon-sur-Saone. To make the image, Niépce dissolved light-sensitive bitumen in oil of lavender and applied a thin coating over a polished pewter plate, which he then placed in a camera obscura. He left the camera near a window, and after exposure to sunlight, the bitumen hardened. The low light sensitivity of bitumen required an exposure time of several hours. The bitumen Niépce used was called asphaltum, which comes from the Greek *asphaltos*, to make stable or secure. In the heliograph, the bitumen makes the image stable—to a certain degree.

Niépce called his image the heliograph, or sun writing, centering the role of light. By renaming Niépce's heliograph process as the petrograph process—oil writing—Cariou shifts attention from the role of light to the role of fossil fuel. Redirecting our awareness from the sun to bitumen deposits below the surface of the earth, this work enacts a reorientation of vision, bringing the pervasive reliance of contemporary culture on petroleum into view. In doing so, Cariou's image destabilizes and defamiliarizes one way of seeing the world—a way of seeing that naturalizes the expropriation of land and the extraction of materials that provide the energy that fuels contemporary life, a way of seeing I have been calling the *extractive gaze*. Accepting this way of seeing allows us not to reckon with the costs and consequences of extraction. Cariou makes the connection between extraction and representation tangible and, in doing so, prompts the question: What does it mean to see through oil? The invitation to see through oil also highlights the dynamic play between transparency and opacity, as "see through" can describe both using something to mediate our vision *and* perceiving something intended to deceive or conceal. Here, to see through oil is an invitation to see beyond the image to the fossil fuel industries that have shaped our world in general, and photography in particular. The image reveals something about how our world is made and the complex ways we move through it, pointing to what Cariou describes as the pervasive "knowledge but not acknowledgment" of the costs of extraction.[5] In this chapter, I read Cariou's photographs not just as objects but also as metaphor and method. The shift in seeing they invite enables us to see the elusive ubiquity of petrochemicals.

Niépce's heliograph is the canonical site of origin for photo historians, and while the presence of bitumen as the enabling medium at this primal scene is provocative, I am not primarily interested in tracing genealogies or lineages.[6] Rather, I seek to turn Niépce's emphasis on light on its head, to allow us to consider the social implications of the omnipresence of mined materials like bitumen in the history of photography. Bitumen's very tactile materiality—sticky, pungent, toxic to work with—sits uncomfortably within the mystique of the medium of photography. Metaphors of unmediated transparency abound in photographic discourse: window, mirror, lens. Walter Benjamin describes how photographs transcend their material origins to become pure representation through the circulation of reproduced images disconnected from their source and, by extension, their materiality.[7] The concept of a bitumen image throws this immaterial framework into question. Its materiality challenges the beholder to acknowledge the world that makes the photograph possible.

Bitumen-based processes never became widely popular because developments in chemistry allowed early photographers to employ more stable and reliable substances. As George Mathiot reflected in *The Philadelphia Photographer*, a popular nineteenth-century trade journal, "If bitumen had been the only sensitive substance, it might have been said, here is both the beginning and end of photography." Luckily, there was an "abundance of the materials for improvement"—notably, Daguerre's experiments with silver, which rescued photography from premature death.[8] However, while a bitumen photograph may be a novelty today, bitumen refined into crude oil is omnipresent in our lives. Oil is deeply enmeshed in artistic culture: petroleum products are used as raw materials in art-making (paints, plastics, film), while fossil fuel corporations fund almost every major museum worldwide and countless individual exhibitions.[9] Despite the geopolitical and economic importance of oil—and the reality that oil underpins much artistic production materially and financially—fossil fuels themselves have been underexplored in cultural production.[10] The disjuncture between the use of bitumen in the first successfully fixed photograph and the relative cultural invisibility of oil suggests that we can learn a great deal by directly considering the representation of extraction.

Cariou's petrographs point us in this direction, asking what new histories surface when photography begins underground? This chapter draws from Cariou's and Niépce's material explorations of bitumen to propose a reorientation of vision from light to minerals. First, I outline bitumen's material history, focusing on its use in early photographic practices and as a photomechanical printing method within the geological survey. While bitumen was quickly abandoned as a photographic material, oil became central to artistic production while remaining culturally opaque, as I discuss through a case study of Tsēmā's lightbox photographs of bitumen. To consider how oil does and does not appear in mass culture, I return to Niépce's heliograph to consider the interplay between light, fossils, and time to show the constitutive links between bitumen and photography and the stakes of their ongoing imbrication today. Finally, I consider how bitumen photographs nuance the visual culture of oil by reading Cariou's very material photographs alongside work by Edward Burtynsky and Allan Sekula, two very different documenters of extractive disaster. I conclude by returning to Cariou's petrographs to argue that Cariou makes a case for seeing as a crucial potential angle of anti-extractive world-making, reworking relationships between peoples and the land. The petrographs transform bitumen from a fuel into a photographic medium, contesting the inevitability that this material must be burnt for energy. Such a reframing demonstrates the potential of critical photography to shift how we see, know, and value the nature and social relations that make it possible.

Bitumen

Petroleum is a broad category of fossil fuel encompassing both liquid and solid forms of hydrocarbon. One of these, of course, is bitumen, the tar-like substance extracted in large mines in the Athabasca region. The composition of bitumen itself, however, varies greatly: ranging from a viscous fluid to an ore, a solid deposit. As a material for human activity, bitumen has a long history. Bitumen gathered from petroleum seeps has been found on tools used by Neanderthals in Syria, dating back some seventy thousand years. Herodotus cites the use of bitumen in building towers and walls in Babylon. Bitumen was used in Ancient Egyptian mummy embalming, and Arabic observers believed that bitumen, or *mum*, caused the dark color of mummies and, thus, called the bodies *mumia*.[11] In China, oil was used as a fuel and transported through bamboo pipelines as early as 400 BCE.[12] The discovery and extraction of unrefined oil is referenced in one of the oldest known books, the Chinese divination text *I Ching*. In the Athabasca region, Métis, Dene, and Cree people waterproofed canoes with bitumen gathered from the land surface. Medicinally, bitumen was used in Europe and the Middle East as a salve for cuts, bruises, bone fractures, ulcers, and tuberculosis.

Despite these many uses, the material was not mined on a large scale for much of its human history. In antiquity, people gathered bitumen of Judea from the surface of the Dead Sea. By the nineteenth century, bituminous rocks were commonly sold in lumps, just as chemical experimentation was revealing its potential as a primary energy source. The Canadian geologist Abraham Pineo Gesner synthesized kerosene in 1846, and the Scottish chemist James Young distilled paraffin from coal and oil shales in 1847. These innovations launched the modern oil industry, presaging decades of dizzying developments in petrochemicals. In 1858, James Miller Williams dug an oil well in Oil Springs, Ontario, Canada. Edwin Drake drilled the first modern oil well in the United States in Titusville, Pennsylvania, a year later. Large-scale bitumen extraction, however, did not emerge for a full century after this. The state-sponsored development of mines and refining capacity in the Athabasca tar sands in the 1960s, on the heels of years of government subsidies of foundational research to unlock the material's industrial capacities, marked a global shift to large-scale bitumen extraction.[13]

Before Niépce experimented with photo engraving bitumen on polished metal plates in the early nineteenth century, however, its more modest aesthetic applications predominated, as bitumen was used as a paint additive and in etching processes. Bitumen was particularly effective for producing shadows, imparting a "shimmering gloom."[14] Mummy brown—a pigment between burnt and raw umber—is the most notorious bituminous pigment. A favorite of the Pre-Raphaelites, the pigment was effective for glazes, shadows, shading, and flesh tones. Originally, mummy brown was made from white pitch, myrrh, and the ground-up remains of mummies, both human and feline; as many mummies were embalmed in bitumen, the pigment was, by extension, bituminous.[15] Bitumen's chemical instability, however, causes decay over time, with a tendency to crack the paint and buckle the canvas, while the ammonia and fat particles it contains can unexpectedly darken other colors. Bitumen-induced degradation is visible in canvasses by Sir Joshua Reynolds, Henry Fuseli, Sir Thomas Lawrence, and Théodore Géricault—another way materials stubbornly make themselves known in representation.

Experiments with the effect of light on bitumen date back to 1798, and it is possible that Niépce was familiar with these experiments.[16] Niépce first used bitumen as a photographic agent in 1814. Living as he did near mining regions in northeast France, where solid and liquid bitumen were available, his eventual success in setting an image followed years of experimentation with various materials, techniques, and chemicals.[17] In 1816, he produced "points de vue": light-captured, fleeting images that were not fixed on the plate. In 1829, shortly after his initial success with permanent image-capturing, Niépce entered a formal partnership with Louis-Jacques-Mandé Daguerre, who continued experimenting with the process after Niépce's death in 1833. Daguerre introduced iodized silver plates and latent development using mercury vapor, significantly reducing exposure time. Due to the "slowness and uncertainty" of bitumen, Daguerre quickly turned to silver as the light-sensitive material in the heliograph process, which took on his name in the daguerreotype.[18] Indeed, one of the challenges bitumen posed in photographic processes was the "most contradictory" results documented by people working with "this very variable substance."[19] The chemical properties and light sensitivity of bitumen samples varied considerably, posing significant challenges to any kind of standardization in photographic processes. As one photographer lamented, "It is impossible to lay down fixed rules for exposure or working, and the only way of ascertaining the photographic properties of any particular specimen is by actual trial."[20] The stability of bitumen, then, is not chemical but rather its quality for use as a binder or sealer: it makes other things secure while remaining unpredictable itself.

The success of the heliograph process depends on the quality of bitumen, which was difficult to assess due to considerable variation in its composition, in turn shaping the possibilities of the photograph. *The Photographic News* analyzed the photochemical properties of asphalts in the 1880s, concluding that bitumen from Syria (or "Judea") and Trinidad were the most appropriate for photographic use.[21] Trinidadian asphalt was gathered from Great Pitch Lake, where the large pitch deposits were hot and fluid. Well over a century later, Cariou collected bitumen from the Athabasca tar sands, where the tar is thicker and less suitable for photography, requiring a significantly longer exposure time. The slow exposure time of tar sands bitumen is one of the reasons why Cariou's images are contact prints made from digital negatives. To replicate Niépce's direct positive process and expose the plate through a camera would require days or weeks to set the image. Thus, in Cariou's petrographs, a digital negative—itself the product of a web of extractive activities—undergoes a secondary exposure to set it in bitumen.

In 1839, photography was announced to the world by Daguerre, who renamed the heliograph the daguerreotype. Daguerreotypes became one of the two dominant forms of early photography, alongside William Henry Fox Talbot's calotype process. While silver replaced bitumen as a light-sensitive material in Daguerre's process, photolithographers continued to use bitumen to mechanically reproduce photographs using a stone or metal plate.[22] In the late nineteenth century, James Waterhouse wrote a comprehensive study of bitumen in photographic processes. Waterhouse was a British photographer who headed the photographic department of the Survey of India and contributed to the ethnographic study *The People of India* (1868 and 1875).[23] Waterhouse wrote extensively about bitumen as it related to his experiments on photomechanical printing, and he contributed fifty chapters to *The Photographic News* between 1882 and 1885. In photolithography, bitumen-based processes enabled sharper lines and superior clarity, making it well-suited for fine reproductions. Bitumen was commonly used in typographic etching on zinc to produce blocks for printing with a letterpress.[24] In light of the unpredictability of the material, Waterhouse conducted many trials that adapted photography to India's climate and used locally available ingredients, highlighting the importance of place, weather, and accessible materials in photographic processes—as well as the centrality of photography to colonial surveys seeking to make conquered land "useful" according to a particular vision for development.[25] Indeed, a combination of highly variable environmental conditions and high demand for cheap, reproducible images made the geographic survey a key site of innovation in photographic reproduction and photomechanical printing.

In fact, geology became a professionalized science in the same decade that photography emerged. In 1839, the year Daguerre announced his method at a French Academy of Sciences meeting in Paris, the Reverend William Whewell proclaimed the "heroic age" of geology to the Geological Society of London, where he served as president. Whewell celebrated the geologists who had "slain [the earth's] monsters and cleared its wildernesses, and founded here and there a great metropolis." Whewell concluded that geologists must "extend her dominion over the earth, till it becomes, far more truly than any before, a universal empire," directly connecting the study of the earth's composition to imperial designs.[26] As this summation reflects, geology as a way of seeing the world and as a field of study was founded as an extractive discipline rooted in expansionist colonial and capitalist projects.[27] Colonization, modes of surveying and depiction, and mineral extraction have profoundly coconstitutive histories. The processes of surveying and cataloguing minerals turn place into property, implicating geology in extraction. The narrative of detached, scientific observation hides the violent social relations and the expropriation of land that lie at the heart of geology.

While photography is not necessarily well-suited to conveying nuanced geological information, it was adopted by geological survey projects quickly and comprehensively, as I discuss further in chapter 2. The mass reproduction of detailed images was especially valuable to the geological survey, enabling the storing and accessing of visualized information in the multiple offices administering colonial programs, highlighting the central role that colonial conquest played in innovations in image-making. For example, the India Survey, which employed Waterhouse, was established to consolidate the territories of the British East India Company. In 1870, Waterhouse meticulously summarized innovations in photomechanical printing by topographical and survey departments across Europe. Several of the processes Waterhouse describes use bitumen, including the 1858 heliogravure process developed by Niépce de Saint-Victor, the cousin of Nicéphore Niépce, who continued to refine Niépce's process. Throughout the late nineteenth century, the mutual development of modes of extraction, projects of empire, and the camera solidified: surveying projects identified and cataloged natural resources, documenting them photographically and reproduced with the products of extraction, and then circulated to stimulate further surveying projects, extending imperial conquest.

Oil's In/Visibility

While Niépce and Cariou work with bitumen as raw material, this is rarely the cultural form in which we encounter bitumen in daily life. Seeing through oil likewise demands that we consider its cultural opacity. Despite the centrality of fossil fuels to contemporary life, as energy humanities scholars have shown, oil as a raw material "has been hidden in plain sight in cultural representation."[28] This is the realm into which Tahltan artist Tsēmā Igharas intervenes, putting bitumen on display in her photographic work. In *Emergence: Bitumen* (2018), the shimmering blackness of a bituminous rock is presented on a velvety black backdrop (plate 4). The glimmering ore is enchanting and otherworldly, seemingly doused in electric blue glitter. Up close, the rock has visual affinities with the twinkling stars that coat the night sky. Though the black background threatens to obscure the equally dark object, the subtle refractions of light accentuate the shiny conchoidal fracture of the ore. Visually, the most striking thing about the image is its saturated blackness. Black is associated with carbon—the fundamental element of life—and therefore represents "the initial, germinal stage of all processes."[29] Mythologist J. E. Cirlot suggests that symbolically, black is linked to "the transmutation of carbon-based life forms" into fossil fuels, which are "hidden as avernal [i.e., underworldly] pools . . . obtained by generative penetration" like the pistoning of the pumpjack that pulls oil from deep underground to the surface. Cirlot argues that there is a deep symbolic link between the material and its extraction.

Tsēmā's digital print is backlit in a lightbox, introducing a luminous glow that evokes luxury advertising and draws attention to bitumen's seductive, charismatic allure. The emanating light and imposing scale of the lightbox accentuate the material presence of the photograph. Backlighting is a practice associated with the photoconceptualism of the Vancouver School (starting in the 1980s), which was first done by N. E. Thing Co. (IAIN BAXTER& and Ingrid Baxter) but was made famous in the practice of Jeff Wall. Backlighting evokes advertising—for example, backlit photographs in bus shelters—but the process also imparts to photography a scale and aura more typical of fine-art painting. The lightbox's multivalent associations with photography, cinema, painting, propaganda, and technology charge Tsēmā's critique of capitalist spectacle. In *Emergence: Bitumen*, the contradictions inherent in a dark lightbox expose the paradoxes in our use of bitumen—simultaneously omnipresent and out of sight.

Despite oil's ubiquity, unrefined bitumen is largely invisible. It is extracted from remote mining sites, transported in pipelines that snake through landscapes, and refined into crude oil. An infographic produced by the International Association of Oil and Gas Producers shows the oil all around us: textiles, cosmetics, heart valves, paint, tires, food preservatives, vegan leather, computers, televisions, and, of course, cameras. The infographic moves between the hospital, the farm, and the home, among other sites, highlighting both obvious and unexpected products that contain oil. Focusing on the commodities petromodernity makes possible—and not crude oil itself—has been a trope in oil industry advertising since the 1950s, when the industry first framed the saturation of oil in society as the precondition for a Western way of life centered on freedom, health, and domesticity.[30] Many objects created from and with oil are featured in advertisements, but these advertisements never show us the oil itself. Oil is hidden in plain sight. The slick, simple infographic neutralizes the ideological point hiding behind the image: it defends the ongoing extraction of fossil fuels in a period when calls for decarbonization have become increasingly loud. It is worth noting the windmills in the background. The image not only implies that extraction could continue in harmony with the increasing use of green energy but also speaks to the contemporary reality of the use of petrochemicals and fossil fuels in green energy production. The infographic identifies cameras and smartphones as oil-based commodities, vivid symbols of the image's central message: we see the world through oil. To give the oil advertisers their due, a noncomprehensive list of the applications of oil in both analog and digital photography includes plastic in roll film, polyester film stock, petroleum-derived photo paper, plastic casing in film packaging and cameras, petroleum-based printing inks, and plastic lenses. The digital infrastructure that allows us to edit and store digital photographs is likewise itself energy-intensive, which, in most of the world, means burning fossil fuels, as discussed in chapter 6. Thus, as art historian Heather Davis reminds us, contemporary photography is "soaked in oil."[31] Tsēmā brings to light how oil, a material integral to the production chain of digital images, hides behind the photograph.

Tsēmā's critique responds to a long history of oil extraction being concealed by the very ubiquity of its product. Photography is an appropriate medium to initiate this critique as the effacement of labor in narratives of photography parallels that of oil, whose mythology has long promised to ease human labor. Capitalism has long functioned by harnessing external sources of energy to lubricate "free" waged work—coal, water, enslaved or indentured workers, underpaid immigrants, or labor-saving machines—but oil's low labor inputs and high energy outputs enabled incredible material transformations. Over the course of the nineteenth century, petroleum became a significant industry, gradually displacing coal as the industrializing world's primary energy source and becoming "nature's legal-tender for the comfort and convenience of mankind," as one oil promoter described it, reducing oil to a form of currency and monetary value.[32]

In cultural imaginaries, then, oil seems to transcend labor. Extracting crude oil is relatively simple and not particularly labor intensive, especially when compared to another hydrocarbon, coal. Coal symbolically represents the industrial working class in many cultures—and their subjugation, made tangible in the coal dust that coats workers' bodies, featured so prominently in labor photography. The brutal labor conditions of coal mines were instrumental in many countries in solidifying the working class itself, particularly in its understanding of its potential mass power at the point of production, where organized workers could exercise their indispensability and shut down operations decisively.[33] In

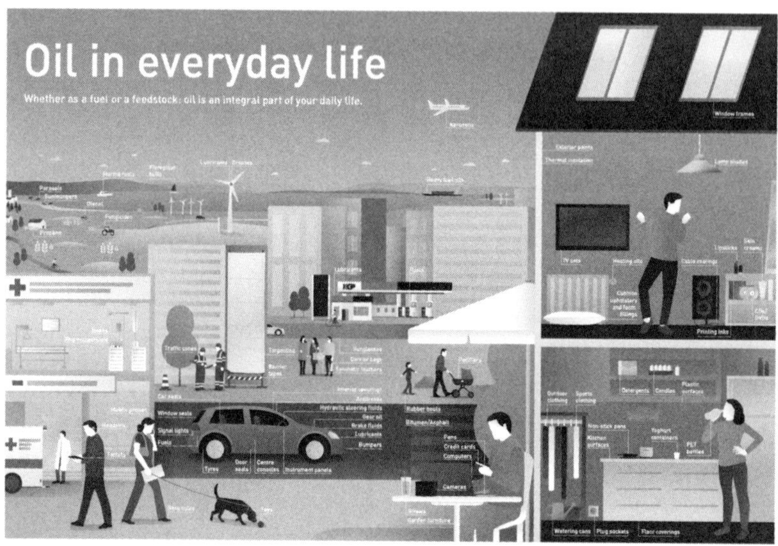

contrast, oil's more decentralized production chains often gave workers less immediately available power to strike in any one location—let alone to visualize the industry's dependence on their labor. Unlike the immiserated mineworker coated in coal dust, the petroleum worker is rarely made visible. The visual culture of oil centers on the oil derrick and the pumpjack, whose slow, rhythmic pistoning into the earth drew crude oil to the surface. The image of densely populated pumpjacks or derricks dotting the oil field is one of petroleum's few tangible, enduring visuals in the cultural imaginary. Aerial view photographs of the Signal Hill oil field in Long Beach, California, are disorientingly natural: the oil derricks resemble spindly trees dotting the urban landscape.

A closer look reveals a more unsettling reality. In 1921, Shell Oil drillers in Long Beach hit a gusher that sprayed crude oil one hundred feet in the air, a narrative that epitomizes the mythology of the lucky oil strike. Oil exploration was a primary driver of growth in Los Angeles, and the Long Beach oil fields were among the richest oil finds in American history. Unlike many other forms of mining, coal included, the oil spilling out of

1.1   International Association of Oil and Gas Producers, "Oil in everyday life."

the ground during this phase was practically refinery-grade and *easy* to extract; it simply had to be gathered. However, there was a slight complication: the land it was found on was already subdivided into a neighborhood of residential homes. Both companies and individuals installed derricks and pumpjacks on parcels of land marked for housing. The photograph shows the densely packed extractive infrastructure dotted amid homes. Visually, production infrastructure dwarfs houses, cars, and actual trees. The uncanniness of the scene is more obvious in colorized postcards of Signal Hill, where derricks and pumpjacks surround the cheerful red roofs dotting white houses framed by neat flower bushes (plate 5). The juxtaposition of domestic and industrial infrastructure functions to symbolically domesticate oil: its extraction does not register as a scene of labor. Indeed, oil derricks and pumpjacks epitomize oil's labor-saving promises in its production. A symbolic split thus surfaces between coal and oil, a fissure that centers on work and automation.

Unsurprisingly, promoters framed oil as a replacement for labor, a force able to end class warfare and gender inequality.[34] As one promoter wrote in 1898, "It saves wear and tear of muscle and disposition, lessens the production of domestic quarrels, adds to the pleasure and satisfaction of living. . . . If it not be a blessing to humanity, the fault lies with the folks and not the stuff."[35] While it is easy to dismiss these statements as naïve boosterism, as historian Dipesh Chakrabarty puts it, the "mansion of modern freedoms stands on an ever-expanding base of fossil-fuel use."[36] The cheap energy facilitated by oil made the expansion of the US middle class possible, as affordable access to oil undergirded so many aspects of modern consumer society.[37] Economic growth in the twentieth century was directly rooted in fossil fuels: a sixteen-fold increase in economic output depended upon a seventeen-fold increase in energy consumption.[38] To explain this in material terms: taken as a global mean, the average human today uses thirty-four gigajoules of energy annually, an amount that has tripled since the Second World War.[39] The majority of this is energy from fossil fuels. If it were measured in units of human labor, it would be the equivalent of sixty adults working twenty-four hours a day for each person. In affluent countries, the number rises to 200 to 240 laborers.[40] The average American family of four draws on more energy than Louis XIV at Versailles.[41] The pumpjacks interspersed with single-family homes in Signal Hill concretize the link between middle-class prosperity and cheap energy while highlighting how the domestication of oil naturalizes the extraction of resources central to settler colonialism.

Signal Hill exemplifies the tension between the hyper visibility and out-of-sightness of fossil-fuel extraction. At this point, I want to return briefly to Tsēmā's *Emergence: Bitumen*. Given the cultural narratives of free or cheap energy built around oil, what are the implications of putting bitumen on display? By emphasizing the striking beauty of the bituminous rock, Tsēmā makes two conceptual moves. The first, as we have seen, shows how petrocapitalism seduces us with the promise of ease by shifting our attention from extraction and production processes to the socially isolated consumer product. Secondly, the redirection to the material draws an implied link to the material and human consequences of extracting and burning fossil fuels. She explains that she isolates the material to connect "materials to mine sites and bodies to the land."[42] The homophonic title of the larger series, *Ore Body*, draws attention to the shared geological origin between human bodies and bitumen:

> *Black Gold is the seductive nature of bitumen.*
> *Black Gold is the destructive potential of bitumen.*
> *Black Gold is our dependence, willing and unwilling, on bitumen.*
> *Black Gold is about the nature of bitumen.*
> *About its energy.*
> *About our energy.*
> *Ore bodies.*
> *Our bodies.*[43]

By linking *ore bodies* to *our bodies*, and *its energy* to *our energy*, Tsēmā proposes enmeshment with fossil fuels, not simply as consumer goods or energy but on a more molecular level, pointing to porous boundaries between bitumen and bodies. While metals and minerals pose challenges to human understanding because they are inert—perceived as outside of life—fossil fuels are ancient remains of life. This is significant, for as Stephanie LeMenager reflects, "The centuries of work we've done as modern humans to immerse ourselves in oil means that, in fact, we are loath to disentangle ourselves or our definition of life from it."[44] Oil is central to economies, and on a more psychological dimension, oil is linked to our cultural understanding of life (or at least a good life liberated from a great deal of toil and hardship).[45]

Tsēmā plays with the tension between oil's transparency and opacity to explore bitumen's desirability and sensuous entanglement in our lives. Critically, the desirability Tsēmā highlights is not the allure of consumer goods—that which oil can make easy or cheap—but in the material itself. Or perhaps, through the link between *ore body* and *our body*, in the *relation* we have to the material. This argument is made aesthetically. Visually, the image is elegant, refined. Materially, the bitumen is raw, crude. Crude oil must be refined before it is valuable to capitalism's consumer economies. *Refining*, an industrial term, has moral connotations, describing the transformation and improvement of both materials and people. Through Tsēmā's lens, the crude material shimmers and seduces. It has value. In this context, the return to the material itself suggests that the value of oil might not lie in what oil does for us but, instead, in land relations. Here, Tsēmā pushes beyond many of the critiques of oil culture to propose another way of understanding the material is not only possible, but right in front of us. By showing the otherworldly beauty of bitumen, Tsēmā doesn't simply make bitumen visible, but makes a case for its inherent value in its raw, unrefined form.

Light, Fossils, and Temporality

Considering how oil does and does not appear in mass culture, let us return to Niépce and Cariou's petrographs to tease out the reorientation of vision that this chapter charts. It is spatial (from the atmosphere to the underground) but also temporal: bitumen redirects our attention from the single moment preserved within the photographic frame to the layered histories of deep time that enabled it. While, at first glance, the ancient timescales of fossils have little in common with photography's instant flash of light, they form a surprising unity on closer observation.

What of light and fossils? Niépce named his photographic process "sun writing," which evocatively tapped into cultural connotations of solarity, drawing on Apollonian tropes of the sun as rational, truthful, and the source of creative energy.[46] Niépce was not alone in emphasizing the autogenic power of light: for good reason, light is central to ontologies of photography and, in many ways, defines photography as a medium.[47] The word *photography* was proposed by Sir John Herschel, among others, and comes from the Greek *photos* (light). Writing on photography has often invoked the sun to make broad claims about the medium's unmediated relationship to the truth. Oliver Wendell Holmes linked photographic truth to the "honest sunshine," describing photos as "permanently recorded in the handwriting of the sun himself."[48] Notions of photographic truth rely on a conception of light as a force that illuminates and clarifies the world around us, though in reality, light just as often distorts and obscures.[49] Photography scholar Geoffrey Batchen has shown that the emphasis on light in thinking about photography was not inevitable. Niépce explored alternatives to *helio*, including the Greek *phusis* (nature).[50] Other nonheliocentric terms were proposed as potential names for the new medium, including *sciagraph* (shadow writing) or *calotype* (beautiful impression). Herschel preferred *actinism*, from the Greek *actino* (ray), to account for the unknown rays that accompanied sunlight. While Daguerre is said to have exclaimed "I have seized the light" upon making a daguerreotype, he renamed the heliograph process after himself, breaking from the emphasis on natural phenomena. While some of these terms are used to describe specific processes, they were almost immediately grouped under the broader umbrella of photography,[51] and light remains central to photographic theory and practice.[52]

The evocation of the sun often functions to obscure human labor. Talbot described photography thus in *The Pencil of Nature* (1844–46): "The plates of the present work are impressed by the agency of Light alone, without any aid whatever from the artist's pencil. They are the sun-pictures themselves."[53] Art historian Steve Edwards situates autogenic conceptions of photography within a larger nineteenth-century discourse that subordinated artisan skill in favor of machine automation, promoting labor's cultural invisibility.[54] The centrality of labor to histories of photography is continually effaced in narratives of photographic production, ranging from the *Pencil of Nature* through Kodak to the smartphone. These discourses reverberate today in narratives of dematerialization and automation. Cariou's intervention works against this tendency, reframing photography as fundamentally material by emphasizing the mineral origins of the medium. Shifting from the immateriality of light to the very material bitumen likewise foregrounds questions of labor—though the petrograph complicates some standard notions of photographic labor, as petrography requires bright sunlight, in contrast to seeing the darkroom as the primary sight of human labor in photographic processes.

The conceptual link between light and the immaterial is often central to the general association of photography's apparent lack of material mediation and its attendant claim for truth. Yet as Cariou reminds us, light is always in dynamic, material interaction with the mined materials within the photograph. The view from the underground reminds us of this fact, and of the fundamental importance of human labor in generating the supposedly sun-given representation. It also allows us to see the connection between the abstract time of the frozen moment in the photograph and the concrete history of subterranean depths and deep time, tied explicitly by bitumen to the temporalities of fossil fuels. In 1785, James Hutton, a prominent figure in the Scottish Enlightenment, first named the deep time of geology. His core intervention in geology was to historicize the apparently stable and eternal natural world as the product of tremendous histories over unfathomable scales that were constantly changing. Geological time asserted that Earth had evolved over thousands of millions of years and carried records of these prior existences in the landscape. Through careful visual observation of rock formations, Hutton theorized that Earth was constantly being formed, deformed, and reformed, a history in which Hutton could "find no vestige of a beginning, no prospect of an end."[55] Heat and pressure initiated chemical reactions while processes like erosion, sedimentation, and volcanic action transformed Earth. Hutton challenged the popular theory that the biblical flood shaped the form of a six-thousand-year-old Earth—biblical time—introducing timescales beyond human comprehension to scientific thought about our world.

The geologist Charles Lyell's *Principles of Geology* (1830) popularized Hutton's theories. Media studies scholar John Durham Peters reflects that Lyell understood Earth as "a recording medium": the geological record formed an imprint of the past.[56] Just as Earth's mantle contains compacted traces of past epochs, so does photography record light that touched its subject—a parallel of indexicality. In photographic theory, the index refers to some kind of direct, often physical, relationship between the signifier (the index) and the signified (the object). For instance, footprints (the index) signify feet (the object) because they are the latter's physical impression. Philosopher Charles Sanders Peirce defined "an Index as a sign determined by its dynamic object by virtue of being in a real relation to it," and the idea of the index as being in "real relation" is the foundation upon which many of photography's truth claims rest.[57]

In this framework, a painter can depict a scene, but a photograph can record it. As Batchen summarizes, it is not photography's claims to capture an authentic reality that defines it as a medium, but rather "proof of that thing's *being*, if not of its truth."[58] The redirection to *being* is productive, illuminating why the physical relationship between the object and its trace lends itself to meditations on time. The flash of light that sets the image in the photographic emulsion reduces time to a singular moment that Roland Barthes reads as "a reality that once existed, even though it's a reality one can no longer touch."[59] In this way, indexicality overlaps with instantaneity: the momentary flash of light that fixes a singular moment, but one that has always already passed. The reality of photographic processes is often more complicated, as prolonged exposure times and composite images undermine the idea of the single moment. However, this temporal paradox is at the core of Barthes's famous observation that the photograph renders what was *there-then* into something *here-now*.[60] As Steve Edwards observes, Barthes's reading isolates the present from the past, preventing the image from acting in the present. While the photograph introduces something as *here-now*, it is infused with loss and nostalgia for a moment that cannot be recovered. In this reading, the photograph becomes "*without future.*"[61] However, the paradox that Barthes highlights in photographic time has something in common with climate breakdown, for while climate catastrophe is *here-now*, it was caused by something that was *there-then*. As climate crisis has materially demonstrated, the traces of the past never entirely disappear, and the future is always prefigured in aspects of the present.

Both photography and geology promise access to traces of the past. Bitumen, formed by fossilized life, has obvious parallels with the common metaphor of image-as-fossil in photographic theory.[62] An 1839 photograph of fossils and shells by Daguerre draws this link between fossils as preserved ancient life (a process that took tens of thousands of years) and Daguerre's invention, which did the same work of fixing something for eternity (or until the photograph degraded) much more quickly.[63] Many photography theorists have taken up this metaphor. Fossils were described by photographer Hiroshi Sugimoto as a form of "pre-photography," as both fossils and photographs are "time recording devices": imprints or traces of preserved life.[64] The negative-positive process of many analog photographs echoes the formation of fossils, for as Sugimoto summarizes, "When you split the strata, the layer on top is the negative image, while the fossilized life form appears as the positive image."[65] Other scholars have linked fossilization and photography as light-powered processes, both partaking in nature's cycles of growth and decay. We can consider Batchen's reflection that "in one sense photography has indeed always been; there has never not been a photography. What is photosynthesis, after all, but an organic world of light writing?"[66] Or Eduardo Cadava's suggestion that there has "never been a time without the photograph, without the residue and writing of light."[67] The idea of the photograph as a fossil also surfaces in Herschel's thinking about photography. As historian of science Kelley Wilder writes, the photograph was not simply a copy for Herschel but "more like a fossil—a surviving imprint from a past time."[68] For media studies scholar Joanna Zylinska, the suggestion that photographs are fossil records reframes them as "a material record of life rather than just its memory trace," suggesting that fossilization is not merely a metaphor but, rather, photography as a method of temporal impression "has always been there, in cosmic deep time."[69] In Niépce's and Cariou's photographs, the conception of photographs as fossil records is not simply symbolic but materially descriptive (see figure 1.2). Unlike silver-based processes where the image emerges when silver halides darken, bitumen photographs don't change color—they *harden*, a process of fossilization.

The ancient timescales of bitumen extend beyond the standard temporal paradox introduced by the photograph. Bitumen is formed from ancient organic matter transformed by heat and pressure into a thick, viscous, tar-like substance: fossils are material traces of the past. These pasts are geological but, in the case of bitumen, also solar. As the astronomer Samuel Pierpont Langley wrote in an article reprinted in *The Photographic News*, the coal that fueled "the whirling spindles of the factory or the turning wheel of the steamboat" was, in reality, "energy gathered in the sunshine of the primeval world."[70] As Langley observes, it is the radiant energy of the sunbeam that fixes carbon within the plant. Over centuries,

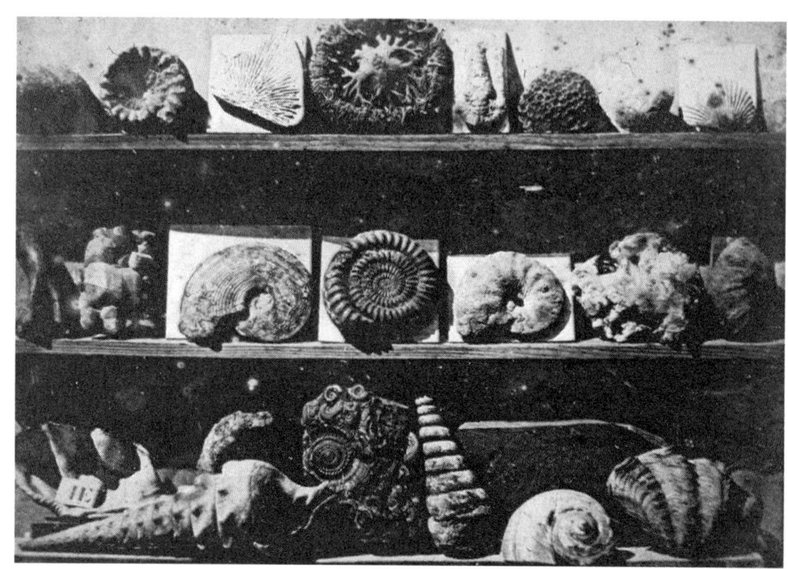

plants and trees are formed into hydrocarbons, imprisoning this energy and moving it up the carbon-bearing food chain. All energy expended by fossil fuels is initially solar energy. A reviving of the dead is revealed to underlie photography's writing with light. Langley's evocative utterance charts a much longer interplay between light and bitumen: rather than a binary opposition brought together in photographic processes, they have been in a multilayered dialectical exchange, shaped over millennia. This directs our attention to the unity of above and below: the atmosphere and underground.

Cariou's petrographs thus gesture to deeper levels of temporality, both through the use of fossil fuels and the return to Niépce and this early moment of chemical exploration. But what can Niépce's initial heliograph tell us about photographic time? Due to its remarkable light sensitivity, silver is the only material capable of instantaneously fixing an image, whereas bi-

1.2 Louis Daguerre, *Shells and Fossils*, 1839. Daguerreotype.

tumen requires several hours of exposure for the image to be set. Niépce's image needed an estimated exposure time of at least eight hours, a delay visible in *View from the Window at La Gras*, as sunlight from a moving sun hits the building from both sides of the frame (see figure 1.3).[71] This creates a scene that the eye could never see—far from the photographic index as a fixed moment in time. As film theorist André Gaudreault observes, the heliograph is not a singular image but rather the accretion of hundreds of overlaid images. This results in a "startling relationship to time in this singular image."[72] The flow of time later captured in moving film is thus present within Niépce's still heliograph. As art historian Kaja Silverman describes, "The elusive image hidden in the illusionistic depths of a shiny pewter plate" is an image that "evolved slowly, through the gradual accumulation of marks," revealing the ongoing development of the image that denies the fixity of the index.[73] The destabilization of the index occurs in the slow setting of the image but also in its viewing, as the pewter plate underlying it is a reflective surface. When the collectors Helmut Gernsheim and Alison Gernsheim acquired the image in 1952, they initially thought that the photograph was a mirror because of the low level of exposure retained on the photographic plate. The image of the view from a window only came into view when the picture itself was angled by a window. The reflective surface slows the viewing process, because the image can only be seen in its entirety as the light moves across the image. This brings the object into the present, for as Elizabeth Edwards reflects, "If the photo-object engages with the body, it also re-temporalizes and re-spatializes the photograph."[74] It is through the extended physical encounter with the image-object that its meanings emerge.[75]

Moreover, the idea of photography's frictionless mechanical reproduction is complicated by the materiality of the heliograph as an image-object constantly in slow motion.[76] Gernsheim had considerable difficulty reproducing Niépce's heliograph. Technicians at Kodak produced a copy, but Gernsheim was disappointed in the results and prohibited its publication until 1977. The Gernsheims added pointillist watercolor dots over one of the prints to make it appear closer to how the heliograph appears in person. This over-painted photograph, in fact, was the primary reproduction circulated for twenty-five years, a startling mixing of mediums that undermines the idea of a photograph as both fixed and reproducible.[77] The

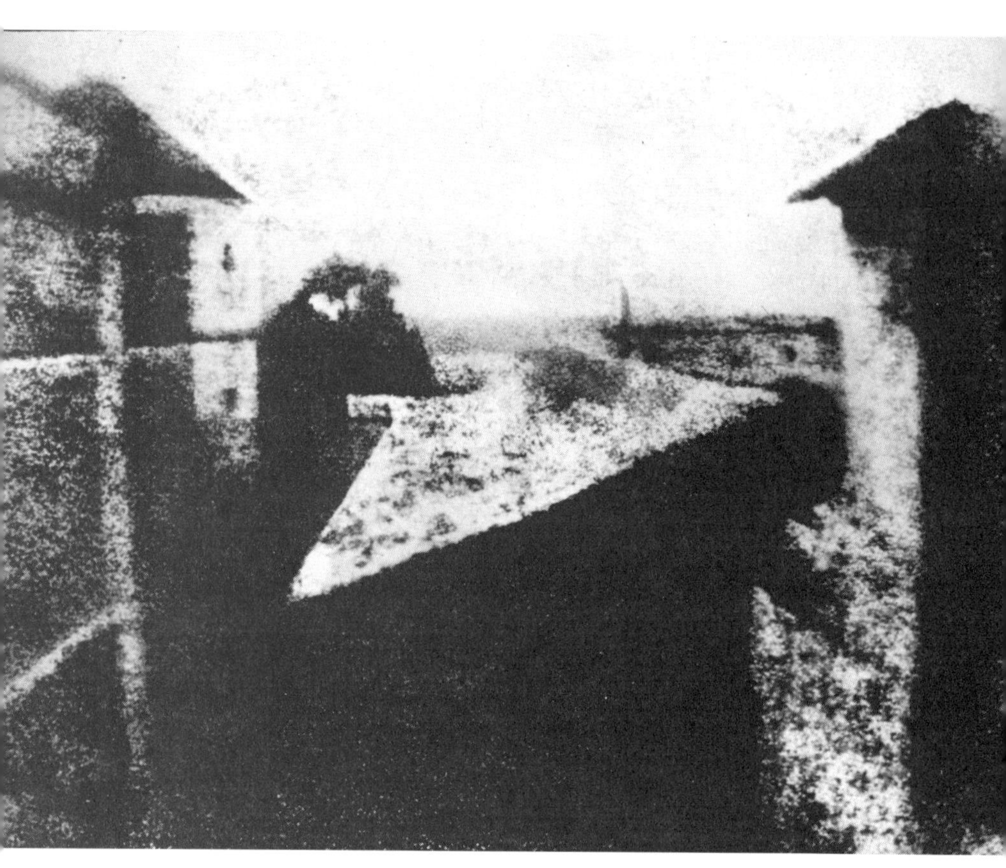

Getty Conservation Institute eventually reproduced the image with newer techniques, though the quality remains unsatisfactory.[78] The poor quality of the reproduction slowed the spread of the image. Like oil itself, this image resists coming into view. The photographic process developed by Niépce and Daguerre introduces a paradox in early photography that breaks with narratives of mechanical reproduction. While Niépce was interested in reproduction, Daguerre was more interested in duration.[79] Ultimately, the first photograph was a singular image whose materiality is, paradoxically, not well suited to reproduction. In *Le Gras*, the sensitivity of the image to changing light conditions tends to unfix and unsettle the surface.

1.3   Nicéphore Niépce, *View from the Window at Le Gras*, enhanced version by Helmut and Alison Gernsheim, 1952.

Photographic time is marked by simultaneity: the multiple temporalities of experience within the photographic frame and through the haptic or optical encounter with the photograph as a material object.[80] The slow overlaying of images that formed Niépce's heliograph draws to mind the imperceptible accrual of geology. The geological record is often framed as forming linearly, one stratum giving way to another temporally and materially distinct layer. However, the sedimented layers of Earth are juxtaposed, joined together, intermingled, and discontinuous, much like the scene captured by Niépce. If the geological layer forms an index of human activity, it is rarely linear. Recognizing this temporal complexity shows us that the past is always active in the present and the future, though the forms may be materially transformed and the consequences removed from causality. Using oil as a stand-in for modernity, Bertolt Brecht wrote, "Petroleum resists the five-act form; today's catastrophes do not progress in a straight line but in cyclical crises."[81] Here, we can recall art historian Matthew Hunter's observation that the kinship between photography, fossilization, and fossil fuels extends into the research of the medium's early experimenters, as research into combustion engines was a common interest among nineteenth-century practitioners of photography, including Niépce and Talbot.[82] Prior to successfully fixing an image in bitumen, Niépce spent a decade experimenting with his brother Claude on an internal combustion engine powered by coal mixed with resin.

A shift from the photograph as a pure representation to the photograph as a material object thus has implications for photography, certainly, but also for broader habits of seeing. Materials, whose extraction and transformation into commodities and infrastructure is often out of sight, underpin our world. Photography's emphasis on the immaterial, seemingly magical ability of light to capture and fix an image aligns with what Marx described as the mystical nature of commodities, which, by manufacturing consumer desire, transcend their banal materiality and metamorphosize into dazzling, magical things, obscuring the more complex social relations of production that brought them into being.[83] The use of bitumen, then, invites the viewer into an ecology of pictures, an invitation to see beyond the image to the complex histories that made it possible. Within the reflective surface of the bitumen image-object, these multilayered pasts—both solar and geological—are brought into relation with the present.

Scale

Thinking about geology foregrounds scale on temporal (the epoch) and spatial (the planetary) dimensions. The scalar tension between the local and the planetary has animated environmental thinking for decades.[84] As Niépce's heliograph demonstrates, time is hard to show photographically, so the desire to *see* the Anthropocene through the camera often leans toward spatial representation. As art historian Jennifer Roberts has shown, however, scale itself is also challenging to see.[85] To think through the potentials and limitations of scale in making climate crisis visible, I return to the Athabasca tar sands pictured in Cariou's petrograph. While we have considered the tactility of these self-consciously material photographs and the role of bitumen as light-sensitive material and energy source undergirding the history of photography, we now turn to its representation *in* photography—remembering, of course, that oil is an essential foundation of all technological and artistic mediums in the contemporary world.

Scale defines the Athabasca tar sands. A single mine in the tar sands moves thirty billion tons of sediment annually, double the sediment moved by all the world's rivers combined.[86] While all extraction has environmental consequences, the Athabasca tar sands are particularly destructive. The bitumen is buried deep underground beneath layers of rock, clay, muskeg, and boreal forest. It is often mined in massive open pits, if deposits are near the surface, or in capital-intensive in-situ drilling operations when buried deep in the ground, displacing tons of land and habitat. While oil is often associated with relatively cheap energy, the tar sands are an unconventional, not easily drillable oil deposit: refining this particular bitumen out of the massive material it's dispersed in and into crude oil requires an incredible amount of energy and requires a high market price for oil to maintain profitability. As a result, the tar sands mining procedure releases at least three times the CO_2 emissions of standard oil production—and this is before the product itself is burned.

In light of the scale of production in the tar sands, it is not surprising that the region is often represented through aerial photography. For example, in *Alberta Oil Sands #6, Fort McMurray, Alberta, Canada, 2007*, the Canadian photographer Edward Burtynsky documented the tar sands in large-scale renderings characterized by saturated color, geometric form, and striking patterns (plate 6). The question of scale is central to Burtynsky's images (they are dramatically large) and his subject matter. He explains, "I have always been concerned to show how we affect the Earth in a big way. To this end, I seek out and photograph large-scale systems that leave lasting marks."[87] His tar sands photographs are part of a larger body of work called *Oil* (1999–2010), which explores society's conflicted relationship with the substance. He reflected on the series, "The car that I drove cross-country began to represent not only freedom but also something much more conflicted. I began to think about oil itself: as both the source of energy that makes everything possible and as a source of dread, for its ongoing endangerment of our habitat."[88] *Oil* preceded his more recent *The Anthropocene Project* (2018), a collaboration with filmmakers Jennifer Baichwal and Nicholas de Pencier, which captures the infrastructure of the Anthropocene, linking sites of large-scale human intervention spanning six continents. The project documents sites across the globe that the Anthropocene Working Group (AWG) identifies as having the most important environmental impacts. Burtynsky's large-scale, luxuriously printed photographs capture the abstract geometry and the bright pops of colors of lithium mines in the Salt Flats of the Atacama Desert, snaking highways carved into California's desert landscape, and stone quarries chiseled out of the earth for cement in Dongying. The juxtaposition of extractive sites is provocative, drawing attention to the myriad ways humans have reshaped the planet, often with devastating human and environmental results. Burtynsky's selection of sites to photograph based on the AWG research categories indicates that the photographs are intended to function as evidence of the changes humans have made to the earth's dynamic systems.[89] His approach reflects Aldo Leopold's observation that "we can be ethical only toward what we can see" and reinforces the common narrative that visibility is an essential step in climate action.[90]

The stunning *Oil Bunkering #1, Niger Delta, Nigeria* (2016) reflects the formal beauty that Burtynsky finds in sites of raw extraction, as the iridescent glaze of oil shimmers on the surface of the Niger River. Toxicity rendered sublime. His trademark aerial view shows the scale of development, but the aerial view also seduces through an aesthetics of abundance and the abstraction of place into formalist geometries. The tension between seduction and revulsion is unresolved in Burtynsky's work, and the locus of responsibility is unclear. A promotional poster released by the Art Gallery of Ontario for his *Anthropocene* exhibit in 2018 made this explicit: "Come See the Art You Helped Create" was the slogan over the image of oil bunkering on the Niger River. The advertising slogan implicated viewers in climate change, addressing a universal "you"—an imagined Western consumer experientially disconnected from sites of extraction. Burtynsky's ambiguity on the question of responsibility points to a larger societal issue, as his works show both the destruction caused by the oil industry and a world that cannot imagine a future without the abundance that oil makes possible.[91] His photographs may be an indictment of industrial capitalism—tinged with fascination and wonder—but they do not move beyond ambiguous critique. Here, we encounter the current state of much Anthropocene discourse. In many ways, Burtynsky is the emblematic photographer of the Anthropocene. As the Anthropocene as a framework emphasizes scale over specificity, critiques (with awe) the oversized impact of humans, and is ambiguous on the question of responsibility, all tensions reflected in Burtynsky's ambitious body of work.[92] The ambivalence of Burtynsky's work means that the photographs can illustrate oppositional viewpoints: corporations on the level of form and environmentalists on the level of subject matter.[93]

This is not to dismiss Burtynsky's practice, for his documentation of the massive impacts of development is a significant intervention into contemporary visual culture. Burtynsky's photographs are a diagnosis of key dimensions of the contemporary moment. However, the sheer spectacular power of their depiction can stymie the growth of political attention to the more subtle and less photogenic consequences of climate change, which are equally important. Allan Sekula cautions that the use of photographs to understand history or politics often "suggests that significant events are those which can be pictured, and thus history takes on the character of spectacle."[94] The focus on what we can see—visually spectacular destruction—overlooks some of the most profound changes in ecosystems, revealing something about the scale of transformation but less about its full significance. This is work that photography struggles to do, particularly in the form of the singular, isolated image. The emphasis on enabling us to grasp large-scale visible phenomena in the Anthropocene is important, but something is lost when a focus on the planetary—the view from the atmosphere—obscures historical changes to specific eco-systems. The politics of exploiting and protecting the earth that happen locally are no less essential to see.

This does not suggest that the local should supplant the planetary in ecocritical thinking. As Amitav Ghosh has reflected, one of the challenges in writing about oil lies in the "intrinsically displaced, heterogenous, and international" world of oil.[95] Oil crosses borders, moves through pipelines, and is transformed into energy, commodities, and profit. Scale also directs our attention to how materials move through space, entangling geographies through commodity flows. While Burtynsky's work shows us the current state of the Anthropocene—the symptoms—Sekula's photographic practice shows us how we got here. For example, Sekula's *Fish Story* (1989–95) provides a visual model for following the flow of oil and capital. Like the *Anthropocene Project*, *Fish Story* places several distinct sites into dialogue over seven chapters while incorporating photographs, text panels, and slide projections. *Fish Story* focuses on shipping and commercial ports, drawing provocative links between shipping containers full of goods crossing the Atlantic, people queuing at a Polish unemployment office, and the economic migration of workers from the Global South. Sekula traces these connections between deindustrialization in the Global North and migrant workers from the Global South. Sekula grew up in San Pedro, California, the port of Los Angeles, which shaped his suspicion of narratives of dematerialization: "A certain stubborn and pessimistic insistence on the primacy of material forces is part of a common culture of harbor residents," for "this crude materialism is underwritten by disaster. Ships explode, leak, sink, collide. Accidents happen every day."[96] While Burtynsky focuses solely on sites of material extraction, Sekula considers both the material realities and the processes of abstraction enacted by extractive capitalism. I use *abstraction* here not in an art-historical sense but in the framework of political economy, thinking of abstraction as the material process of commensurability imposed on concrete labor processes and their products by a global market in commodities—innumerably diverse labor, things, and lifeways treated as though all simply represented a quantity of economic value. Sekula's historical-materialist project shows the integral links between extraction and abstraction—which, as Imre Szeman and Jennifer Wenzel remind us, come from the same Latin root, *trahere* (to drag, to draw out).[97] The implications of the links between extraction and abstraction are significant, for, as Rob Nixon observes, capitalism has a propensity "to abstract in order to extract."[98] This form of abstraction can be found in the photograph, which separates processes that are part of interconnected systems while enacting a universalization that obscures historical and cultural specificity. As Kevin Coleman and Daniel James show, "Capitalism and photography operate according to a shared logic of abstraction, alienation, and the conversion of use value into exchange value."[99] Sekula's integration of text and image, and his carefully constructed sequences of images, works against the loss of photographic meaning. What sets Sekula's critical documentary practice apart from many other projects that use text and sequence to nuance photographic meaning is Sekula's insistence on a materialist critique of the systems that structure and naturalize the relations the text and images depict. Sekula nuances the narrative that petrocultures are culturally invisible by emphasizing that we struggle to see networks that link geographies, materials, and labor. Thus, Sekula connects economic and environmental justice, placing responsibility with corporations and governments prioritizing profit over people and ecosystems rather than the Art Gallery of Ontario's imagined "you." *Fish Story* shows the material conditions that underlie the global exchange of commodities, the reshuffling of geographies of production and consumption, and the forms of fixity and flow these logistics "pipelines" produce and destroy.[100]

Sekula, like Burtynsky, has turned his practice toward the symptoms of ecological catastrophe, but in a manner that deeply integrates it within the human societies that produce and suffer from its consequences. The distancing implicit in many visual representations of extraction finds a parallel in how mass culture experiences ecological disasters, which are often mediated by the visual. The experience of watching far-off ecological catastrophe is often marked by horror, fascination, and experiential distance. Peter Galison and Caroline Jones highlight how news photographs of the oil spill from the Deepwater Horizon, an offshore oil rig operated by British Petroleum in the Gulf of Mexico, neutralized its violence. The "surface images (seafloor surface, ocean surface, and shorefront)" limit the scope of vision by excluding the impacts on marine life.[101] The images obscure as much as they reveal. In contrast, visceral materiality characterizes Sekula's series *Black Tide/Marea Negra* (2002–3), which documents the aftermath of a spill off the coast of Spain in 2002 from the tanker *Prestige* (plate 7). The spill was at the time the largest anthropogenic disaster in European waters. *Black Tide/Marea Negra* consists of twenty photographs in ten frames and a libretto for an opera set in 2032, the estimated timeline for the damage from the spill to recede. Sekula thus draws attention to the prolonged temporalities of ecological damage, which, in this case, extend thirty years into the future. The project extends photographic time to consider the temporality and spatiality of environmental disasters in the deep time of social life.[102]

Sekula juxtaposed photographs of the sea and those of volunteers in white overalls performing the Sisyphean task of cleaning up the oil. Unlike Burtynsky's bird's-eye view, Sekula returns to a human scale to situate the viewer at sea level. In doing so, the series does not primarily emphasize the spectacular nature of the spill. Rather, Sekula trains attention on the quixotic futility of trying to clean oil from the ocean, as the viscous, liquid materiality of the oil blends into the waves, as volunteers attempt to remediate the symptoms of a global capitalist system that continues to grind on in the background. The emphasis on the individual literally up to his elbows in oil grounds Sekula's critique of government responses. After a storm punctured a hole in the side of the tanker, the French, Spanish, and Portuguese governments refused to allow the ship to dock for repairs while it was still possible to mitigate some of the damage. Instead, a salvage company towed the tanker offshore, extending the spill's range. The ship ultimately split into two and continued to leak oil after sinking. Twenty million gallons of heavy fuel oil poured into the sea. In a sense, this stands as an allegory for climate crisis, as states and corporations shift responsibility while ignoring possibilities of mitigation and adaptation that exist in the present. In the face of government inaction, thousands of volunteers contributed to cleanup efforts organized by the Spanish company TRAGSA and the environmental justice organization Nunca Máis.

The dark, slick oil that stains the white TYVEK coveralls of the activists reflects a tactile engagement with the problem of fossil fuels—a striking contrast to both the distanced contemplation of Burtynsky's aerial view and the way governments washed their hands of the problem. In the visual emphasis on mess, Sekula proposes the desire to remain aloof is a fallacy: there is no outside of the damage caused by oil extraction and consumption. The volunteers model a practice of what Donna Haraway calls "staying with the trouble": acting ethically requires accepting our implication in damaging systems to work from a place of response-ability and relationality to the natural world and other humans. For Haraway, the importance of moving beyond either despair or hope is a process of "learning to be truly present, not as a vanishing pivot between awful or Edenic pasts and apocalyptic or salvific futures, but as mortal creatures entwined in myriad unfinished configurations of places, times, matters, meanings."[103] In *Black Tide/Marea Negra*, we are given a tangible example of what that means as the volunteers slowly work against the effects of 81,000 tons of oil—and the economic system that made the disaster happen. The final piece in Sekula's accompanying libretto, *The Song of Society against the State*, reminds viewers of the stakes: the forces of global capital swamp the individual and the collective—made material in the sticky, toxic oil— but people do the work anyway. The laborious process of scraping oil off the rock is a tangible reminder that facing ecological devastation is not and will not be quick or simple—addressing environmental crisis requires a fundamental change to global economic systems. As Gabriele Mackert reflects, *Black Tide/Marea Negra* is a "speculation on the conditions of living and working in uncertainty": the task is Herculean, and the outcome is in doubt.[104] At the same time, the scene is somewhat peaceful: the worker seems contemplative, framed by the gentle undulations of the sea. This meditative framing marks *Black Tide/Marea Negra*, pushing back against the tendency to spectacularize ecological catastrophe. If the bird's-eye view of Burtynsky's formalist experimentation abstracts, Sekula's work confronts the viewer with the materiality of oil and the material conditions of a society that prioritizes economic growth over sustainable and just futures. What those futures may look like, and where the thinking enabled by critical photographic practices could point us, serves as this chapter's conclusion.

There-Then, Here-Now, and Yet-to-Come

Cariou's petrographs reroute our vision from the seemingly immaterial atmosphere of light to the very material bitumen of the tar sands, from the flat temporality of the index to the ancient histories of deep time. The use of bitumen moves viewers beyond the surface, which typically defines the photograph, to encounter subterranean depths in space and time. Due to their extended exposure times as images, and the ancient timescales of the hydrocarbon that they depict and that makes them possible, the complex relationship to time inherent in bitumen photographs further nuance this perceptual transformation. What implications might this emphasis on materiality and time have for thinking through climate breakdown? If fossils direct our attention to the past, burning fossil fuels directs our attention to climate breakdown in the present and future. Similarly, photographic time is not simply linked to memory and history; it contains multiple temporalities that can look to the future. Photographic time is also *yet-to-come*, which proposes a futurity enacted in the photograph, though what that future looks like is not fixed.

Climate science has conclusively shown that burning fossil fuels poses an existential threat to life on the planet. Despite this, an extractive future premised on burning fossil fuels remains our most likely outcome at the moment of writing. Governments and corporations are actively building an extractive future in the tar sands and beyond. Provincial and federal governments subsidize the energy sector in Canada an average of $14.3 billion per year, either through direct investment or uncollected taxes.[105] In 2018, the Canadian government under Prime Minister Justin Trudeau spent $3.4 billion to buy the Trans Mountain oil pipeline, which spans a 715-mile route from Alberta to the British Columbia coast. This pipeline has been subject to resistance from Indigenous land and water protectors, opposition that has been met with state violence, cementing this carbon-intensive path in economic development plans and government revenue projections. This vision of extractive futurity is also enacted culturally. This process relies in part on the continued experience of what Nicholas Mirzoeff calls "Anthropocene visuality," which allows us to ignore the reality that the tangible manifestations of climate crisis—extreme weather events like fires and floods, climate migration, higher costs for basic goods, and mass extinction—are both here-now and yet-to-come. According to LeMenager: "Environmental damage yet to come, without (current) aesthetic dimensions, does not stir up alarm or activate an ethic of care."[106] Creating another yet-to-come is urgent: a postextractive imaginary. Another yet-to-come, I argue, is what we find in another petrograph by Cariou, titled *Prayer Tree* (2017) (plate 8).

A tree trunk is wrapped in cloth, a practice done in Cree and Anishinaabek territories to honor the trees and mark the site as a space of ceremony or contemplation. The fabric is an offering to the tree and the earth. The prayer tree stands in contrast to the extractive practices of marking and wrapping trees in forest management practices, in which loggers use paint or colored ribbon to mark trees for removal. Logging is central to oil extraction, as corporations seek deposits below the Boreal Forest. Industrial production dams rivers, clear-cuts forests, and removes everything its extractive gaze reduces to the status of overburden. The tar sands are being deforested at a rate second only to the Amazon rainforest basin.[107] Rather than showing us the destruction caused by extraction in the tar sands, Cariou diverts a small amount of bitumen out of the energy industry and into a medium of representation to show us a parallel history of the tar sands. Marking the tree for prayer is, in a sense, an act of diversion—taking the tree itself out of the capitalist system and assigning it an entirely different form of value, as the petrograph does for bitumen.

The intimate scale of Cariou's petrographs draws the viewer into a relational encounter with the image, a stark contrast to the aerial view of Burtynsky's *Alberta Oil Sands #6*, which situates the viewer well outside of the scene while suggesting a sense of immediacy, a disconnection heightened by the overwhelmingly large size of the print and its display on the gallery wall. Some of Cariou's petrographs use the aerial view, but the effect is fundamentally different. Due to the small size and the reflective surface, there is an embodied experience of looking at the object that changes through the act of looking. Scale and reflectivity draw the viewer into the scene, an invitation heightened by the glimmer of the image. The sheen calls to mind Roland Barthes's reflection on "the shimmer": the "integrally and almost exhaustively nuanced space . . . whose meaning, is subtly modified according to the angle of the subject's gaze."[108] The ability of the viewer to modify—however subtly—the image introduces an agency and, perhaps, response-ability in viewing the image. Following Barthes, the possibility presented by shimmer is not didactic or directed but uncertain and invitational. The subtle glimmer that Cariou utilizes directs the viewer to another yet-to-come by immersing the viewer in this landscape scene, as the wrapped tree is a reminder that the tar sands are a sacred place, a boreal landscape of waters, more-than-human species, and the traditional territories of Métis, Dene, and Cree people. Cariou foregrounds the relations of settler colonialism and extractive capitalism that have worked to deterritorialize land and transform it into a commodity or resource. At the same time, *Prayer Tree* refutes the territory's abstraction to an industrial area, mine, or extraction zone. This connects bitumen as a material in tangible, nonabstracted ways to land, people, and nonhumans.

The extraction and burning of fossil fuels have turned the tar sands landscape into an environmental sacrifice zone, exposing life in this region to toxicity and contaminants. Red River Métis anthropologist Zoe Todd describes how industrial processes transform oil into a pollutant: the extraction and burning of oil "weaponises" bitumen.[109] Todd asks what would shift if we thought of fossils as our kin. Fossil fuels are the remains of ancient life, and coal combustion releases this ancient energy. To excavate and burn subterranean deposits of coal—to "conjure" coal—is to disturb the dead. By reframing fossils as kin, Todd asks what responsibilities we might have to materials like bitumen and other human and nonhuman species that make up the Athabasca region. Todd's framing also reveals that the language we use to describe oil—fossil *fuels*—suggests that oil's natural role is to serve as an energy source, locating an extractive futurity in the word itself.[110] Todd invites us to imagine how we could know and relate to bitumen differently. This is a complicated and unresolved question in settler societies, which have largely severed these kinship ties between the human and nonhuman worlds. Still, Todd highlights that oil itself is not the problem. It is the large-scale extraction of petroleum for profit that is damaging. This is a critical distinction: it's not the material; it's the system.

Or, following Michif-settler scientist Max Liboiron, it is the *relations*. Liboiron argues that pollution and toxicity are fundamentally problems of colonial land relations: environmental damage is a symptom of colonial violence that prioritizes economic growth over bioflourishing worlds. In contrast to this extractive understanding of land, anti-colonial methods place land relations that are "specific, place-based, and attend to obligations" at the core of all engagement.[111] Cariou's photographic practice aligns with what Liboiron calls "different Land relations" that oppose the colonial logics that causes pollution. The doubled indexicality of Cariou's image—gathering the oil from this territory to document this territory— roots these images in place. Cariou's emplaced photographs reflect his connection to this territory. Cariou is from Meadow Lake, Saskatchewan, located near the infrastructure of the tar sands. The bitumen gathered from the ground becomes the ground of image-making. The site-specificity of Cariou's practice introduces a commitment to the nonabstraction of land, employed to fight against the tendency to view land as the backdrop of human activity.

Contemporary scholarship on photography has explored its relational qualities, and Cariou frames his approach to photography as "gathering" photos instead of "taking" them.[112] In this, Cariou seeks to break from the often-extractive nature of photographic practice. In the language of photography—recall aim, shoot, trigger, capture, take—we see a convergence between the ideology of photography as a medium and the drive to accumulate that underpins extractive capitalism. By shifting the relations between the photographer, the subject, and the materials used in the process, Cariou moves toward "different Land relations" within photographic practice. Gathering is a way of being in relation to what is gathered, initiating a relationship that consciously tries to move away from extractive relations.[113] This reframing of photography resonates with Ursula Le Guin's carrier bag theory of fiction, which argues that the gathering bag predated the spear in human society, placing gathering, holding, and sharing at the center of human relationships to each other and land.[114] In contrast to extraction, which sees nature as resources to be accumulated for producing wealth, Cariou's practice approaches nature as a gift, not a commodity.[115] Robin Wall Kimmerer (Potawatomi) has shown how gifts "establish a particular relationship, an obligation of sorts to give, to receive, and to reciprocate."[116] The offering also is part of Cariou's practice:

> I also think of the bitumen as a kind of medicine, one that can cause damage if used improperly, but one that can also provide valuable gifts if it is approached with the proper respect. This is why I leave an offering of tobacco or sweetgrass wherever I gather the bitumen I use in my petrography, in keeping with Métis cultural teachings about reciprocity, gratitude, and maintaining good relations with the land.[117]

Through the focus on bitumen as a material and the landscapes from which it is extracted, Cariou introduces an alternative way of understanding the potentials in the materials we have come to call fossil fuels.

In his petrography, Cariou engages materially with a substance that has come to define the region, while at the same time proposing a different relationship to the material. Cariou reflects on the intimate nature of making petrographs: "As I spent more time with the bitumen, I came to understand it not only as a source of potential danger but also as a creative collaborator, helping to reveal new ways of seeing the world."[118] This new way of seeing is a postextractive imaginary. Here, I argue, we have one final reorientation of vision: from an extractive gaze to a relational encounter. In Cariou's petrographs, the relations that bind land, material, and bodies come into view. Before its use as an energy source, bitumen was used in medicine and art. It healed and created meaning. As one views the image, the land is inseparable from the material and, by extension, the body superimposed in the picture. This encounter offers a brief glimpse into a future where our relations to land have shifted. Through the reflective surface, the photograph invites the viewer to imagine themselves in this landscape, to enact a different relationship to place. The uncertainty of the unfixed surface—which shifts as the light changes and is altered every time someone enters into relation with it—marks its continuous motion, its ever-unfolding future.

Cariou establishes a new relationship with bitumen, one that considers simultaneously the damaging and desirable nature of the oil industry and the economy that it underwrites. Once it is set in the photograph, the bitumen becomes a vehicle for environmental justice activism and Indigenous sovereignty. Cariou subverts the mutual implication of romantic and extractive gazes through a relational engagement with the material itself, to center questions of land, ethics, and relations. This is a profound gesture in our current moment, as bitumen extraction in the tar sands reshapes climate globally. Through the complex entanglement of material, process, and meaning, Cariou responds to the conceptual and material challenges of the Anthropocene by proposing new ways of relating, making, and seeing. Cariou's practice shows us that bitumen is a material that exists outside of extraction and that the tar sands landscape and other extractive zones have value outside of the wealth that is physically taken out of it. In doing so, it counters the extractive gaze by showing us other forms of relating to this land that existed before, during, and after extraction in the tar sands. Cariou's petrographs introduce a more complex understanding of environmental damage and the ongoing possibilities of life in regions slated for environmental sacrifice. The following chapters turn to different metals and minerals, but the modes of seeing proposed by Cariou will remain with us as we move through other materials, their processes and histories, encouraging us to see the relational and antiextractive potential latent in much of their environmentally and socially destructive uses.

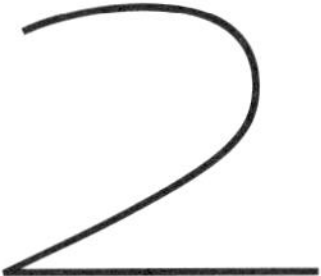

Silver and Scale

"Silver," *Kodak Photo Magazine* (1947)

SINCE THE POPULARIZATION OF PHOTOGRAPHY, the primary light-sensitive material in analog photographs has been silver. The light sensitivity of silver halides allows for short exposure times, providing for generations the only viable way of capturing a negative instantaneously in the camera. Over the twentieth century, photography used an estimated 25 percent of the silver produced worldwide.[1] Much of this silver was consumed by Eastman Kodak Company, the company that spearheaded photography's transition to a standardized, mass-produced (and mass-consumed) industry. Early photography was labor-intensive and required knowledge of chemistry, limiting the medium's widespread take-up. A series of innovations by George Eastman's Eastman Kodak Company transformed photography from a medium practiced by skilled professionals and autodidacts experimenting with photographic chemistry into a mass cultural practice. In creating a new mass market for amateur photography, Kodak likewise stimulated the demand for more mined materials. Silver was critical to their operations. While Kodak vertically integrated most aspects of its production process in-house to ensure a steady supply of high-quality materials, procuring silver posed a distinct challenge. Silver occurs rarely in pure form, with most of our supply mined as a byproduct of other metals, including lead, copper, and zinc, a fact that introduces significant unpredictability to supply chains seeking silver alone. By the late nineteenth century, most of the world's major silver deposits had been discovered and mined. Ensuring its steady availability and managing its cost were ongoing concerns to the company, for, as they explained: "Silver goes into all film, all sensitized paper; without it, there would be no photography."[2]

Given the metal's outsized role in Kodak's operations, the silver vault at the Kodak Park plant in Rochester, New York, was often reproduced in their advertisements. A 1945 advertisement in *Life* celebrated the "treasure behind your snapshot," disclosing that Kodak's "normal reserve, to insure continuous manufacture of film and paper, is about $3,500,000 worth. Two tons are one day's supply" (plate 9).[3] Superimposed over the photograph is a magnified insert, revealing the filamentary threads of pure, metallic silver entangled on the paper. Here, the "treasure behind the snapshot" is directly depicted: we see the photograph *and* the silver that forms the print, the image and its material basis. Elsewhere, Kodak reflected, "To dig metal out of the dark earth, flash light on it for a tiny fraction of a second, and obtain a picture marvellously rich in detail and texture—there's a touch of the miraculous in it, isn't there?"[4] The emphasis on the miraculous nature of photography directs attention to photography's (al)chemical transformations, which transform pure silver bullion into the silver halides that form the blackened traces of silver in the photograph.

The history of photography is, in one important sense, a history of the mine. Silver has to be dug out of the "dark earth" before it can be used in photography. In the Kodak ad, however, we see the raw material, not the labor that extracts or transforms it. While the quote names a chain of photographic production—the mine, the darkroom, the print—the focus is on the "magic metal" rather than the labor required to process and transform the raw material. Indeed, Kodak required incredible amounts of materials and labor to meet the burgeoning demand for home photography. The massive chains of work that make photography possible were rarely made visible, subsumed under the heading of "the rest" in Kodak's famous slogan: "You press the button—we do the rest." To scale up the production of photography, Kodak swept all the work that made it possible out of the viewfinder.

Silver brings into view the impact of photography as a mass industry, introducing the themes of scale, abstraction, and labor. Chapter 1 focused on a method of making photographs that relied on bitumen, an originary photographic process that lacked market scalability. But the mineral history of photography writ large is soldered to silver. As a result, this chapter is slightly different than the others: while the other materials under consideration have sharply specific histories vis-à-vis photography, silver's significance is its ubiquity. Photographic silver is thus a complex story to tell; in some ways, the difficulty *is* the subject. Scaling up production changed photography—not just in quantity but in what it made visible and invisible, possible and impossible. This chapter embraces the complications of silver's presence throughout photographic history by thinking through another significant challenge in representation: the problem of labor. Paradoxically, the work of photography becomes obscured at the moment where it is the most present.

Underpinning my exploration of labor is an emphasis on scale and abstraction, two themes introduced by silver. Scale is central to capitalism's functioning, which maximizes profits by scaling up production. Somewhat counterintuitively, scaling up often makes things less visible, as a large enough scale can make it challenging to grasp the whole.[5] Silver allows us to track the dynamics of scale in terms of expanded production and in terms of visual proportion. Visually, scale often abstracts, but it also connects: tracing silver as it moves through economies of scale allows us to see complex production chains that link the mine to the image, reminding us of the centrality of labor to histories of photography. The shift between scales also reveals tension between the material realities of extraction, refining, and making commodities with the more abstract workings of capital. Abstraction is the second animating theme of this chapter and another context where silver's history complicates a straightforward material history. For most of silver's history, the metal circulated as currency. Its dual status as raw material and marker of value makes it distinct from the other materials charted in this book, as silver was a mediating form in capitalism's system of abstract value, rendering work processes and commodities commensurable on the market. As the economist Herbert M. Bratter states, "What is meant by 'silver'? A commodity? Money?"[6] By the time photographic industries required large amounts of silver, it had largely fallen out of use as currency. But as a material, it retained dual signification. It is both tangible and abstract, glimmering and self-effacing. In silver, we locate an uneasy dance between the concrete and the abstract. Accordingly, the chapter scales up and down, toggling between silver as a global commodity, a marker of value, an object of speculation, and a specific material with particular qualities that enable its use in photographic industries.

This chapter traces a long temporal arc and moves throughout the Americas to think through labor as it relates to scale and abstraction. I begin by introducing silver as a material, engaging Karl Marx's writing on metals and value. I then turn to the early modern world in Potosí (now Bolivia) to trace the historical roots of the global silver economy, arguing that the development of global pathways of colonization, forced labor, and intensive mining around the pivot of Potosí surfaces like the coloration of silver halides materially underlying contemporary photography. From there, I turn to the second-most significant silver find in history: the nineteenth-century silver deposits in the western United States, with a focus on Timothy O'Sullivan's photographs of Comstock Lode, Nevada, taken for the US Geological Surveys—the material used to take the pictures needed to strategize where more workers could be sent to produce more of that same material. O'Sullivan's images are an entry point to trace silver's complex public and private supply chains. This period saw an increase in silver mining and the demonetization of silver as currency, flooding the market with the cheap silver that made photography as a mass practice possible, setting the stage for the rise of Kodak and the modern consumer economy of home imaging. Throughout, I endeavor to bring work into view, often with a degree of struggle. I conclude with a focused meditation on how labor does and does not come into view photographically. A focus on silver brings into sharp relief the scale of extraction required for photography as a mass medium and the knotty interrelations between labor and capital. Following silver from extraction to the representation of extraction directs our attention to the material labor upon which capitalist production, even so-called cognitive and creative work, rests.

Silver

Silver (Ag, atomic number 47) is a precious metal and was one of the seven known metals in antiquity. In the hierarchy of metals in ancient Greece, silver, associated with the moon and the goddess Artemis, ranked second only to gold. Metals designated precious are ascribed inherent value based on their beauty, relative scarcity, and the significant human labor required for their extraction.[7] Historically, the primary use of silver was in coinage and decoration. For most of economic history, silver has been traded as currency and was central to the development of global capitalism. In *Contribution to the Critique of Political Economy*, Marx deconstructs why silver and gold came to stand in for value. In part, it was due to their scarcity and their glittering, "sensuous splendor" or aesthetic appeal.[8] Critically, however, the relative uselessness of silver and gold as instruments of production made them potentially valuable as a medium for currency. Gold and silver are malleable and soft, which makes them ill-suited to be melted down into tools or used in building projects but well-suited to consolidation in differing quantities. Thus, vulnerable to neither rust nor moths, silver and gold are "the solid, palpable and glittering form of exchange-value, to exchange-value in the form of the universal commodity as distinct from all particular commodities."[9] This does not at all suggest that the value of silver is necessarily stable. Only 25 percent of silver extracted is from silver ore, leaving the supply highly erratic depending on its output as a byproduct from smelting other metals.[10] As a result, mining silver—unlike gold, which is usually found in its natural form—requires "the aid of capital, association, and machinery," further complicated by the difficulty in refining the metal itself.[11] Silver is thus "capricious" in its value.[12]

As silver's fluctuating value highlights, it is fluid and fungible in multiple senses. As a raw material, silver lends itself well to abstraction: it is relatively light and can be easily melted down, making silver well-suited to stand in for standardized, exchangeable value in differing amounts. Currency is the "fluid form of money," and this fluidity—the ease of circulation—is key to linking money metals to value.[13] Marx's theory of money shows how use value (the physical, tangible nature of the commodity that allows it to meet needs) and exchange value (its social function as a representative form of an amount of socially necessary labor time) are entangled. As Marx writes, ease of circulation is critical to the value of these precious metals:

> As means of circulation gold and silver have an advantage over other commodities in that their high specific gravity—representing considerable weight in a relatively small space—is matched by their economic specific gravity, in containing much labour-time, i.e., considerable exchange-value, in a relatively small volume. This facilitates transport, transfer from one hand to another, from one country to another, enabling gold and silver suddenly to appear and just as suddenly to disappear—in short these qualities impart physical mobility, the *sine qua non* of the commodity that is to serve as the *perpetuum mobile* of the process of circulation.[14]

The mobility of silver, a commodity with a high ratio of labor intensiveness to mass and volume, facilitated the easy and standardized circulation that is key to the movement of capital. Upon contemplation, Marx's summary of silver's material suitability as currency also evokes photography.[15] By the nineteenth century, silver circulated as currency *and* in photographs.[16] Physical mobility is the sine qua non of photographs themselves, which, as Walter Benjamin would later reflect, function as a perpetuum mobile, objects in ceaseless circulation.

Before the invention of photography, however, the transnational exchange of silver from the sixteenth century onward facilitated the growth of a global economic system. In *The Wealth of Nations*, Adam Smith singled out the remarkable global reach of the silver trade as one of the principal exchangeable commodities linking the Americas, Europe, and Asia.[17] The story of silver as it related to the transnational flow of materials, labor, and currency takes us to Potosí, Bolivia. In the sixteenth century, Potosí was part of the viceroyalty of Peru, an imperial province within the Spanish empire. In 1545, silver was discovered on the traditional territory of the Charcas and Chullpas people. It was the richest silver find in history. Extraction undertaken by the Spanish empire initiated an extraordinary movement of capital, knowledge, technology, and forced labor, forming a commodity frontier that would reshape systems of value on a global scale. The extracted silver was melted down into coins that could be easily transported and standardized, which rapidly increased the amount of money in circulation.

Between 1500 and 1800, 85 percent of global silver was extracted in Mexico, Bolivia, and Peru. The silver extracted from veins in the Americas was largely taken to the Pacific Coast, where it was transported to the port of Acapulco, then to Manila, the trading post between the Spanish and Chinese empires. China purchased about 40 percent of that silver. The drive to increase extraction stimulated the triangular trade between Europe, Africa, and the Americas and likewise fueled the transatlantic slave trade as enslaved Africans were brought by the Spanish to mine silver. The colonial administration introduced a form of forced labor, called the mit'a, for people indigenous to the lands now claimed by Peru and Bolivia, which came to form the backbone of the Spanish imperial silver-extraction complex.[18]

This model of mining silver through forced labor had ancient roots. In ancient Greece, enslaved people mined silver bonded to highly toxic lead.[19] Spain itself was the primary site of extraction in the ancient world under the Roman Empire. The political economist Alexander del Mar writes that "Spain was to the ancients what Mexico and Central and South America became in later ages to the Spaniards; the Dorado . . . the place where, after the country was plundered, the metals gold and silver were found in the greatest abundance, and where they could be procured by the forced labor of captives and slaves," resulting in the diversion of waters, deforestation, soil erosion, and widespread poverty along the way.[20]

The *scale* of extraction at Potosí, however, was unprecedented and it caused global environmental and economic shifts. By supplying silver, which circulated as currency, Potosí became a source of the means of capital accumulation, used to buy and sell other profit-producing commodities, a process that fueled the early modern European economy. Emperor Charles V, Holy Roman emperor, called Potosí the "treasury of the world."[21] In *Don Quixote*, Miguel de Cervantes coined the expression "Vale un Potosí": to be worth a Potosí, to be priceless. Between 1492 and 1600, the supply of precious metals in Europe increased eightfold due to mining in the Americas. This influx of wealth, channeled through agents of expanding global trade, shifted power from the aristocracy to the merchant and banking classes, which had seismic economic impact. Mining in the Americas also had significant environmental impact. The Columbian Exchange—the movement of plants, animals, people, and viruses between Europe, Africa, and the Americas—alongside the extraction of raw natural resources and the decimation of Indigenous populations due to disease, intensive exploitation, and violence produced a change large enough to register in the geologic layer. In light of this, the geographers Simon Lewis and Mark Maslin propose that the Anthropocene started in the early seventeenth century with the Columbian Exchange, suggesting that climate breakdown is integrally connected to patterns of accumulation through dispossession and extraction developed in the process we are examining.[22] This argument locates the ecological disruptions of the Anthropocene in a set of extractive relationships imposed by colonialism and extractive capitalism that terraformed ecosystems, displaced populations, and severed relationships to lands, waters, plants, and animals.[23]

The circulation of currency, labor, and capital investment was mirrored in the circulation of images of extraction in Potosí. By 1553, a woodcut of *Cerro de Potosí* by Pedro de Cieza de León became widely known, and Potosí entered the global image world (figure 2.1).[24] Reproductions of the "rich mountain" circulated in maps, paintings, and engravings, particularly in European histories and travel narratives from the mid-sixteenth century onward. The reproduction of images stimulated interest and investment in Potosí, another example in the long history of using the visual to promote extraction. One such image is an eighteenth-century painting by Gaspar Miguel de Berrio (plate 10). With its bird's-eye view, the painting captures the dramatic scale of mining in Potosí. The mountain towers over the city, as the Andes rise in the background. The city became one of the largest in the Americas following the discovery of silver: by the early seventeenth century, 160,000 people had accumulated there. The rapid growth of population and industry transformed the ecosystem, resulting in deforestation, while mercury, lead, zinc, and other toxic byproducts entered the air, water, soil, and bodies. As artist Harun Farocki argues, de Berrio's image speaks through its silences: what it hides tells us something about both colonization and capital circulation.[25] The visual emphasis on scale enacted through the aerial view of de Berrio's painting abstracts the infrastructure of production and tells us very little about the socioecological consequences of extraction at Potosí. While these histories would be hard to locate in any form of visual production, the distanced framing of the aerial view—far from naturalistic in its pictographic inventory of parts of the social structure—shows nothing of the labor and violence that underpinned production. At the same time, one does not have to look far to find ambivalence underlying the celebration of extraction. Potosí, for instance, is featured prominently in Peter Paul Rubens's and Gaspar Gevaerts's 1634 ceremonial arch *The Arch of the Mint*, designed for the arrival of Philip IV at Antwerp. The frontispiece celebrates the wealth created for the Spanish empire through extraction, but the back ties the wealth of the mint to Cerro Rico. Steven Mullaney deconstructs the ambivalent messages embedded in the arch, which, while unlikely a deliberate critique, offer a "shadowy glimpse" into the "violence, theft, labor, and death" that lay at the heart of mining.[26] The contradictions at the heart of extraction are impossible to smooth out.

When we read the history of photography through the lens of silver, Potosí is a precondition for the emergence of photography. Photography theorist Ariella Azoulay argues that photography does not find its origins in the 1830s and industrialization but rather in the fifteenth century, with the colonization of the Americas. She explains that the arts are underwritten by "invasion and plunder" and, in turn, come to affirm this violence as progress.[27] For photography to emerge in the 1830s "the centrality of imperial rights on which photography was predicated had to already be accepted."[28] Azoulay draws attention to the reality that photography is enmeshed within and often reinforces structures of power that are premised upon the dispossession of peoples and the extraction of resources and labor: "Photographs cannot be understood outside the context of their production."[29] At the same time, Azoulay's reorientation to the relations that enable the broader event of photography brings into view how the photograph is an encounter that can enable "new possibilities of political action and forming new conditions for its visibility."[30] Azoulay's

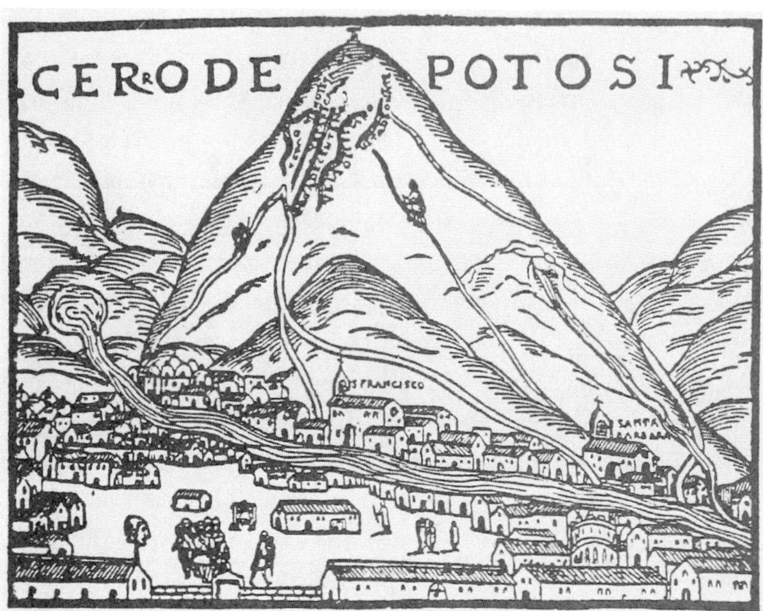

proposed start date of 1492 directs our attention to the colonialization of the Americas, which saw the expropriation of land for extraction and the genocide of Indigenous peoples. Marx situates this history ironically as "the rosy dawn of the era of capitalist production," as capitalism was made possible by primitive accumulation rooted in the expropriation of land and Indigenous and African labor.[31] Once invented, then, photography was applied to produce the visual archives of colonization and used as a facilitating technology of further conquest.[32] Mining, of course, plays a central role in this history.[33]

While Azoulay locates colonization as a prerequisite for photography, Marx grounds photography within the speed and growth of the nineteenth century, which saw the annihilation of space by time, a process made particularly tangible by the movement of photographs. The expropriation of silver in Potosí shifted the centers of silver extraction to the Americas. When photography was announced to the world in 1839, silver extraction was powerfully ensconced in the landscapes of the Americas.

2.1   Pedro de Cieza de León, *Cerro de Potosí*, 1553. From de León's work *Crónicas del Perú*. Woodcut.

By the time photography became a mass medium toward the end of the nineteenth century, the United States and Mexico were responsible for over 50 percent of global silver production. The photographic images documenting the new ways of life being made on colonially occupied and metropolitan land from that period were set in silver extracted in settler-colonial states across the Americas. The hemispheric circulation of knowledge, labor, and capital shaped the mining industry while images symbolically affirmed the value of extraction and the capitalist production of wealth.

Photography emerged from alchemists' experiments with light, heat, and photochemical reactions, many of which centered on the product squeezed out of the earth in these colonial silver mines. Of particular importance was the identification of the chemical substances silver nitrate, silver chloride, and ferrous salts, which would form the basis of experiments that led to photography as we know it today.[34] In 1717, the German polymath Johann Heinrich Schulze discovered that silver nitrate darkened when exposed to sunlight. Schulze's experiments produced impermanent impressions, but he did not succeed in fixing an image. In 1757, the Italian physicist Giovanni Battista Beccaria found that light turned silver chloride violet, and Swedish chemist Carl Wilhelm Scheele published similar results on the sensitivity of silver chloride in 1777.[35] Thomas Wedgwood and Humphry Davy continued to experiment with silver nitrate, succeeding in printing impermanent contact prints in 1802, but they, too, were unable to fix the image for any length of time. It would be another quarter century until experiments by Niépce, Daguerre, and Talbot successfully fixed a photograph.

For use in analog photographs, raw silver ore is refined to create bullion. Silver bullion is dissolved in nitric acid, producing pure silver nitrate crystals, the starting material in the film-developing process. Silver nitrates are transformed into halides: chemical compounds that form between silver and halogens. Halogen means salt-producing materials, and halogens react with metals to produce light-sensitive salts, or halides. Once exposed to light, the halides darken, thus drawing the light not blocked by the photograph's object in an image in silver. Other metals and minerals are used in photographic processes, but silver halides are the most common in film photography.

According to art historian Robin Kelsey: "The emergence of silvery images in the darkrooms of photography has therefore always had a grim counterpart in the extraction of silver underground."[36] If, as Marx suggests, silver "appear[s], in a way, as solidified light raised from a subterranean world, since all the rays of light in their original composition are reflected by silver"—a description that vividly evokes photography—what might silver reveal about the relationship between photography and extraction?[37] If photography promises light and clarity, the mine—which functions symbolically in the cultural imagination as a site of darkness, peril, and riches to be unearthed—can be read as the allegorical inverse of photography. As Agricola observed, however, the mine is also the very possibility of art itself, linking extraction and photography on the material plane. While Agricola's *De Re Metallica* predates photography, one can imagine that this new pictorial art would have provided another powerful argument in his artistic defense of mining.

Labor and Speculation in the Comstock Lode

Because of how rarely silver appears in large amounts in its natural form, there have been very few significant silver finds in history.[38] Following the discovery of the sixteenth-century mines in Peru and Mexico, the next major find was Comstock Lode, in what is now western Nevada, in 1859. Nineteenth-century silver mines reflect both the scaling up of the settler state's territorial reach and extractive intensiveness, and of speculation, the inflationary scaling up of value in anticipation of future growth. In the nineteenth-century United States, as in the British Empire, the geological survey marked the critical convergence between photography, geological knowledge, and extractive activity. Photographers participated in geological survey projects funded by the US government, building photographic archives documenting the geological features, natural resources, and communities of the expanding settler state's western territories. Visual records of survey expeditions were widely distributed in reports, journals, albums, and prints. The prominent role of photographers within the survey process shows the importance of the visual documentation and representation in the execution of these projects. Photography was not necessarily the most crucial tool for scientific purposes compared to other methods of measurement, but photographs were particularly effective in promoting the work of the survey and legitimizing broader narratives of national expansion.[39]

In 1859, prospectors discovered silver on the Comstock Lode, an ore body under the eastern slope of Mount Davidson on the Virginia Range, in what would become Virginia City, Nevada. Miners en route to California's gold fields discovered small amounts of gold trapped in crumbly blue dirt. The blue rock turned out to be an almost solid mass of silver sulfide. The Comstock Lode was the first major silver discovery in the United States and one of the largest silver discoveries in the history of the world. The ore body formed in the Tertiary Period from volcanic rock over millennia. Fault fissures in the volcanic vents filled with heated water risen from deep levels in the earth's crust. Once cooled, the water deposited precious metals and quartz into the rock. Precious metals were deposited into the rock alongside quartz as the water cooled. Silver and gold typically run through the earth in narrow cracks which are, at most, a few feet wide. One hundred eighty feet below the surface at Comstock Lode, the silver ore vein was over forty-five feet wide. The bonanza captivated the attention of speculators and prospectors, turning Virginia City into a major center of exploration. By 1863, more than $12.4 million of silver had been extracted. San Francisco's first stock exchange was founded in 1862 to trade Comstock Lode shares, and the market value of Comstock Lode stocks was $40 million at its peak. For comparison, the assessed value of all property in San Francisco at the time was less than $50 million.[40] Quickly, the Comstock Lode attracted investors, industrialists, geologists—and image makers.

Survey photographer Timothy O'Sullivan's enigmatic photographs of the Comstock region, taken for Clarence King's US Geological Exploration of the Fortieth Parallel (1867–72) and George M. Wheeler's US Geographic Surveys West of the One Hundredth Meridian (1872–79), captured the development of mining in the West during this period.[41] O'Sullivan documented geology, landscapes, the work of the survey, Indigenous communities, and mining operations. The survey traveled by pack mule, and a drawing of the photographer's mobile studio, reproduced in *Harper's* in 1869, shows the cumbersome nature of survey photography in the wet-plate collodion era, requiring a portable darkroom tent, several bottles of chemicals, and the camera itself.[42]

The first published volume of the King expedition's official report to the government was devoted to the subject of the mining industry. A significant amount of the report details the mineral body and the mining operations at Comstock Lode. Maps and diagrams illustrate the volume, but only one of O'Sullivan's photographs of Comstock Lode appears: opening the report is *Savage Mine, Curtis Shaft (Nevada)* (1867–68) (plate 11). The scene departs from the stark but striking landscapes captured elsewhere by O'Sullivan. The image is stiff and controlled, lacking the effortless grace of his outdoor landscapes. Six men gather at the surface of the shaft. In the center, a man stands behind an ore car. The scene is slightly uncanny as, upon first glance, the miner appears to be inside the car. All figures except the man in the center are slightly blurred, their clothing muddied. Here, the emphasis is on *imagining* the work of the mine, not seeing it. As Allan Sekula observes, the photograph that was circulated showed "a neatly symmetrical view that presented a disciplined work crew about to descend into the depths."[43] The labor processes within the mine are recounted in considerable detail, but the labor, the actual work of human hands, is not made visible within the report.

Why might this be? Once within the depths of the mine, the orderly and structured surface presented in O'Sullivan's shot of the work crew gave way to darkness and a complex network of narrow tunnels blasted into the rock. Initially, mining the silver and gold at the surface was relatively simple, as the soft ground could be mined with a simple shovel. Once the surface veins were depleted, underground mining began. The deep underground mines were a far cry from placer mining, the surface form of panning in water that defined the California Gold Rush.[44] In underground mining, teams of miners followed rock veins through deep shafts with high temperatures and cramped conditions. The high rates of accidents and death underground belie the orderly image of the surface. O'Sullivan captured the particularities of the underground worksite in a shot of a miner framed by blasted rock in *Miner Working inside the Comstock Mine, Virginia City, Nevada* (1867–68). This image, however, was not included in the report. The miner—without any visible safety gear— chips away at the rock face, which dwarfs him. Another worker's arm reaches into the scene from the far right, seemingly swallowed by the earth he excavates (figure 2.2). O'Sullivan's tight framing of the workspace emphasizes the claustrophobia of the mine, while the structure holding up the shaft feels precarious, out of joint. The tunnel carved out of the earth barely fits the workers who carve away at the rock face, drawing the viewer's eye into the receding darkness of the shaft. The image reveals the intimate labor of hard rock mining, as the miner follows the narrow veins in search of silver. The workers' tools are perched precariously on the left side of the frame: a hammer, shovel, and a lit candle. The candle illuminated the workspace but also posed a fire risk, a danger heightened by blasting with black powder (gunpowder).[45] Two years after O'Sullivan went underground, thirty-five miners died in the Yellow Jacket fire at the nearby Gold Hill mine. Caused by a methane explosion, it was one the most significant mining workplace disasters in the United States up to that point.

The geology of Comstock introduced particular challenges to under-ground mining. Traditional methods for reinforcing the mine shaft against cave-ins were insufficient due to the remarkable size of the ore body and the softness of the rock. Instead, heavy timber was stacked in giant cubes, known as square-set timbering, which is visible in O'Sullivan's photo-graphs. The rock was quite soft: the walls were primarily clay and the quartz was "crushed and water soaked."[46] When the rock face was pierced, the clay often swelled, causing shafts to collapse, while on other occa-sions, steam and scalding water that reached temperatures of 170 degrees Fahrenheit poured into the tunnel from repositories in the ground. The temperature of the shafts was uncommonly hot: it could reach 120 degrees, and the air was filled with sulphurated hydrogen.[47] Poor conditions also characterized above-ground labor as processing the ore was challenging. Ore was shattered by hand with sledgehammers before being pulverized into dust by machines. This dust was mixed with water and mercury to refine the ore, as the mercury attracted the particles of precious metals. The mercury—a highly poisonous metal—evaporated in the heat, leaving pure silver and gold. Here is yet another parallel with film-development processes, as using mercury vapor as a development agent was also com-mon in early photography. The exposed photographic plate was treated with heated mercury, which extracts the unexposed image and reveals the latent image in silver, much as mercury-refined silver in mining processes. At Comstock, fifteen million pounds of mercury were used to extract gold

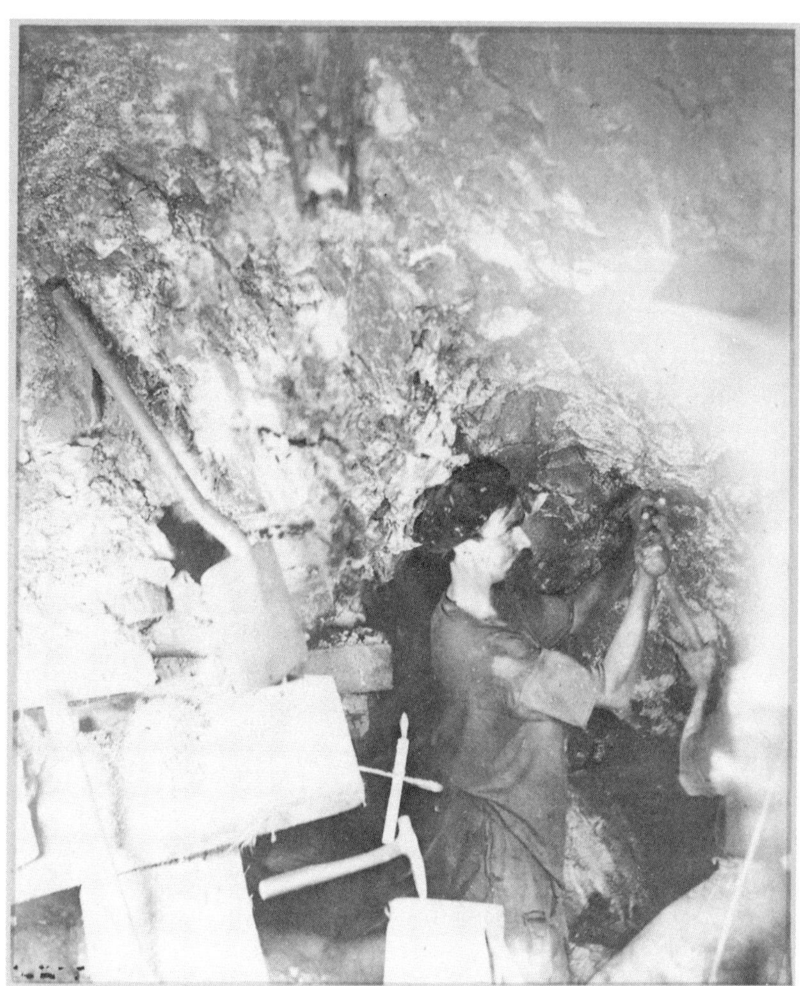

and silver from the ore. Today, the region has one of the most severe cases of mercury contamination in the United States, highlighting the long temporalities of legacy pollution.[48]

The brutal conditions in silver mines radicalized workers.[49] At Comstock, miners were a wage-earning industrial workforce, initially paid three dollars and fifty cents per day, though workers' organizing later pushed this to four dollars.[50] In 1863, the first miners' union west of

2.2   Timothy H. O'Sullivan, *Miner Working inside the Comstock Mine, Virginia City, Nevada*, 1867–68. Albumen print.

the Mississippi was formed at Comstock, though military force ordered by Governor James Nye broke their first strike. Subsequent organizing solidified labor's power, which became a significant force in Nevada.[51] Labor struggles were central to shaping the region, and O'Sullivan's images reflect a distinct awareness of class politics. O'Sullivan's laborcentric framing of the mine as a worksite did not align with the mandate of the War Department of the US Army Corps of Engineers, which was far more interested in promoting settlement and further economic exploitation of the land than in documenting the brutal labor conditions such exploitation entailed. As a result, this subset of his images were not widely circulated, and only one image—the miners about to descend underground—was published in the official report. Robin Kelsey argues that the visual choices made by O'Sullivan to emphasize the oppressive nature of the workspace reflect his awareness of his class position.[52] As an immigrant of Irish descent, O'Sullivan occupied a subordinate position within the hierarchy of the survey despite his technical skills. While Sekula argues that the photographer is often a small-scale entrepreneur and thus positioned in a contradictory position between capital and labor, Robin Kelsey shows that O'Sullivan's awareness of his status as a worker created the possibility for an identification with the miner as a fellow worker. The omission of the underground images in the report reflects a discomfort among the report's management with the labor relations that these photographs made visible. To see the mine in this way is to confront how our world is made.

The limited circulation of O'Sullivan's groundbreaking photographs raises questions about how we see—or don't see—the mine as a worksite, questions that are heightened by comparison with O'Sullivan's photographs taken above ground. In *Snow Peaks, Bull Run Mining District, Nevada* (1871), taken for the Wheeler survey, O'Sullivan flattens the mountainous landscape into planar, sweeping geometric forms (plate 12).[53] Scale is the defining formal characteristic of the image. The smooth whiteness of the snow-coated mounds is contrasted by the linear forms of tall, narrow pinenut trees, which cast shadows onto the bright white snow. The tiny trees provide a sense of the relative immensity of the geological formations. The overall effect of the visual forms is a graceful downward slope, smooth and still under the frozen ground. A small trail of footprints are imprinted in the snow, emphasizing the sublimity of nature in contrast to humans' apparently faint—and fading—impact. The image appears outside of time, nearly untouched by human history.

Snow Peaks, Bull Run Mining District, Nevada, opens an 1874 album published by the Wheeler survey, where the stillness of the image is nuanced by the descriptive legend that accompanies it:

> A summer view of the snow-capped peaks of the Bull Run Mountains, about 9,000 feet in altitude, which form the eastern limit of the broad valley of the Owyhee, in Nevada. In these mountains are mines of silver which were examined by a portion of the expedition in the early part of the season of 1871. This detachment came from Carlin, a station on the Central Pacific Railroad, which was the ground of the first rendezvous camp of that year. In crossing this range, great difficulty and danger in the deep fields of snow which were yet remaining, although it was then the month of June, was encountered. Observe on the picture the line which runs to the right obliquely up the nearest hill; this is the trail made by the marching party, whose animals were obliged to labor and flounder through snow which lay at a depth of four or five feet. In the cañon which crosses the foreground of the picture the mining camp, which was their objective point, lies hidden from view. Reaching this they found the miners inactive, awaiting the more complete disappearance of the winter snows in order that work might be resumed, impeded greatly, in these remote mountain districts, by the rigorous and enduring cold of the winter, and by the inefficient means of transportation to the outer world.

The descriptive legend describes rich ore bodies waiting to be mined underneath the snow. Why, we might ask, would they choose to show not the processes or infrastructure of extraction but instead the *suggestion* of it? Without text, the photograph is a striking mountainous landscape that contains no visible traces of extraction. The combination of image and text suggests latency, something existing but concealed—not yet developed. Much like the latent traces of silver waiting to be revealed on the print in the darkroom, the survey directs the reader to imagine the future extraction of this material. The dual function of the survey is evident in this context. Surveyors created an inventory of nature and industry framed through a lens of contemplation, a tradition that bridged scientific objectivity with aesthetic appreciation.[54] Yet despite the scientific lens of the surveyor, there was also an imaginative component to the survey. Its speculative dimension surfaces here, inviting the viewer to imagine the potential wealth extracted from this place without the viewer being forced to confront the harsh labor and environmental realities of extraction. The striking mountain landscape is unsullied by extraction. The image encourages the viewer to imagine an untouched and abundant nature: an endless storehouse of natural resources that would continue to fuel economic growth without the industrial withdrawal of those resources scarring the landscape.

Photography is thus particularly effective in capturing the speculative gaze of extraction. In his analysis of O'Sullivan's photographs, Kelsey observes the close relationship between the visual-geological search for natural resources and how the documentation of these resources fueled economic investment in both survey projects and extraction itself.[55] The survey images invite the viewer—the intended audience being particular actors within the white settler state—to see and appreciate the potential for futures of extraction and settlement. In the process, the prospector's gaze eclipses histories of land use by Indigenous people and other settlers involved in nonmining economies. Within the survey, the abstract value of extractive capitalism is mirrored by the abstraction of the photograph, which renders land as a potential site for production through the distant, aesthetic gaze of the surveyor. Kelsey concludes that the images are "both a mining and a minting."[56]

Indeed, a discourse casting photography as a system of value dates back at least to Oliver Wendell Holmes. Recall that in the early 1860s, Holmes described carte-de-visite photographs as "greenbacks," or paper money, though the more glaring connection for our purposes is that with metal currency.[57] The Kodak ad that opens this chapter utilizes the perceived natural value of metals to sell Kodak products, but the invocation of metals in photographic advertising dates back to the medium's early years. For instance, in the 1840s and 1850s, the Connecticut-based Scovill Manufacturing Company, which made daguerreotype plates, produced currency-shaped advertising tokens for several photographic studios. Given that this was a period when paper advertisements predominated, the choice to make metal tokens indicates an engagement by daguerreotype companies with then-contemporary debates on the supposedly natural value of metals in contrast to paper money's apparently more abstract value.[58] These advertisements aligned metal photographs with metal currency, suggesting they have authentic value and truth, in contrast to paper money and competing paper photographic methods. By 1866, Scovill was supplying the United States Mint with blanks for copper, nickel, and bronze coins. This is not merely a provocative coincidence. In both the cultural mythology of advertising and the material realm of production, the link between means for producing photographs and means for the exchange of commensurate value is particularly concrete.

For the Wheeler survey, O'Sullivan purchased his photographic materials from Scovill.[59] Where did Scovill source silver? As we have seen, silver has been a large-scale global enterprise since the sixteenth century, complicating any simple recounting of its supply chains as the production and exchange of silver has expanded worldwide.[60] But the discovery of the Comstock Lode catapulted the United States among the leading producers of silver, and worldwide production rapidly accelerated alongside mining in the western United States: in 1861, global silver production was 35 million ounces; by 1900, annual production reached 173 million ounces.[61] By 1870, the United States supplied 30 percent of global silver, followed by 20 percent each from Mexico and Germany.[62] Photography's development into a major industry is inextricable from the rapid increase in the international supply of silver: as much silver was extracted in the thirty-year span from 1870 to 1900 as was produced in over 450 years between 1403 and 1869.[63]

Where did the silver extracted from Comstock end up? The paths it followed could be fairly complex. The financial pressures of the US Civil War made Comstock's silver a vital source of funds for the Union. In 1863, Congress passed a law establishing the Carson City Mint in Nevada to provide money to finance the Civil War. However, the war ended before the mint began production in 1870 and the mint was only operational for twenty-one years, turning only a portion of Comstock silver into currency.

The remaining silver was sent to New York, the largest international commodities trading market near Comstock, and in New York's metal exchanges and harbors, the majority of US silver was processed, sold, and shipped.[64] The trade in metals was mediated by bullion brokers, the largest of which, Handy & Harman, began as producers of silver harness fittings until, in 1867, they switched from working the metal to selling it. Handy & Harman became prominent suppliers for jewelers, silversmiths, and photographic goods companies.

Being melted down into currency or purchased for industrial uses were two possible outcomes for Comstock's silver. However, most of the silver that passed through New York was sent to London's silver market, which used established preferential trading ties to primary silver markets in India and China to establish itself as the center of the global silver trade. London's "temple of silver"[65] was controlled by four trading firms: Mocatta & Goldsmid, Samuel Montagu & Co., Pixley & Abell, and Sharps & Wilkins, while Johnson Matthey and N. M. Rothschilds & Sons handled the refining and assaying of the silver that reached the London market.[66] Importantly, silver was commonly purchased en bloc before it was refined and smelted, a process that could take months and, before which, the full yield of silver was unknown.[67]

In metal markets, silver as a piece of the earth metamorphized into silver as a commodity and, when further abstracted by financialization onto the stock exchange, into a derivative, or speculative, object: a tradeable title to income streams should certain things happen to the value of the commodity it referenced. Silver was thus bought and sold as a material commodity and as an *idea*, tying financial gains to the dematerialized abstractions of money—differences between shifting prices—rather than the value of the material itself. As Marx quipped, political economists theorizing about currency "forget that in order to calculate values in terms of gold and silver, neither gold nor silver need be 'present.'"[68] The seemingly indexical relationship between the material and its natural value was shown in its price to be speculative and unfixed: transactions were completed in London and New York's silver bullion market that traded a grand total of silver that significantly outstripped the standard amount of actual silver available.[69]

Back at one of the origin points, however, similar mismatches wreaked economic havoc for some: though an incredible amount of actual silver was extracted from the Comstock Lode, ultimately, more money was invested into the mining region than profits were produced:

only five mining companies eventually paid out dividends that exceeded the amount of capital invested.[70] As Mark Twain summarized, "It takes a gold mine to run a silver mine."[71] The form of speculation economically necessary for even the simplest kind of future-oriented investment in production, hoping to cover costs and turn a profit, reflects, upside-down as in a camera obscura, the baroque forms of speculation on a commodity-futures market.

Paralleling the rise of photography as a mass medium—and the rise of silver mining in the West—was the decline of silver as a form of currency in the United States. In 1873, the United States eliminated bimetallism, which fixed the dollar's value to a specific quantity of gold and a specific quantity of silver, thus also fixing their relative value to each other and established the gold standard, whereby the dollar expresses a claim on a fixed quantity of gold.[72] The exchange rate between gold and silver had remained relatively stable in Europe and the Americas for centuries, and the demonetization of silver precipitated a significant decrease in the market value of silver bullion contained in existing silver dollars. As the United States was a primary silver producer in the nineteenth century, a surplus of silver piled up following its demonetization. The switch to the gold standard was contentious: politicians from the silver states like Nevada lobbied to prevent the introduction of the gold standard, and its demonetization became known there as the "Crime of 1873." The controversy over silver's monetary status would last until the end of the nineteenth century when the gold standard officially became law in 1900. With its formal demonetization, the value of silver to gold plummeted further to 30:1.[73] Silver's market value declined between 1873 and the twentieth century with "practically no interruption."[74] The price shift had significant social and economic consequences unevenly spread across regions and industries. The relative value of gold somewhat softened only after the discovery of gold on Klondike Creek, Canada, which led to the Yukon Gold Rush in 1896.

For our purposes, however, this development was crucial: the scaling up of photography, as a practice and as an industry, into a mass-participation medium found in demonetization an important material surplus and increasingly cheap supply of one of its now-indispensable components. As silver became less important in currency, the precious metal took on a new role as a system of value in relation to extraction, as photography became central to cataloguing ore bodies and promoting further extraction in survey and corporate photography. In a sense, photographs made with silver become a substitution for silver currency, for photographs, like coins, circulate and literally contain value. This directs our attention to a third possible path for silver extracted from Comstock: to be set in a photograph. Some Comstock silver was purchased directly by companies like Scovill on New York's metal exchanges and ended up in photographs like O'Sullivan's *Snow Peaks, Bull Run Mining District, Nevada*.

Silver beyond the Mine; or, The Modern Consumer Economy of Photography

O'Sullivan's images disclose the scalar tension between actual mining labor and the global exchange of minerals, revealing how the interplay of these activities near at hand and across the planet transform silver from a material into a commodity and, eventually, an anointed representative of value—only to be thrust back among the common commodities with the end of bimetallism. To view the convergence of silver and mass photography from another angle, however, we can return to the Eastman Kodak Company to focus on silver as a means of production. By the end of the nineteenth century, the labor-intensive processes that defined the work of photographers like O'Sullivan gave way to much more widely and cheaply reproducible methods by means of standardization. While working as a bank bookkeeper in 1877, the twenty-three-year-old George Eastman, founder of Kodak, began experimenting with photography. After founding Eastman Kodak Company (originally called the Eastman Dry Plate Co.) in 1881, he introduced a succession of simplifications in photographic technology, including dry plates in 1880, transparent film in 1885, preloaded roll film in 1888, nitrocellulose film base in 1889, daylight loading in 1891, and daylight developing in 1902. With the introduction of the one-dollar Brownie Camera in 1900, the "effortless abundance" Kodak promised was within reach of middle-class consumers who eagerly adopted the medium, having been exposed to its more expensive boutique applications with increasing frequency.[75] When Kodak released the first branded camera, the company promised ease of use: "You press the button—we do the rest." Kodak reduced the highly skilled artisanal work of photography to taking the exposure and sending the film back to Kodak to be developed.

Kodak maintained tight control over its supply chains, production, and corporate image. The opacity of mining labor shown in de Berrio's painting echoes the often hidden work of making photographs. In the early years of photography, Oliver Wendell Holmes observed that with the rapid growth of the "business of sun-picturing," most people would find "'No Admittance' over the doors of their inmost sanctuaries" or the factory.[76] Should this sound familiar, Marx too narrates in *Capital* his entry into "the hidden abode of production, on whose threshold there hangs the notice 'No admittance except on business.'"[77] Holmes highlights a significant limitation in documenting the work of the mine or factory: limited access to corporate sites often makes them less visible, both in strictly visual terms and conceptually. The images of labor that Kodak circulated conveyed precision, expertise, and corporate benevolence, buttressing Kodak's desired reputation as a compassionate employer.

In 1890, the Kodak Park plant opened in Rochester, New York. The scale of production at Kodak Park was spectacular, as evidenced in a 1922 aerial-view photograph. The soaring bird's-eye survey communicates the ambition of Kodak by centering the image on two billowing smokestacks, signifying advanced industrial production. The plant is orderly, neat, and contained; small buildings cluster around the towering forms of the smokestack. Shot from above, the roads running through the plant have a geometric precision lined by clusters of trees. Kodak widely reproduced the image, including postcard reproductions of the image decorated visitor passes to Kodak Park (figure 2.3). Here, we see how the use of the aerial view communicates scale representationally, a useful formal device as definitive assessments of scale are especially complicated within photography. Photography theorist Olivier Lugon shows that analog silver prints are objects without a fixed scale: scalar choices are made when the image is taken *and* in the darkroom; prints can be enlarged or miniaturized.[78] Regardless of what size it was printed in, the Kodak Plant photograph exemplifies what scholars have termed the "technological sublime," a vision of awe-inspiring industry and innovation at a scale that dwarfed even the natural world.[79] By the 1880s, massive industrial infrastructures represented a potent crystallization of the ambitions of a world centered on human ingenuity as the marker of authority. Corporate photography sought to capture this sense of scale while isolating the factory from the world around it—or, in the case of Kodak, building an entire world around the factory.[80] Kodak Park functioned more like a city than a manufacturing complex and employed tens of thousands of workers who worked on a complex spanning thirteen hundred acres linked by an intraplant railroad.

The regimentation of the factory—and Kodak's vertical integration of most aspects of production—enabled a rapid scaling up of its labor force and its production of profit. As early as 1861, Marx had identified photography as a hallmark industry of the new era of production.[81] In 1863, Oliver Wendell Holmes observed the division of labor in photographic sectors, noting that at the E. & H. T. Anthony & Co. factory in New York, "a young person who mounts photographs on cards all day long confessed to having never, or almost never, seen a negative developed, though standing at the time within a few feet of the dark closet where the process was going on all day long."[82] This protoassembly-line subdivision of labor enabled rapid growth that escalated over the second half of the nineteenth century. The Kodak Park plant reflects the grandiose ambitions and industrial efficiency that characterized capitalist production at the time. For instance, by the mid-1950s, the Kodak Park plant had, by the company's description,

> more than 116 major manufacturing buildings; Operates 5,200 telephones in a dial system; Maintains a complete fire department; Makes enough refrigeration in its own machines to supply all the homes in New York City, or the equivalent of 1,500,000 mechanical refrigerators; Uses 22,000,000 gallons of water a day; Recovers about $375,000 a month in silver from scraps of discarded film and paper; Runs a fleet of 200 trucks and other vehicles over 12 miles of intraplant roads; Turns out enough perforated film (mostly amateur and professional movie types) a year to make a strip 660,000 miles long—enough for a round trip to the moon and several times around the world![83]

The raw materials, chemicals, and labor required to run Kodak Park and fill the ever-increasing demand for photographs were indeed immense.

Here we can see how the deep interdependence between silver as a photographic material and silver as money and commodity enters our story again. With rising production in photographic industries, by 1920,

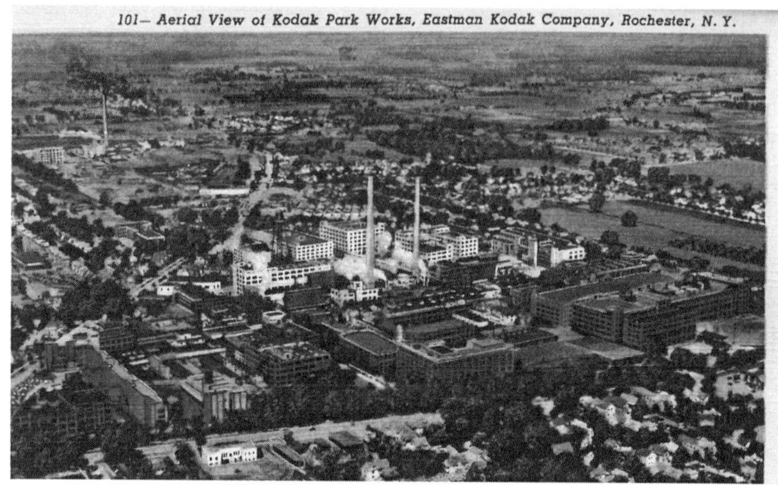

101— *Aerial View of Kodak Park Works, Eastman Kodak Company, Rochester, N. Y.*

Kodak became the largest consumer of silver in the United States outside of the US Mint.[84] By 1998, photography accounted for 40 percent of all silver used in the United States.[85] In a trade article written for Kodak employees entitled "Silver—Metal, Money, and Mixture," the author reflects that silver "finds a great industrial use in the photographic industry. And because of its importance to our industry, its use as money, and the world's money policies[,] are important to us."[86] Industries using silver as inputs, and industries producing, trading, and moving silver, found their industrial futures and bottom lines intimately tied up with the supply of metals and with the political decisions setting parameters for the US monetary system. Kodak's sourcing thus had to be nimble enough to follow shifting supplies and avoiding potential geopolitical obstacles. By the mid-twentieth century, the mint purchased most American silver, as silver continued to be used in dimes, quarters, and half-dollars despite the gold standard, so the majority of Kodak's silver came from foreign sources.[87] This shift in sourcing also reflects the declining status of the United States as a leading producer of silver: by 1930, 42 percent of global silver came from Mexico, reducing the United States' output to 21 percent of global supply.[88] Almost 20 percent of global silver was used

2.3   A postcard of an aerial view of the Kodak Park Works, Eastman Kodak Company, Rochester, New York, 1946.

in the arts and industry in North America.[89] Kodak's consumer pricing was thus highly sensitive to fluctuations in the price of silver: in 1976, Kodak increased the price of films, papers, and chemicals by an average of 6.5 percent, citing the rising cost of silver, which had approximately doubled in the previous year.[90]

Silver's capricious supply and price was challenging to a company that sought to control as many variables as possible. To introduce more stability into supply chains, Kodak recycled silver, turning the factory floor into a kind of secondary silver mine. The unexposed silver halides can be recovered and reused in the developing process. This practice predates Kodak: methods for recovering silver were a common subject in nineteenth-century photographic journals. For example, *The Photographic News* argued that "besides the individual gain, it is a saving, in the gross, to the commercial world, as in the case of waste, so much silver goes, practically out of existence, and is lost to the world."[91] Indeed, 10 to 15 percent of the silver used across industries annually is recycled.[92] Kodak described its recycling operations as "probably the most efficient 'silver mine' in the world."[93] Given the scale of production at Kodak, recycling silver gave them a buffer to minimize the cost and continuity effects of a volatile supply chain. Interestingly, as digital photography rose to prominence in the twenty-first century, Kodak would seek to minimize the perception that the cost of analog photography was tightly tied to the price of silver. In a report published in 2015 that outlined the enduring significance of silver halide paper, they wrote:

> Yes, silver is an expensive precious metal, and the price fluctuates in the commodities markets. But this has little bearing on its use as a light-sensitive component in photographic paper.... This serves to minimize both costs and environmental impact. More detail is needed about this because there are "myths" being promoted in the industry that the price of silver is directly linked to the cost of using a photographic color paper.... The "myths" go on to say that because of the high cost of silver, the cost to the processing lab (and ultimately the end consumer) must be higher than using other technologies. This is simply not true. Silver is merely an intermediate step in the process of making a print.[94]

In this self-narration of an industry in transition, the significance of materials is subsumed as a mere auxiliary to the more meaningful production processes, a shift in emphasis from the supply of raw materials to the production of the final commodity.

As much as Kodak directs our attention from the mine to the factory, the consequences of extraction continue. The environmental impacts of Kodak did not end with the initial extraction of silver. Transforming silver into silver halides for film emulsion also required the extensive use of chemicals. For instance, in the 1945 *Life* ad about how silver "gets the picture," the text explains that the metal is "freed from its chemical partners" by chemicals such as nitric acid and iodide.[95] Other chemical solutions were used to increase light sensitivity, add color, or function as hardening or spreading agents. As mentioned above, in the mid-twentieth century, Kodak Park produced over 350 different chemicals for use in photography—a single day's production would fill eighteen tractor trailers—and more than four thousand research chemicals.[96] Kodak was, more than anything, a chemicals company, and the use of chemicals in production processes has had a substantial environmental impact.

Communities around Kodak Park were exposed to high levels of chemical contamination as emissions and pollutants—including methylene chloride, acetone, methanol, and dioxin—were released into the local ecosystem. By the end of the twentieth century, Kodak Park drew between thirty-five and fifty million gallons of fresh water daily from Lake Ontario. After being used in production processes, the now-carcinogenic water was dumped in the Genesee River.[97] Richard Maxwell and Toby Miller conclude that "the competition effect of capitalism—its paradoxical tendency towards monopolization—propelled environmental exploitation and despoliation in the raw film business as much if not more than the rising commercial demand for film stock."[98] As they suggest, it was not simply consumer demand that drove environmental degradation but specifically the firm's competition-induced demand for higher profits, many of which could be captured through economies of scale in the use of chemicals—with tremendous environmental externalities. Between 1980 and 2000, Kodak produced more toxic emissions than any other company in New York State.[99] Kodak's environmental impact on the landscape and workers' bodies was incredibly damaging. Anthropologist Elizabeth Povinelli describes Kodak photographs as having a "toxic unconscious," for "those billions of stilled memories and moving fantasies" resurfaced as "the material afterimages of those Kodak moments emerged as fibromyalgia, neuropathy, and primary biliary cirrhosis."[100] The intensity of Kodak's production made the pollution immense. Scaling up any interaction with the earth, which is what production is, always has ecological consequences.

Considering these histories through the lens of the cumulative and successive environmental crises that characterize our present moment, scale is a useful analytic. The drive to create a new consumer market for photography required more and more materials. As Kodak drove consumer demand by making photography more accessible, the extraction, refining, and processing of the materials used in industrial production left their mark on the local and distant environments. The view from the corporation abstracts, doing active work to place these histories outside of the field of visibility. The aerial photograph of the precise and tidy plant hides pollution's seeping, drifting, toxic legacy.

Photography and Labor

Thus far, the chapter has considered the scaling up of silver mining and photographic production, but the latent theme running throughout this story is labor. By way of conclusion, I want to scale down to consider photography's ability to depict work by turning to the earliest known photograph to represent labor, *Boulevard du Temple, Paris, 3rd Arrondissement*. In 1838, Louis Daguerre photographed a Parisian street from an elevated position. The boulevard winds around the left side of the frame, and, unusually for a busy boulevard, it is empty. Daguerre exposed the image on a small, silver-plated sheet of copper—13 by 16 centimeters—and the detail is remarkable (figure 2.4). The cobblestones of the boulevard are visible in the foreground, as are the chimneys on buildings stretching into the background. The high contrast further accentuates the details. Like *View from the Window at Le Gras*, the image is a unique print. Daguerre continued Nicéphore Niépce's experiments with iodized silver plates, introducing latent development using mercury vapor, which reduced exposure time, while using hypo (sodium thiosulfate) to wash away unexposed silver salts prevented silver halides from continuing to darken after the initial exposure.[101] These innovations resulted in an exposure time of roughly ten to twelve minutes, significantly shorter than the estimated eight hours of Niépce's exposure. The scene is an image of work: it captures two figures in the distance, a shoeshine and his client. Portraits posed a significant challenge to the daguerreotype method, for as a contemporary wrote, the slightest movement results in "a confusion of light on the plate."[102] The slow exposure time eliminated most activity on the busy street, as the carts and people in motion disappeared from the photographic frame before solidly registering. The shoeshine exchange, however, had the customer fixed in relative stillness long enough for Daguerre to capture him on the plate. In contrast, the shoeshine becomes blurred, for he, unlike his client, is in motion. In a second exposure of the same scene taken at midday, the boxes for brushes and polishes are visible, but the human activity has disappeared. As Allan Sekula asks of this image: "What is celebrated? The static moment of consumption, the fashionable pose. What is obscured, denied, disavowed? The productive moment, the energetic blur of that other body, unacknowledged, the working body, the invisible shoeblack."[103] A "dual erasure" is enacted in this shot: the camera hides the labor that made the photograph possible, and the labor of the shoeshine is erased within the photograph.[104] The camera and the photograph collude to hide traces of work.

Photography's complex relationship to labor extends beyond Daguerre's photograph into other contemporaneously developed methods. For example, William Henry Fox Talbot created the calotype, a reproducible process printed on paper. Talbot's origin story for his invention of what he called "photogenic drawing" centers on his inability to draw, insisting that with his invention, the "picture is what makes ITSELF."[105] Talbot narrates his desire to fix the image as emerging while drawing on Lake Como in Italy, a scene rooted in leisure rather than labor. The narratives that displaced human agency from the scene of production in factories find echoes in descriptions of photography as "autogenic": "the pictures make themselves; the apparatus draws the scene; objects draw their own likeness; the sun makes the picture; the self-acting apparatus makes the commodity."[106] This framing echoes bourgeois conceptions of money, which, Marx writes, locates the value of metals in nature, not labor, or the fetishism of commodities, in which the interaction of people as economic actors "assumes here, for them, the fantastic form of a re-

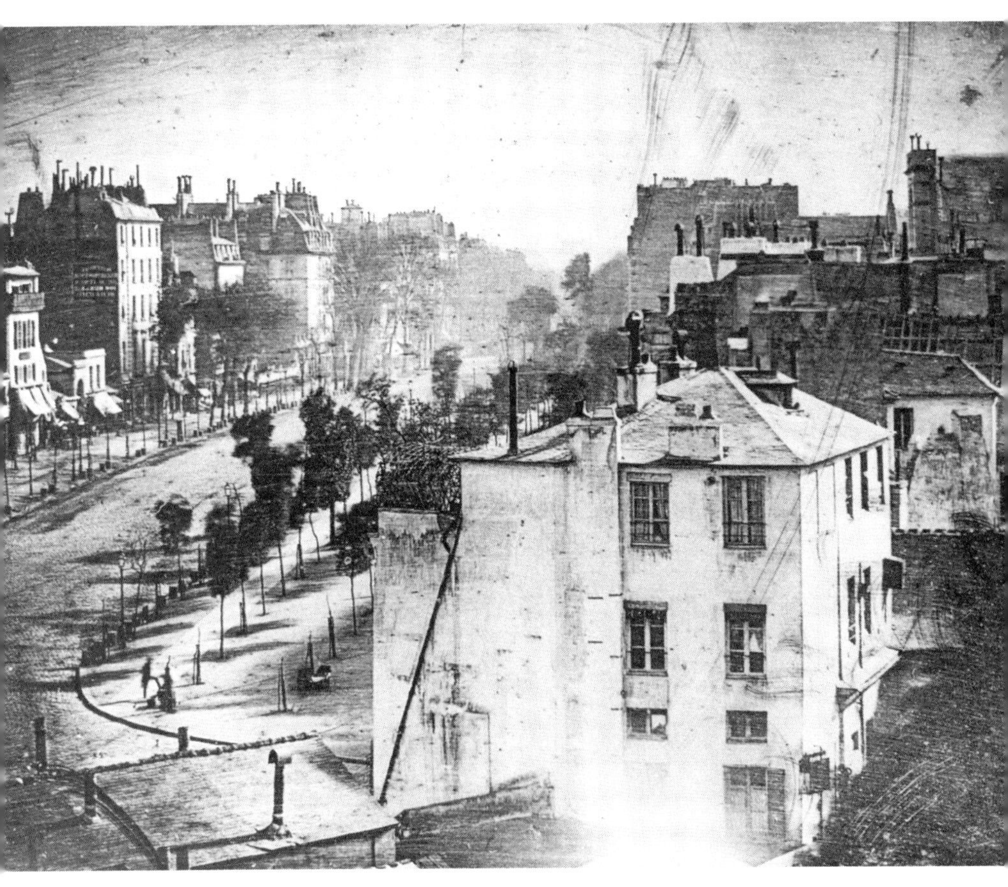

lation between things."[107] Despite these narratives, early photography still required specialized knowledge of photographic chemistry that separated skilled photographers from lesser artists, let alone lay consumers.[108] As representatives of a craft, artisans of early photography express a complex relationship to labor, which colors early photography in its narration and its practice.

I want to suggest that this, too, is a problem of scale: a tension between the individual and the collective identity of the worker. Historians Kevin Coleman, Daniel James, and Jayeeta Sharma ask: "Can photography depict the worker as both an individual and part of a larger political and economic process? If a photograph manages to represent labor, does that

2.4   Louis Daguerre, *The Boulevard du Temple, Paris, 3rd Arrondissement*, 1838. Daguerreotype.

entail abstracting out the lived experience of a particular historical subject?"[109] Considering the miner as a general class fraction necessitates an
abstraction of the concrete person who does mining labor. This problem
also underlies this chapter: there can be a loss of meaning as one shifts
scales between thinking concretely about labor and thinking about industry as a whole. Rather than attempt to reconcile the two scales—and in the
process, flatten them out—I want to hold them in tension, despite their
discontinuous implications. Both have something to tell us, but perhaps
they cannot be neatly reconciled.

Sekula's critique of the erasure of labor enacted in Daguerre's photograph develops an analysis that dates back to Bertolt Brecht's and Walter
Benjamin's observations of photography's tendency to aestheticize and
isolate, flattening complex networks of power and economy in the process.
Framing and cropping an image removes the scene from its surroundings,
cutting it out like ore from a mountainside. Brecht and Benjamin draw
links between the surface impression captured in the photograph, the lack
of context enforced by photography's tendency to crop scenes from the
world they are a part of, and photographic aesthetics in general. These
phenomena all point to photography's potentially superficial engagement
with the world. Brecht polemically argued that because photography exists in the world of the surface, it cannot reveal "reality":

> The situation is now becoming so complex that a simple "reproduc-
> tion of reality" says less than ever about reality itself. A photograph
> of a Krupp factory or the AEG says practically nothing about these
> institutions. Reality itself has shifted into the realm of the func-
> tional. The reification of human relationships, such as the factory,
> no longer betrays anything about these relationships. And so, what
> we actually need is to "construct something," something "artificial,"
> "posed." What we therefore equally need is art. But the old concept
> of art based on experience is invalid. For whoever reproduces those
> aspects of reality that can be experienced does not reproduce real-
> ity. For some time reality has no longer been experienceable as a
> totality.[110]

Brecht argues that the social and economic relationships structured by
capitalism are not visible within photography: the photograph of the

Krupp factory cannot show how the factory is placed in the functioning of capitalist social relations. Here, Brecht points to a profound problem of scale, noting that reality can no longer be grasped as a whole. Benjamin uses similar language in an analysis of the New Objectivity style of the photographer Albert Renger-Patzsch. The aestheticization in modernist photography, Benjamin believes, strips meaning from the image, with the result that photography "is now incapable of photographing a tenement or a rubbish heap without transfiguring it. Not to mention a river dam or an electric cable factory: in front of these, photography can only say 'How beautiful.'"[111] Benjamin suggests that the photograph as mere direct representation turns activism into aesthetics, politicized moments into objects of consumption. This disconnection from context means that deeper meaning cannot necessarily be gleaned from the surface of the photographic image.

However, O'Sullivan's underground images situate the miner in context, which refocuses attention to the mine as a worksite. If we compare the miner's labor to the photographer's, how can we understand occupational portraits? Consider a tintype by William Cook from 1868. Cook poses, sitting and mixing liquid "to use in the picture business." The image is an occupational self-portrait of a photographer who ran a studio in Anoka, Minnesota, from 1860 to 1867.[112] Studio portraits were a popular genre in the latter half of the nineteenth century, and people often posed with the tools of their trade. In O'Sullivan's occupational portrait of the miner, he excavates rock using the tools of his trade. In the tintype, we see the chemical stage in the labor of photography: the mixing of chemicals to make the image appear, a trope also drawn on by Kodak (figure 2.5). The photographer looks at his materials, engrossed in the task, evoking the alchemist. Underneath a jacket with visual resonances to a lab coat, he wears a suit, vest, and bow tie: signifiers of professional respectability (figure 2.6). The constructive role of photography is applied here for the self-presentation of skilled labor. Tintypes, meanwhile, were an affordable means of image production invented in the 1850s as a cheaper alternative to daguerreotypes. The dark, muted tones that are typical of tintypes are shown in the portrait, as the photographic emulsion was applied to a thin sheet of metal, typically iron, coated with a dark lacquer—itself a step on the road to the deskilling of the chemical aspect of the labor of photographic production.

The controlled space of the studio stands in marked contrast to O'Sullivan's portraits, which confront the viewer with the labor conditions underground at the hidden abode of production itself. Compare, for instance, O'Sullivan's images to studio daguerreotype portraits from the California Gold Rush, in which young men used portraiture to construct new identities rooted in the mythology of the lawless frontier.[113] Prospectors posed with tools, weapons, and gold. In the only definitively identified self-portrait of the photographer Carleton Watkins, he poses not as a photographer but as a miner.[114] In *Primitive Mining, the Old Rocker*, from 1883, Watkins poses with rudimentary mining tools, wearing a white collared shirt and vest (figure 2.7). In the background is Watkins's traveling darkroom. The image is tinged with nostalgia. Here, the mine labor is placer mining, a surface form of panning with very different work conditions than the underground mining at Comstock Lode, largely those of quasi-independent contractors. Watkins's self-portrait reveals an identification with the miner—perhaps a playful one—and shows how important mining was to his photographic practice, as the mining industry was an important client. What sets O'Sullivan's portraits apart from these other occupational portraits of the miners is the immersion within the site of the mine. This gesture redirects the occupational portrait from an honorific form of self-construction to a more materialist consideration of labor under conditions of alienation and restricted autonomy.

The desire to make labor visible and the technical limitations of photography continue as themes throughout the medium's history. The first motion picture film crystalizes this tension: Auguste and Louis Lumière's *Workers Leaving the Lumière Factory in Lyon* (1895). Shot on 35-millimeter film, the Lumières used a Cinématographe—a device that could shoot, project, and copy film—to film the forty-six-second movie. As with most analog film, the base of the motion picture film is silver, giving its name to the silver screen while likewise evoking the connotations of glamour and wealth associated with precious metals. In the short film, workers leave the photographic supply factory—owned by the Lumière family— in Lyon-Monplaisir through two gates, pouring out of the frame on both sides. The film shows workers in photographic industries, though they are shown leaving work.[115]

While the slow exposure time of the technical process used to make Daguerre's photograph removed movement from the frame, the Lumière

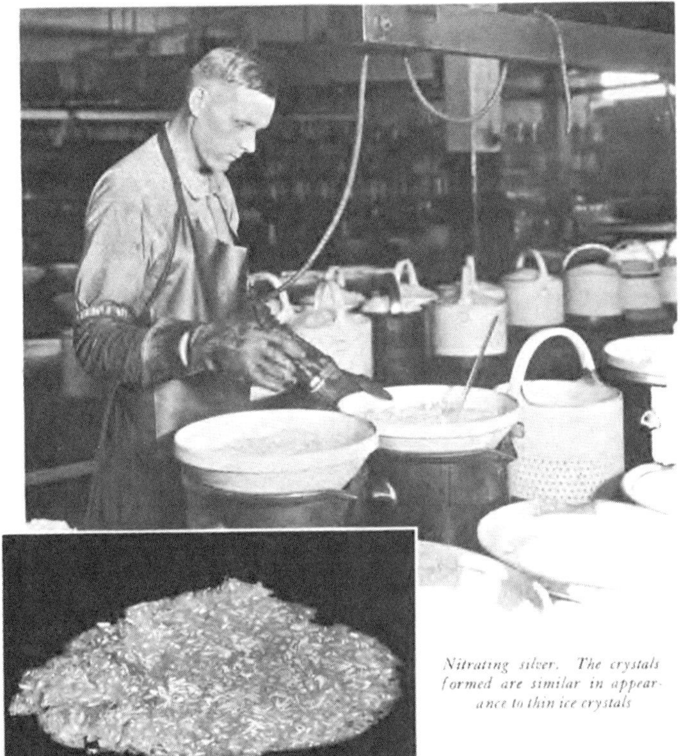

Nitrating silver. The crystals formed are similar in appearance to thin ice crystals

brothers' focus is on motion. Symbolically, the cascade of workers leaving work marks their transition from the collective identity they share as workers inside the factory. As they disperse, they become individuals again, although this subjectivity is located outside the motion picture frame. As Harun Farocki reflects, the scene is not a literary trope but rather a symbol "taken from reality. It is as though the world itself wanted to tell us something."[116] Here, we can return to Brecht's observation that a photograph of the Krupp or the AEG factories "tells us nothing" and some form must be constructed, acknowledging the limits of photography—or the surface—to reveal the larger structures of power that shape the world. Farocki suggests that the Lumière brothers were using a new artistic and technological form not to replicate signs and tropes from lit-

2.5   Nitrating silver. Reproduced in *The Home of Kodak,* 1929

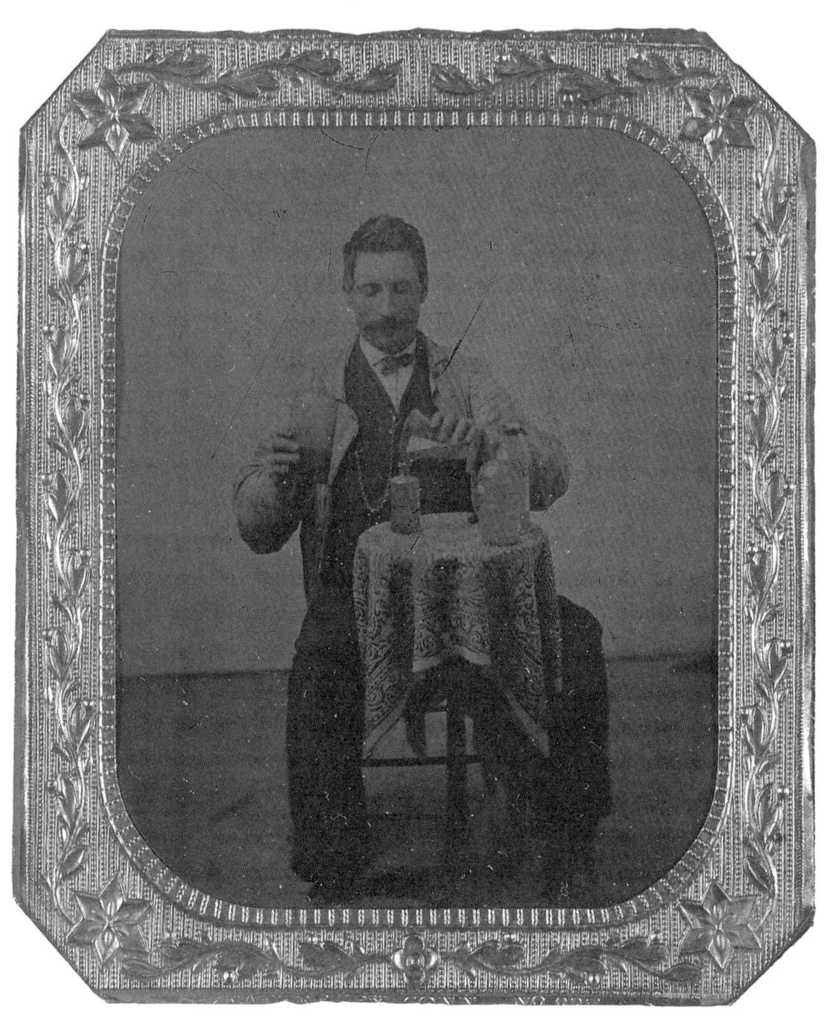

2.6 William Cook mixing materials for use in the picture business, 1868. Tintype. Courtesy of the Peter Palmquist Cased Photographs Collection, Yale Collection of Western Americana, Beinecke Rare Book and Manuscript Library.

erature or art but to represent the real. The workers stand in for the larger system they are part of, "the visible movement of people is standing in for the absent and invisible movement of goods, money, and ideas circulating in industry."[117] Farocki also draws attention to the problem of labor and representation introduced by Sekula: "The first camera in the history of cinema was pointed at a factory, but a century later it can be said that film is hardly drawn to the factory and is even repelled by it. Films about work or workers have not become one of the main genres, and the space in front of the factory has remained on the sidelines."[118] The relative disinterest in concrete labor processes as a theme within cultural production has long been in evidence, such that the exceptions are particularly notable. Media studies scholar Sean Cubitt's reading of Don McCullin's *Early Morning, West Hartlepool* (1963), which documents the West Durham coalfields, outlines what photography *can* show us regarding labor and capital:

> We cannot see in that image what brought about the conjuncture it captures: the coal and steel industries, the careless poisoning of the environment, the class structure and class struggles, the forms of industrial capitalism or the National Health Service that picked up its wounded, the gendering of employment and the monoculture of North-East England in the 1960s. But we can see what none of those grand social and historical forces can: the horror and waste they visit on this specific landscape, this specific man, on this specific day, and equally significantly, the transfigurative power of the photograph to imply that another world was possible, even in this dark wasteland.[119]

Here, the photograph pulls us back to Earth: from vast abstractions of politics and economy to the individual worker. The subjective, anecdotal power of photography can testify to a singular moment in time, to mark a moment or a person as significant. Literary scholar Jacob Emery argues that photography is the medium best suited to reveal "the raw edge between a legible scene and the larger world from which it has been wrested."[120] This tension between the concrete and abstract shows us something about the tense but necessary relationship between the part and the whole. Here, we can locate photography's political potential in the context of the global interplay of labor and capital.

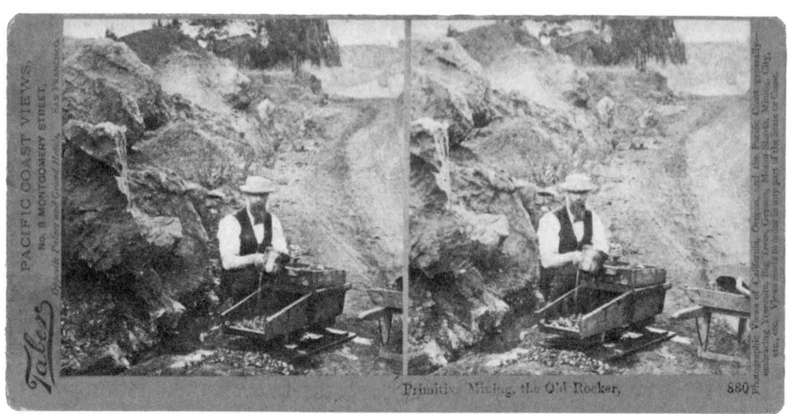

Despite these early foundational ties between mining and photography and the identification shown by both Watkins and O'Sullivan with the labor of the mine, narratives of photography and photographic production continually efface the centrality of labor to these histories. By engaging with early histories of photography and film represented by Daguerre and the Lumière brothers, Sekula and Farocki highlight the unresolved tension in photography: between the world of the surface and the labor that makes the world, labor that is often underground or out of sight. At moments, however, something singular and specific comes into view, like O'Sullivan's miners at Comstock Lode, that lead an attentive viewer into a more complex history. O'Sullivan's underground photographs are not contemplative or distant. Instead, they look unflinchingly at what it means to extract metals from the earth, metals that enable us to see things like this labor process itself. The complexities and contradictions in our cultural relationships to extraction come into sharp focus through such images.

Silver's remarkable material properties made the scaling up of photographic production possible. Mass-producing photographs would not be possible with other materials—bitumen, as we've seen, is too slow and variable, and platinum, as we shall see, is too expensive and rare. However, grounding the history of photography in silver causes its own distortions. We can recall Marx's observation that in economies, silver can

2.7   Carleton Watkins, *Primitive Mining, the Old Rocker,*
1883. Albumen silver print.

"appear and just as suddenly disappear." Silver appears and disappears in the photographic developing process: its capacity to move forms the print. Trying to ground photography in the depths of the mine, such as in O'Sullivan's portraits, is an incomplete gesture. To understand mining at Comstock Lode, we have to scale up to the larger, speculative workings of the capitalist mode of production as a whole. An attempt to move our analysis underground thus quickly pulls us up into the abstract valuations of ticker tape on the stock exchange. Likewise, a history of exchange cannot be understood apart from labor. As silver scales up and down, its meanings fundamentally shift, moving from material to commodity, from so-called natural value to speculation. The sites become entangled but not equivalent. Here, the distinction between scalar viewpoints and the ways of seeing they enable becomes significant: capitalism's scale is the aerial view, the global market of commensurable labor processes; the view from the worker's eye reorients us to the body. As we redirect our gaze to the underground, however, we can also reconsider some criticisms of photography's perceived activist potential that emphasize photography's limiting relationship to the surface. Just as silver moves between scales, literally surfacing on the back of labor, how photography flits between the concrete and the abstract destabilizes a singular reading of the image's political role in the context of extractive capitalism.

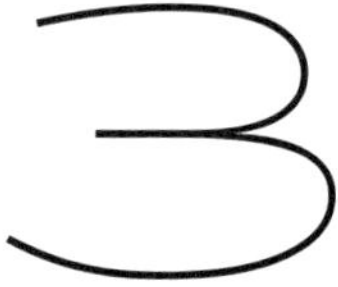

Platinum
and Atmosphere

SILVER IS THE MOST COMMON MATERIAL in analog photography, but platinum is the most luxurious. Appropriate for a metal that is both expensive and rare, platinum prints have a sophisticated tonal range, a matte surface, and a soft expressiveness. These formal qualities are displayed in South African photographer David Goldblatt's *Old Mill Foundations, Tailing Wheel and Sand Dump, Witwatersrand Deep Gold Mine, Germiston, August 1966* (1966, printed later) (plate 13). The mill formations have fallen into ruin, their desolation amplified by the mournful, muted gray tones. The visual emphasis is on the broken tailings wheel that, powered by water and steam, crushed ore to separate gold, silver, and platinum from waste rock. The scene is still, elegiac. Movement is implied in the stormy, smoky sky, where puffs of black smoke give way to muted gray clouds, intermingling natural and industrial atmospheres. The extended tonal range of platinum is utilized to its maximum capacity in the print: the velvety dark grays of the sky form a striking contrast to the stark white sand dump, a white so pure it almost looks illuminated by floodlights. The sand pile is a slag heap, a hill built out of refuse from mining. Slag heaps are the most tangible marker of the distinction between what is valued and what is left behind in mining: as metals and minerals are processed to create value, they create a value differential between the valued mineral destined for the market and the devalued waste, which needs to be discarded in a kind of land it would come to define—the wasteland.

Goldblatt knew mining wastelands well. He was born in the Johannesburg gold-mining suburb of Randfontein; his life was shaped by both extraction and Apartheid.[1] Johannesburg was founded as a mining camp in 1886, and the planning of the city reflects both racial segregation and the spatial demands of mining, including the toxic detritus left behind by industry. South Africa's mining regions famously contain some of the richest mineral deposits in the world, including gold, diamonds, coal, and platinum. The demands of extraction fundamentally transform the landscape of mining regions, but, owing to the social relations and financial structures that govern capitalist extraction, the wealth extracted from the earth is rarely reinvested in local communities' infrastructures or even significantly in workers' wages.[2] The government of South Africa has inherited six thousand abandoned mines—sites of production turned into sites of waste—estimated to cost approximately R30 billion (approximately US$1.6 billion) to maintain and remediate.[3]

Platinum's rarity and cost often dictate a certain status-awareness in its use as a photographic medium, making mine tailings an improbable subject for platinum printing. Irving Penn's platinum prints of cigarette butts provide an apt comparison, as the contrast between the rarity of the materials and waste as subject matter made the "very process of creation the agent of his meaning."[4] The image of the tailings dump is from Goldblatt's series *On the Mines* (1965–70), which was originally printed as a photo book. The photojournalistic style of the silver gelatin prints in *On the Mines* was later supplemented by a limited run of platinum prints printed on Arches Platine paper. Goldblatt reproduced only a few prints from *On the Mines* in platinum: a portrait of a boiler-house attendant Joe Maloney, a miner waiting on the surface, shaft sinking, and a tightly framed shot of a man's torso and tools. Darkroom supplies were costly in

South Africa, introducing a further practical consideration to the limited run of prints.

So why print in platinum? Platinum is the most chemically stable metal, giving platinum prints superior stability as photographs. To print in platinum speaks to a desire for permanence. If the crisp precision of the silver gelatin print suggests an unmediated view of the world, one that testifies to the authenticity of the scene and invokes photographic or photojournalistic truth, platinum reveals the artist's intervention. The soft tones and matte surface of platinum on cotton are expressive and aspire to timelessness—a desire to transcend the singular moment captured in the image through recourse to something more universal.[5] The temporality of the image is extended; the print is at once outside of time and intended to endure into the future. Printed in platinum, the mine tailings become a stand-in for a much longer history. The expense and rarity of the print singles it out as significant, perhaps as something that illuminates the realities and tensions underpinning our world. Here, it marks the mine tailings as something worthy of consideration. Through Goldblatt's lens, the slag heap is transfigured into a thing of beauty and, printed in platinum, a thing of value. What the photograph cannot show us are the elemental entanglements of the tailings landscape—how the fine particles of sand are carried in the air, depositing lead, cyanide, arsenic, and radioactive traces of uranium into water, soil, and human bodies. We can see how smoke darkens the sky, but we can't feel the choking sensations as particulate matter invades the throat and lungs. Within the landscape, the divisions between the atmosphere, surface, underground, and body are troubled, as toxins permeate the ecosystem at every level, a complex process flattened by any photograph. As the Goldblatt image invites us to consider, atmosphere is the conduit through which traces of the mine enter bodies and ecosystems.

Atmosphere has emerged as a key analytic in the environmental humanities for the interplay between media and environmental change.[6] Atmosphere is the gas above a planetary body and it suggests an in-between space: there is an indeterminacy to the phenomenon atmosphere describes. Visually, atmosphere describes the overall impression of a work of art, often with an inflection of perceived emotion. Combining these two meanings, atmosphere suggests that the work of art extends outside its objectness, creating an environment that envelops or invites the viewer: here, we see the world-building potential of the image. Art historian Mark Cheetham proposes that thinking atmospherically invites us to shift how we do art history by bringing the "material realities of planetary phenomena to bear on the understanding of artworks and their contexts of creation," which in turn requires a shift in art-historical methodologies.[7] Thinking through atmosphere thus establishes stakes that are both aesthetic and material.

Inspired by platinum prints' formal and material links to atmosphere, this chapter undertakes an atmospheric reading of photographs. I begin from the premise that, aesthetically, platinum prints excel at rendering atmosphere. Materially, platinum's use in photography is also linked to problems of physical atmosphere: experiments on platinum printing were partly driven by the susceptibility of silver prints to fading or discoloring through exposure to the contents of the air. Platinum is less vulnerable to contaminated air pollution caused by the burning of fossil fuels than silver-based prints. *Photographic News* summarized that "it is a known fact to chemists that metallic platinum in all its states, massive and molecular, is totally unchanged or unaltered by atmospheric influences."[8] Platinum promises stable boundaries—it doesn't react to other things, enhances permanence, and seals the image-as-representation against the elements. However, as we will unravel through an emphasis on atmosphere, the idea of unalterable elements and fixed boundaries is a fiction. Atmospheric haze blurs boundaries pictorially and materially. Atmosphere moves between scales, but it meets on the surface. Accordingly, this chapter stays above ground to consider how atmosphere interacts with land and the body, intermingling three sites we consider distinct. Throughout, I trouble art history's traditional split between figure and ground, pointing to how materials circulate between bodies and the landscapes they inhabit.

After outlining platinum's chemical stability and extensive use in industrial processes, the chapter then introduces atmosphere as an aesthetic concern, focusing on the pictorialists who championed the atmospheric aesthetics of platinum prints. From there, I show how platinum became an important substitute for silver-based processes, partly because of silver's vulnerability to atmospheric pollution in the industrial society that enabled its photographic use. The rise of commercial platinum printing was spearheaded by William Willis's Platinotype Company, which allows us to trace platinum's concentrated supply chains to the mining landscapes where the metal was extracted, documented in Simon Starling's platinum prints of South Africa's platinum mines. I read Starling's prints atmospherically to consider how the dust and particles of the mine drift through the landscape, which introduces questions of embodiment, as the mining landscape enters into the worker's body. Here, I return to David Goldblatt's photographs to think about labor conditions and occupational disease. The chapter concludes with Larry McNeil's platinum prints of coal mining in the western United States for a consideration of atmosphere and futurity, an inescapable theme throughout our consideration of this metal's boutique visual applications.

Platinum

Silver-white in color with a lustrous surface, platinum (Pt, atomic number 78) is among the rarest and most expensive precious metals. Platinum is classified within the platinum group metals (PGM): iridium, osmium, palladium, platinum, rhodium, and ruthenium.[9] In antiquity, there were seven known metals (gold, silver, copper, tin, lead, iron, and mercury), and it wasn't until the eighteenth century that platinum itself was formally identified, making it the eighth metal to receive a name. Naturally occurring in the alluvial sands of rivers, platinum was used in ceremonial jewelry in pre-Columbian South and Central America. It was also used in ancient Egyptian artifacts, including Shepenupet II's casket at Thebes. These early uses of platinum were largely alloys of the platinum group metals, for true platinum has an especially high melting point, making it difficult to isolate in its pure state. In 1557, Julius Scaliger described a metal in Spanish Central America "which no fire nor any Spanish artifice has yet been able to liquefy."[10] This description of a metal that could not be melted is the first description of platinum by Europeans, who initially understood the metal as an impurity in silver. The Spanish naval officer Antonio de Ulloa published the first report on platinum in 1748 after encountering it in gold mining in the Chocó region of Colombia in 1735. He called the metal "platina," or little silver. Roughly contemporaneously, the British scientist Charles Wood encountered samples of platinum from Colombia while in Jamaica. Wood sent specimens to the scientist William Brownrigg, who published a study introducing platinum as a new metal in 1750. A commercial process for refining pure platinum was later developed in the early nineteenth century by the English chemist William H. Wollaston. However, while European scientists are often identified as discovering platinum, people in the Americas were using platinum for centuries prior to this.[11]

Platinum is only found in significant quantities in Canada, Russia, and South Africa, and the vast majority of the mineral's stores on this planet—roughly 80 percent—is found in South Africa, where it is mined as a byproduct from nickel and copper. Platinum is very rare: gold is mined at a rate of fourteen times that of platinum.[12] Imagine an Olympic-size swimming pool (a volume of 2.5 million liters): all the gold ever produced would fill the pool three times over. In contrast, all the platinum ever produced would just reach the ankles of a person standing in the pool.[13] Twice as much steel is poured in one day in the United States than all global platinum production in one year. Platinum is abundant in meteorites—driving a contemporary interest in asteroid mining—and, as a result, is often mined in impact sites. Due to its rarity, platinum is (typically) the most expensive metal, lending its name to such things as the platinum records certification awarded by the Recording Industry Association of America (RIAA) and "platinum" credit cards. Although platinum is often worth more than gold—during the 2014 miners' strike in South Africa, it was twice the value of gold—the price of platinum is incredibly variable.[14] While platinum is a precious metal, it is rarely used in ornamental contexts because its myriad applications in industrial processes weigh heavily on its pricing. As one of the purest precious metals—typically 95 percent pure—and the least reactive metal, it has tremendous advantages in industrial processes requiring high heat. Critically, it is resistant to oxidization, which is significant for histories of photography.

For application in photographic processes, platinum is reduced to salts with particularly useful photochemical properties. Johann Wolfgang Döbereiner first observed the photochemical properties of platinum salts in 1826.[15] Platinum prints, however, rely on a primary reaction with iron. In 1842, Sir John Herschel developed a method of printing photographs that used the light-sensitive salts of iron rather than silver. These iron-based printing methods are known collectively as siderotypes, from the Greek *sideros*, or iron. However, unlike silver, iron(II) oxalate requires a secondary reaction to make a permanent print, either by reducing a noble-metal salt to its metal form, like platinum (platinotype), palladium (palladiotype), gold (chrysotype), silver (argentotype), and mercury (celaenotype), or by forming a pigment such as Prussian blue (cyanotype).[16] The archival permanence of iron-based processes is much higher than silver-based processes, thus introducing stability to the photographic prints they produce. The archival stability of siderotypes was particularly pronounced with platinum prints, given platinum's extreme chemical stability. However, while platinum prints were celebrated for their durability, using this metal posed other challenges. The high light sensitivity of silver halides allows for an instantaneous exposure. In contrast, iron-based processes depend on the photochemistry of iron(III) polycarboxylates, which have a low light sensitivity and thus require longer exposure times. Iron-based processes make contact prints from same-sized negatives, which requires a large-format camera. As cameras became smaller and more portable, silver edged out these more cumbersome iron-based procedures in mass use.

While Herschel's experiments with siderotypes showed the viability of platinum processes, printing in platinum remained challenging, and an effective process took decades to develop. In 1873, William Willis patented the platinotype process and established the Platinotype Company five years later.[17] The prints had a totally matte surface with a tactile quality and a subtle gray tonal scale. The *British Journal of Photography* described how "the tones of the pictures thus produced are most excellent, and possess a charm and brilliancy we have never seen in a silver print."[18] Platinum printing was lauded as a more stable alternative to silver-based photographic processes, whose vulnerability to decay and fading had become a problem. Promoting the process at the photographic section of the American Institute, Willis explained, "The object of this process is to produce photographic prints which shall be permanent (or unalterable) under all atmospheric conditions."[19] Indeed, the National Photographic Record Association, whose mandate was to document English life and landscapes for posterity, insisted in 1897 that all prints were to be in platinum or carbon.[20]

Despite the superior archival permanence of the process, the platinotype was slow to be adopted, as critics perceived it as having an "unphotographic" appearance.[21] The cool gray-black tonality and matte surface of the platinotype diverged from popular taste, which had been shaped by the warm tones and glossy purple-brown finish of albumen prints. While the general public was slow to adopt the cool tones of the platinotype, the process was championed by photographers asserting photography's fine art status, who admired the "fine engraving black without meretricious gloss."[22] The photographer A. Horsley Hinton reflected that "the savage, brilliant and gaudy colours and glittering tinsel [of silver printing] appeal to barbarian taste; the beauty of delicate tones and blended hues is only known to the highly civilised and cultured."[23] A matte finish was considered more elegant than gloss, while the extended tonal range of platinum processes was particularly effective for capturing shadows and luminance.[24] The process became popular with fine art photographers, including Peter Henry Emerson, Alfred Stieglitz, Frederick H. Evans, and Paul Strand. In annual exhibitions of the Royal Photographic Society from 1893 to 1901, between half and a third of the work shown were platinum prints, followed by carbon and silver.[25] Willis adapted the process to meet market demand, introducing sepia-toned and glossy papers. By 1888, several companies began to sell platinotype papers, including Kodak, Ilford, Gevaert, and American Aristotype.

The popularity of platinum prints was short-lived, however, as multiplying industrial uses put pressure on its price. As early as 1858, *Photographic News* lamented that "its scarcity, and consequent high price, make its employment inadmissible in the ways in which it would be of the most use"[26] and in 1866 bemoaned that fifty years prior, platinum was "rather cheaper than at the present."[27] We can recall how Karl Marx reflected that silver and gold were appropriate as currency because they didn't have very many applications—their "individual use-value does not conflict with their economic function"—making the value of silver and gold relatively stable. In this way, silver and gold stand in contrast to platinum. Platinum is used in cancer treatments, fiberglass, flat-panel and liquid crystal displays, computer hard disks, hybridized integrated circuits, and multilayer ceramic capacitors. Platinum was a core component in Sir Humphry Davy's miner's safety lamp, a firesafe lamp for use in coal mines to prevent methane gas explosions.[28]

Perhaps most importantly, platinum is used as a chemical catalyst: a substance that quickens a chemical reaction. In the first decade of the twentieth century, the price of platinum rapidly increased, following the chemist Wilhelm Ostwald's 1902 discovery that platinum could be used as a catalyst for the oxidation of ammonia to nitric oxide—the raw material for fertilizers, explosives, and nitric acid.[29] Platinum became an important catalyst in many other chemical industries, and the cost of the metal began to rise. Platinum printing was banned in Germany in 1901, as the government deemed platinum for filaments in incandescent light bulbs a more critical application of the material.[30] In World War I and World War II, platinum was identified as an essential metal due to its application in wartime industries—as a key material for producing explosives—and its use was banned in photography in England by 1916. Some photographers, such as Frederick H. Evans, gave up photography completely when Platinotype stopped production. A process using palladium, one of the platinum group metals, was adopted, and many twentieth-century platinum prints are in fact made with palladium.[31] Production using platinum group metals started again after the war, but the cost was prohibitive and supply-chain disruptions introduced further challenges to the process. As platinum became restrictively expensive over the course of the twentieth century, attention in photochemical research shifted to refining processes that used silver gelatin photographic paper. By 1937, the Platinotype Company closed, and platinum printing became very rare. The commercial production of platinotype paper stopped in the United States in the 1930s and Great Britain in 1941. When the Alternative Process movement in the late twentieth century took a renewed interest in the method, practitioners had to coat their platinum paper by hand. Ultimately, while luxe platinum prints had many aesthetic advantages for photographers, platinum's industrial use values were too significant to justify its applications in photography: supplying chemical industries—especially during wartime—trumped representational needs. As chemist Mike Ware observes, photography's most refined process arrived late and departed early.[32]

Pictorialist Atmospherics in a Polluted Atmosphere

Platinum printing is primarily associated with pictorialism, the late nineteenth-century movement that advanced photography as an artform, as opposed to a mere documentation of reality. Tonality, composition, and expressiveness marked a successful pictorialist image. Alfred Stieglitz, a founder of the Photo-Secession movement, which promoted pictorialist aesthetics, remarked on the importance of atmosphere:

<blockquote>Atmosphere is the medium through which we see all things. In order, therefore, to see them in their true value on a photograph, as we do in Nature, atmosphere must be there. Atmosphere softens all lines; it graduates the transition from light to shade; it is essential to the reproduction of the sense of distance.[33]</blockquote>

This atmospheric quality came to be found in the particular qualities that platinum prints provided. Hinton described how the intermediate grays of the platinum print were particularly effective for nature photography, "especially of suggesting the idea of 'atmosphere'—that most precious quality of every good picture."[34] The softness of lines lent an air of intensified expressiveness to the images. The pictorialists championed platinum's more suggestive visual form, in contrast to the stark precision of the silver gelatin print, the sharpness of which seemed to erase the hand of the photographer-as-artist.

The hallmarks of atmosphere—mist, fog, steam, and vapor—became key visual devices for the pictorialist interpretation of the world.[35] These could be both natural and industrial, as in Stieglitz's *The Hand of Man* from 1902, which captured a billowing cloud of black smoke pouring out of a train powered by a coal-burning steam engine (figure 3.1). Emissions become a visual device in this pictorialist moment, as the dark smoke pouring out of a train powered by a coal-burning steam engine blends into the billowing clouds. Here we see both natural and industrial atmosphere, swirling together at the edges, dissolving into one another. The softness of the puffs of smoke contrasts the swerving, gleaming steel of the tracks of the Long Island City train yards. The title nods to the human intervention in the landscape, as well as the hand of the photographer. The image was printed as a gravure—a reproducible method of making prints that was popular with pictorialists—in the inaugural issue of Stieglitz's magazine *Camera Work*.

Atmosphere is a formal property but also has a material dimension. As we have seen, silver, the most common material used in analog photography, is reactive to sulfur compounds, which cause silver to tarnish. In the coal-fueled Victorian period, as urban industrial emissions ballooned,

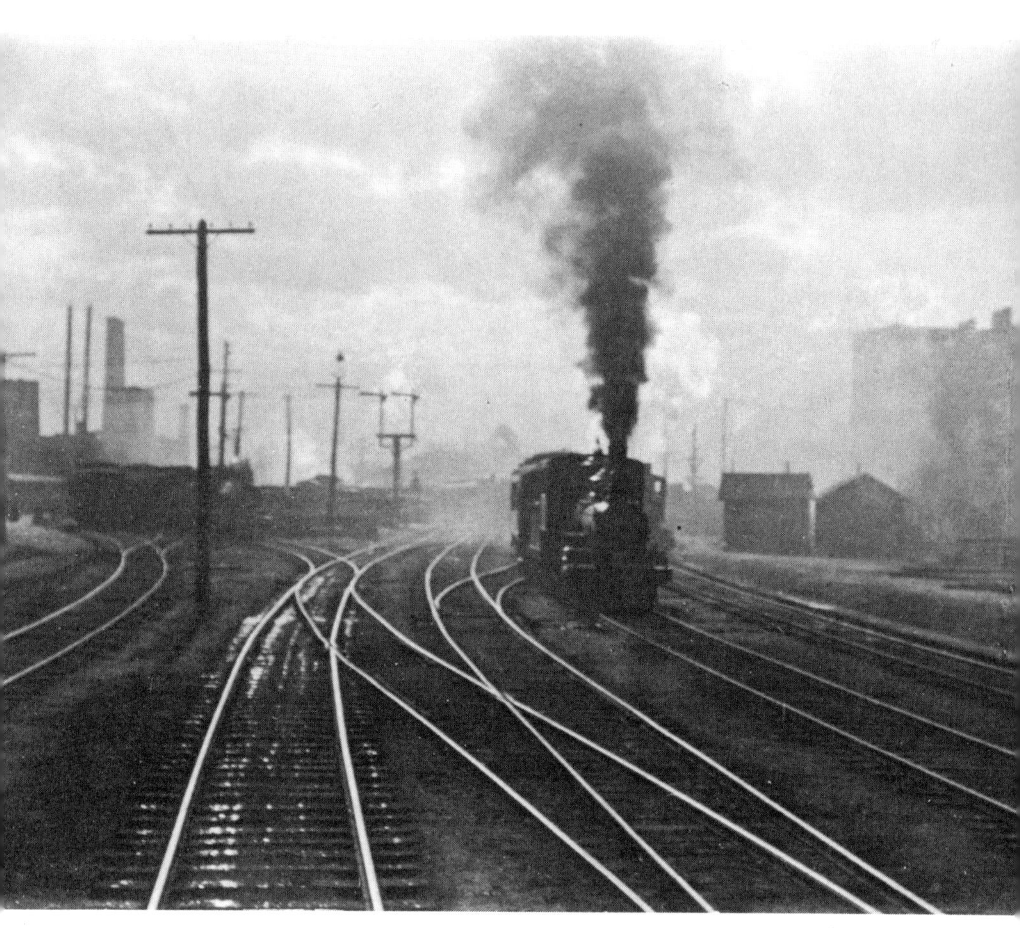

this was a significant problem. Reactions with atmospheric sulfur gases from burning coal damaged silver prints, resulting in the faded, brownish tones that characterize many nineteenth-century photographs printed in the metal.[36] In 1880, *Photographic News* described the "vitiated," sulfur-filled atmosphere of industrial cities as one of the significant challenges facing photographers:

One other cause of fading, beyond that of the mounts, must be borne in mind, however, in discussing the stability of silver prints, and this cause is independent of the operations of fixing, washing,

3.1   Alfred Stieglitz, *The Hand of Man*, 1902. Photogravure.

and mounting: *we mean a vitiated atmosphere. In a coal and gas consuming country like this, the air of our towns and cities must always be more or less contaminated with sulphur.* An ordinary sample of coal contains about one per cent, of sulphur, and coke, unless it is very good, contains the same. Therefore, for every hundred tons of coal or coke consumed in our cities, one ton of sulphur is given off into the air in the form of sulphurous acid, from whose attack few objects are safe. In our apartments, again, the gas burning in dining-room or library contains plenty of the destroying sulphur, and *when we remember that the photograph is, after all, but a thin film of metallic silver, and silver is of all metals one of those most prone to suffer from the action of sulphurous acid,* the picture must indeed be well protected if it is not to sustain injury.[37]

Silver's reactivity to the polluted atmosphere threatened the photographic print's stability, and therefore permanence.

In photography's first decade, there was a growing awareness in Britain of the negative impacts of coal. Prior to the 1840s, carbon and sulfur were understood as air purifiers. This conception that the polluted sky of London was beneficial changed as antismoke campaigns emerged in the 1840s, accompanied by government-initiated studies on the effects of coal combustion on the local environment. Burning coal releases soot, sulfur dioxide, and carbon dioxide into the atmosphere, which was perceptibly causing public health problems, while Manchester's acidic rainfall was linked to its combustion. With this awareness hanging in the air, consideration of the effects of industrial pollution on photographs themselves was not far behind.

As public concern around air pollution grew in the 1840s, daguerreotypists, who relied on light exposure to make images, were likewise very aware of the effects of smog during the exposure process. The smog from burning coal was commonly cited as a cause of the poor-quality daguerreotypes made in England, which were often dark and lacking in definition compared to their counterparts from less intensively industrialized settings. At the 1851 World's Fair, American photographers won three of the five medals awarded to daguerreotypes. A correspondent wrote, "The excellence of American pictures is evident, which is to be accounted for by several reasons. In the first place, American skies are freer from fog and clouds—from bituminous coal not being much used, the atmosphere of our cities is free from smoke."[38] Art historian Kim Beil argues that the success of the American daguerreotypist John Jabez Edwin Mayall in London disproves the smog hypothesis. Instead it was technical and chemical innovations that made American daguerreotypes more precise.[39] Beil concludes that "critics were so alert to the presence of air pollution that they erroneously thought it was affecting their photographs."[40] While smog may not have been solely to blame for the quality of British daguerreotypes, better light conditions did produce higher-quality photographs. Nor is pollution simply a question of light. As the previous quote reminds us: "The photograph is, after all, but a thin film of metallic silver, and silver is of all metals one of those most prone to suffer from the action of sulphurous acid." Air pollution affected prints themselves, especially ones printed on paper. The innovations required to compensate for the smoggy sky also point to the ways in which the interaction between environmental factors and materials drove technological experimentation.[41]

Discussions in photographic journals about the atmospheric effects of air pollution caused by coal combustion became more prominent in the last quarter of the nineteenth century. John Ruskin's 1884 lecture *The Storm-Cloud of the Nineteenth Century* linked the polluted sky to coal combustion in an early attempt to theorize the meaning of the "plague-wind and the plague-cloud," which bring in "peculiar darkness, they blanch the sun instead of reddening it."[42] Ruskin highlights the "uselessness of observation by instruments, or machines, instead of eyes,"[43] referring to scientific instruments. Still, his remark could as easily be extended to cameras, which are limited in their ability to capture atmosphere.[44] The result is "blanched Sun,—blighted grass,—blinded man," linking the polluted atmosphere to a moral gloom and an inability to see.[45] *The Storm-Cloud of the Nineteenth Century* was an attempt to name something that hadn't yet been conceptualized, to theorize the materiality of the elements and make visible the seemingly immaterial changes in light, particles, and atmosphere.[46] As art historians have shown, these atmospheric changes impacted the visual form of nineteenth-century painting. J. M. W. Turner's vivid sunsets are believed to have been influenced by the high levels of tephra—rock and dust emitted by volcanic explosions—in the atmosphere following a series of volcanic eruptions, while the impressionists made the hazy skies of industrial cities an aesthetic trademark.[47]

The technical discussion on photography and atmospheric pollution, however, leaned more directly into new scientific research, which journals noted "has indirectly an important bearing on photography," going so far as to contemplate geoengineering schemes for removing dust from the atmosphere. Practitioners enthused, "What a boon this would be to photographers we need not say."[48] The chemist Harold B. Dixon gave a series of lectures that were discussed in *Photographic News*, titled "On the Use of Coal Gas," which explained that "London gas contains about twelve grains of sulphur per hundred cubic feet," and that "it is probably this sulphur which gradually spoils dry plates which are kept where much gas is burned, and its destructive action on metal work is well known."[49] It was taken as commonplace that the atmosphere of London was "impregnated with soot and vapors" and the "evidence is not wanting that the proximity of the metropolis materially interferes with the photographic, if not the Optical observations undertaken."[50] *Material interference*, by the physical waste of the very industrial society that likewise produced photography's cherished basic materials, lies at the heart of this practitioner's problem.

Fading and impermanence was an ongoing concern in early photography, as prints dulled and lost their sharpness. The *Philadelphia Photographer* lamented that "the art of making pictures by the action of light, was but a beautiful toy, a kind of fairy-like dream, now thrilling us into misery and disappointment, exquisite pleasures from the beautiful images obtained, pain and misery as they faded from our sight."[51] In the *British Journal of Photography*, Cosmo Burton invoked the language of blighted atmosphere, arguing that photographs should be "unpolluted by Silver" if they were to last.[52] Because of the unreliable materiality of silver and the particular challenges of fixing a silver print, alternatives to silver-based processes were actively sought. As Sir John Herschel reflected on the challenges of silver in 1839, "I was on the point of abandoning the use of silver in the enquiry altogether and having recourse to Gold or Platina [platinum]."[53]

In the present, a return to the platinum standard, of sorts, once again directs our attention to atmosphere. Today, in addition to its catalytic properties, platinum is valued as an essential means of combating atmospheric pollution.[54] Platinum collects oxygen atoms (O) and binds them with toxic carbon monoxide (CO), to create the less harmful carbon dioxide (CO_2). Accordingly, one of the main uses of platinum is in internal-combustion engines—one-third of the world's platinum is now used in catalytic converters in vehicles to reduce carbon monoxide, hydrocarbon, and nitrous oxide emissions. Emission-control catalysts significantly impact air pollution: a 1989 test found that noncatalyst cars produced six times more nitrogen oxide, eleven times more hydrocarbons, and thirty-five times more carbon monoxide.[55] Further, platinum electrodes are necessary for alternative energy sources that use fuel cells, which generate electricity without noxious emissions.[56] If the rise of coal enabled a mineral-based economy, platinum is one of the metals essential for an electric economy.[57] Today, the need for catalytic converters to reduce car emissions continues to link the precious metal with atmosphere, though now with a fantasy of clearing its haze. On both ends of the photographic and atmospheric dialectic, the chemical materiality of platinum has enabled its use toward distinct socially defined goals, laying bare the inextricability of industrial life and the materials of cultural production.

Supply Chains and Mining Landscapes

Atmosphere is a nebulous, hard-to-ground material state that, as we have seen, has nevertheless had a tangible impact on the surface of photographs. From there, we can turn to the question of the surface as introduced by the materiality of the platinum print. Silver is suspended in emulsion in silver gelatin prints, forming a barrier between the metal and the atmosphere. The emulsion protects the metal from chemical reactions to atmospheric influences. In contrast, the platinum print image is formed on a single layer of paper, without a binding agent such as albumen, collodion, or gelatin. The platinum is embedded directly in the surface of the paper. William Willis, the inventor of the platinotype process, described how platinum was "left entangled in and amongst the matted fibres."[58] Nothing need mediate or protect the metal from the atmosphere for it to do its photosensitive work.

Willis's Platinotype Company developed and patented the platinotype process, initially cornering the market on platinum print production. The vast majority of platinum prints were produced using materials made by the company. Because of this, it is possible to trace the platinum supply chain in a way that is very difficult for materials like silver, whose paths extend across dense and variegated global networks, mediated by multiple competing firms. The platinum used by the Platinotype Company was sourced in the form of the salt potassium chloroplatinite from the chemicals company Johnson Matthey, a leading supplier of precious metals.[59] Johnson Matthey was founded in 1817 as a gold assayer but quickly established a specialization in platinum: their precious metal products were exhibited in the 1851 World's Fair while the company provided the platinum used in key experiments, including Michael Faraday's research into magnetism and Humphry Davy's work on platinum wire as a catalyst for the combination of oxygen and hydrogen.[60] Unusually for a precious metal supplier in the nineteenth century, Johnson Matthey established a research department, placing their firm on the cutting edge of precious metal refining.

In the nineteenth century, 98 percent of platinum was sourced from Russia's Ural Mountains. During World War I, the Ministry of Munitions in the United Kingdom appointed Johnson Matthey the sole agent and distributor of platinum. The escalating demand for platinum in war industries faced significant hurdles, as the Russian Revolution and the establishment of the Soviet Union disrupted this platinum supply chain, leading to what the industry experienced as a "platinum famine."[61] Johnson Matthey managed to process 1.3 tons of platinum during the war, but the use of platinum in photography was banned.[62] However, in 1924, the geologist Hans Merensky discovered platinum-bearing rocks on the Merensky Reef, a layer of igneous rock in the Bushveld Igneous Complex in South Africa. This was a seismic shift of a find because, as mentioned, 80 percent of produced platinum worldwide today is sourced from South Africa. The ore had an extremely high concentration of platinum, but it was bonded to sulfides which complicated its extraction and processing. The ore samples were sent to refineries worldwide, and Johnson Matthey developed the Powell-Deering process to refine the ore. As such, Johnson Matthey became the sole refiner for Rustenburg Platinum Mines, the world's largest producer of platinum. Extracting the metal in open-pit mines on the surface as well as underground, where the narrow and deep shafts have high temperatures and cramped conditions, is labor- and capital-intensive.

The industry came to shape entire regions of South Africa, with the country's platinum mining landscape becoming the subject of British photographer Simon Starling's *One Ton, II: 5 Handmade Platinum/Palladium Prints of the Anglo American Platinum Corporation Mine at Potgietersrus, South Africa, Produced Using as Many Platinum Group Metal Salts as Can Be Derived from One Ton of Ore* (2005) (plate 14). Starling's platinum print directs our attention to the surface of the picture *and* of the mining landscape. An open-pit mine is shot from an elevated angle, eliminating the horizon line. The framing of the image situates the viewer in the flat grays of the mine. The monochrome subtlety and sharp tones characteristic of platinum prints emphasize texture, drawing attention to the carving away of the earth, revealing the different rock strata. The angled and stepped sides of the pit, built to prevent avalanches, form an abstract geometry indexed in the earth. This is a landscape transformed by mining, but the typical hallmarks of a mining site—tailing ponds, slag heaps, headframes, or other structures—are absent. Instead, the photograph draws attention to deficit, as the open-pit mine is chiseled out of the landscape, leaving behind something like a wound. Within the photographic frame, all that is left are the textures, shadows, and layers of the earth itself.

One of a series of five prints, the platinum set in the print was extracted from the Mogalakwena Mine, the Anglo American Platinum corporation mine at Potgieter, the largest open-pit platinum mine in the world. Anglo American supplies 40 percent of platinum worldwide. In a move parallel to that of Cariou (see chapter 1), Starling repurposes platinum extracted from the mine to document extraction in the mine. The title directs the reader to notice the physicality of Starling's photographic print as a way of making tangible to the viewer the incredible effort required to produce five small prints. The modest size of the prints (29.75 × 27.50 × 1.75 in.) is determined by the amount of platinum in one ton of ore, a striking dissimilarity that draws attention to the high energy demands (both in labor and manufactured energy) of platinum mining. The haul road that snakes through the pit provides a sense of scale, as trucks resembling toys wind around the curve. *One Ton* draws a direct link from platinum mining in South Africa to the platinum print, revealing an inseparable relationship between the process of mining platinum, the image, and the chemical processes underlying what it makes visible.[63] Within the image-object, we see both the specificity of the print and the process used to make it, as well as the industrial structures that make the photograph possible in a larger sense. Starling draws a dynamic connection between the image, the material, and the economies of production, as well as the circulation of energy, minerals, and images.

Extractive landscapes are often in tension with the conventions of landscape as a visual genre, for as Raymond Williams observed, a working country is hardly ever a landscape, as conventional landscape art implies distance and contemplation.[64] Mining, in particular, poses representational challenges, for in such extractive landscapes, the reshaping of the natural world and the historical production of nature are made visible. Mining landscapes testify to how humans and other natural forces have acted upon land. Art historian Rachael Z. DeLue states that mining "manifested as a problem for the environment and as a problem for representation, for it presented the artist with a loss of form, with absence and deprivation, a deficit of material rather than a bounty of things to depict."[65] Open-pit mining amplifies this sense of deficit as the mine is carved out of the earth. As these materials are excavated, trace particles enter the atmosphere—just as when coal combusts. The surface transformations of land invite a consideration of *drift*, which the geographer Deborah Dixon describes as "the surficial detritus of unconsolidated sediments, such as boulders, gravel, sand, silt and clay, that lie atop and obscure the consolidated layers of rock that we refer to as strata." Dixon criticizes the geological turn within the humanities for reinforcing an idea of the earth as a "stratified, lithic archive," instead highlighting how drift is mobile, dynamic, and exceeds the desire to delimit it into clear categories or spaces.[66] Read atmospherically, the instability of the surface landscape—and the way that it drifts into the air—underpins the contained scene.

While photography is limited in documenting the more complex spatial networks of extraction, it can form an evidentiary record of the transformation of territory through such extractive processes. Potawatomi environmental scientist Robin Wall Kimmerer writes, "The story of our relationship to the earth is written more truthfully on the land than on the page."[67] To borrow a photographic analogy, mining landscapes are an index of mining processes and economic systems.[68] The mine shafts, slag heaps, tailings ponds, clear-cut forests, flooded lakes, and open-pit mines form a record of the surface transformations caused by industrial processes, written directly on the landscape. The interplay between the geological register of the photograph and the flash of light that produces the index is thus paralleled in the transformation of land. Both the photograph and the landscape are an index of economic and industrial history, a reality highlighted in Goldblatt's and Starling's platinum prints, which connect the material being extracted from this landscape and the representation of that same landscape of extraction.

Labor and Embodiment

One hundred years after O'Sullivan captured a work crew above ground at a Nevada mine, we find ourselves at the surface of the mine once again. In a platinum print by David Goldblatt, *A Miner Waits on the Bank to Go Underground, City Deep Gold Mine, 1966* (1966, printed later), a man stands with his arms crossed, legs spread wide but planted firmly on the floor (plate 15). He stares at the camera, though the framing is off-center and the image is sharply angled. The hard hat identifies him as a manual laborer, and the signs in the background locate the scene at the City Deep Gold Mine in Johannesburg, where gold, silver, and platinum-group metals are mined. Stark strokes of sunlight form abstract patterns that cut across the image, heightened by the velvety matte surface and sophisticated tonal range of platinum prints. Here, the interplay between light and shadow obscures more than it reveals. A splash of sunlight splits the miner's face down the middle, brightening the left side and hiding the other half. Tension characterizes the image, both in the linearity of the framing and the tautness of the miner's body language. While the miner addresses the camera, he doesn't come fully into view.

The image was originally published in *On the Mines* (1973), a photo book containing fifty-six black-and-white photographs structured by three chapters: "The Witwatersrand: A Time and Tailings," "Shaftsinking," and "Mining Men." The first section explores the landscapes of mining, and very few of its photographs have any people in them. In the second, Goldblatt goes deep underground to document blasting and shaft-sinking. The conditions underground required working with both low light and fast film, resulting in painterly photographs with hazy lighting and ambiguous forms. The third section turns to the broader ecosystem that makes mining possible. Through this range of sites, Goldblatt captured the complexity of life in extractive zones, taking viewers into the corporate offices of white industrialists, crowded buses where workers make long commutes due to restrictive zoning laws, and quotidian spaces like barber shops. This body of work makes visible the Black workers whose labor was central to the city's functioning but whose visibility in key spaces was suppressed under Apartheid's racial segregation. Stylistically, *On the Mines* aligns with photojournalism, amplified by Goldblatt's sequence of images and explanatory captions. For example, in a photo of shovels stacked into a sculptural formation at the Central Salvage Yard, Randfontein Estates, Randfontein, Goldblatt explains: "'Lashing' shovels retrieved from underground. Every grain of sand in the yellow tailings dumps that made the Witwatersrand landscape and every grain of gold that made its wealth, came from a rock off a Black man's shovel underground."[69] The dynamics of racial capitalism underpinning extraction are explicitly drawn out by Goldblatt in the caption, directing the reading of the image beyond the frame to consider the structural factors that shape mining work.[70]

Goldblatt's photographs reveal that the mining landscape cannot be separated from mining labor. South Africa's mines are the deepest in the world, and the ore is comparatively low-grade, making underground mining capital-intensive. In gold mining, the price of gold is relatively fixed, so the costs of extraction cannot easily be passed along to consumers. Deliberately deflating wages below the cost of living was a key strategy firms undertook to make extraction profitable. Over the course of the twentieth century, a series of laws constructed an underpaid Black workforce, most notably the Mines and Works Act (1911 and 1926), which banned Black workers from skilled mine labor. This legislation was combined with punitive taxes, the criminalization of quitting, and spatial and occupational segregation.[71] Platinum mining has been similarly charged, characterized by low wages, high levels of industrial disease, and little job security. This instability for workers is matched by the volatility of the entire platinum market. The US Geological Survey ranks platinum as the highest on its criticality index of all metals, susceptible to supply-chain disruption by economic, environmental, political, and social events.[72] The need for the South African state to contain such volatility spilled over into violence during the Marikana massacre on August 16, 2012, when the South African police killed thirty-four striking platinum mine workers, injuring a further seventy-eight strikers.[73] The material stability of platinum, then, depends on the stable enforcement of the costs of the precarity of platinum extraction onto the workforce that provides it.

Goldblatt's choice to print the portrait of the miner and the mine tailings in platinum directs attention to the bodily impacts of mining. Just as the landscape forms an index of industrial and economic processes, so does the human body. Sociologist Rebecca Altman writes, "Material culture and industrial infrastructure carry the history of their making. What happens when their residues enter the body? Do they transfer that history to us?"[74] Altman directs our attention to the reality that traces of industrial processes in the form of toxins accumulate in human bodies. What, then, of embodiment in South Africa's mines? The mineral formation of the Bushveld Complex exposes miners to several substances that cause industrial disease: silicosis from crystalline silica, interstitial fibrosis from copper and iron, pneumoconiosis from cobalt, stannosis from tin, adult respiratory distress syndrome from zinc or sulfur, and lung cancer from arsenic or nickel.[75] In platinum refineries, too, workers have a high risk of developing platinum salt sensitivity associated with respiratory symptoms of asthma, rhinitis, urticaria, and dermatitis.[76] Industrial disease was also a problem for workers in factories producing platinum papers. So far, in this book, I have focused primarily on the mine. But minerals and metals move through complex chains of production in which they are refined and transformed by large workforces, including refineries and factories where these materials are turned into the base products of photography. Workers at every stage of the production chain differentially share in these occupational health risks, and the collective effect on their bodies is likewise part of the material history of the photograph.

Sir Thomas Oliver's *Dangerous Trades*, a groundbreaking 1902 study of occupational health in England, included miners alongside factory workers producing chemical products for photography, including potassium, sodium dichromate, and mercury, linking the risks faced by working-class laborers across industries. Oliver notes the considerable difficulty in defining industrial disease, which includes both direct and indirect effects. For instance, lead exposure has direct effects, such as colic, which surfaces as sharp, sudden spasms of pain, but kidney disease develops slowly and is a "remote or *indirect* consequence of plumbism."[77] Ventilation, density and duration of exposure, previous health concerns, and poverty are other factors that contribute to the impact of workplace exposure. As historian Jennifer Tucker has shown in her analysis of visual representations of chemical workers in England, chemical factories were some of the most dangerous workplaces in nineteenth-century England, even as the chemical industry products were tied to Victorian middle-class ideals of "civilization."[78]

In 1911, a study of workers' occupational health in forty different photography studios in Chicago discovered that working with platino-type paper was a health hazard. Handling the halogenated salts of platinum caused an allergy to reactive halogen ligands, leading to identifiable kinds of occupational disease.[79] Workers experienced severe irritation of the throat and nasal passages, asthma (wheezing, coughing, running of the nose, tightness of the chest, shortness of breath, and cyanosis), and dermatic skin lesions, including eczema and urticaria. The risk came when platinum was reduced to salts, with dust inhalation posing the highest threat along with direct contact with the surface. Studies on workers in platinum refineries have shown similar patterns of health issues.[80]

More generally, photography has been a "chemical playground," and experiments significantly affected earlier photographers.[81] In 1864, *Photographic News* called disease one of the "special dangers of the profession," documenting the dismissal of the risks of certain chemicals by "high medical authorities" though the "alarm has been sounded over and over again."[82] This language echoes the organizing campaigns by miners who sought to have industrial illnesses recognized, though it was well into the twentieth century before there was any movement on this front.[83] While occupational disease is a significant problem for workers in photographic industries, photographic representation struggles to capture the attritional violence of unsafe workplaces. While there have been photographic studies of workers' health and disease, occupational illness largely evades photographic capture.[84]

In light of this permeability of the working body, I want to return to Goldblatt's image of the miner on the surface. The light that splits his face is a problem of exposure: light deforms the image. But exposure perhaps has a double meaning in this context: exposure to light and exposure to toxicity make the image—and what it itself exposes—possible. What Elizabeth Povinelli calls the "toxic unconscious" of photographs becomes grounded in the worker's body in the photographic supply chain. The workers inhaling dust in the mine and the photographic workers inhaling platinum salts show how the mine and minerals physically enter human bodies, often through dust or other trace elements like radon in the air.

These links are not limited to the workplace, as the mine tailings show: environmental racism permeating the spatial structure of our human landscapes exposes low-income and racialized people to higher levels of industrial disease in their communities and homes.[85] Industrial disease complicates the distinctness of the surface, showing the permeability of the human body as elements move through people, troubling a clear distinction between the human and the natural, between the inside and the outside.

Stability, Permanence, and Futurity

By way of conclusion, I want to turn to a different landscape to consider how the material stability of platinum prints might speak to the future. In a platinum print photograph by Larry McNeil (Tlingit/Nisga'a) titled *Demented Coal Paradox* (2013), a human figure is superimposed over the industrial infrastructure of the Antelope Coal Mine in Wyoming, run by Peabody Coal (plate 16). McNeil gives standard hallmarks of industrial production a dystopian twist, as the human wears a gas mask woven from spruce root. The patterning of the mask mimics the geometric forms of the steel transmission towers that connect the power lines cutting through the scene. The towers evoke the networked world of energy and communications technology, but transmission also names the spread of viruses, the airborne movement of particles. The interplay between the surface and the permeation of the body takes a new resonance here as the figure is sealed off from the atmosphere—a protected but unsettling presence. The superimposition of the figure denaturalizes the industrial scene. Here, it is illustrative to return to Stieglitz's comment that "*The Hand of Man* is an attempt to treat pictorially a subject which enters so much into our daily lives that we are apt to lose sight of the pictorial possibilities of the commonplace."[86] Stieglitz aestheticized the coal-powered train to make it visible; McNeil makes the coal mine strange, uncanny. It is not a scene pulled from nature but from an imagined—albeit likely—polluted future, in which masks are necessary to protect people from everyday particulate matter, raising the question of our transforming atmosphere. The edges of the print fade into darkness, evoking a darkening sky or gathering smoke. Here, atmosphere is encroaching, threatening, stifling—full of smoke and CO_2.

The human figure wears a gas mask woven using Tlingit basketry techniques, suggesting that Indigenous knowledge is needed to protect humans from the devastation caused by extraction, a recognition of the reality that Indigenous peoples, who make up less than 5 percent of the world's population, protect 80 percent of its biodiversity.[87] Spruce-root baskets are an essential tool for gathering in Tlingit cultures, and their continued use reflects the endurance of Indigenous cultural traditions. As Indigenous cultures faced ecosystem collapse, the destruction of traditional foodways, and cultural genocide under the imposition of colonial-capitalist structures, spruce-root basket weaving endured. This reminds us that, from an Indigenous perspective, there is nothing new about climate crisis. As Kyle Powys Whyte (Potawatomi) explains, "Climate injustice, for Indigenous peoples, is less about the spectre of a new future and more like the experience of déjà vu," as the geographies of extraction and the interlocking catastrophes catalyzed by colonialism, industry, and capitalism have disproportionately impacted Indigenous nations.[88] At the same time, however, Whyte draws from Anishinaabe (Neshnabé) intellectual traditions to theorize collective continuance, revealing that adapting to environmental change is at the heart of Indigenous ways of relating to land.[89] In the face of large-scale climate change driven by extractive capitalism, empire, and settler colonialism, Indigenous traditions adapt and continue as ways of living, relating, and making the world.

Why would McNeil produce a platinum print? In part, it responds to historical image-making. The rise of platinum prints at the end of the nineteenth century coincided with an escalation in white settler violence against Indigenous nations. Lavishly printed platinum portraits of Indigenous people by Edward Curtis, Joseph Keiley, and Gertrude Käsebier exoticized the so-called vanishing races of Indigenous nations. The soft expressiveness of the prints heightened the romanticized, stereotypical portrayal. A form of visual salvage ethnography, a premise underlying the portraits is that authentic Indigenous culture had to be documented photographically before it was eclipsed by Western modernity. The archivally stable prints promised to preserve a likeness for posterity in the face of cultural disappearance. Repurposing platinum responds to this historical moment, contesting the superficiality of the portraits, which abstracted Indigenous peoples into types. The reactivation of a historical process by McNeil responds directly to settler photographers' early photographs of

Indigenous people. It suggests an ongoing dialogue between history and the present that denies the fixity or distance of the past by showing how it coexists in the present.

McNeil's print itself underwent stages of material transformation: from Kodachrome film to a digital image to the platinum print. This series of shifts destabilizes the idea of the photograph as fixed or straightforwardly indexical, as the likeness takes multiple forms. This intermedial process responds to static notions of cultural authenticity as fixed in the past, as McNeil's process journeys from analog to digital and back again. At the same time, McNeil utilizes the archival stability of the print to make something with permanence: an alternate vision of the future that centers an Indigenous gaze. As *Indelible*, an exhibition at the Smithsonian National Museum of the American Indian that featured this work, summarized, using the platinum process is an assertion of Indigenous futurity that proposes that in climate breakdown, Indigenous traditions "persist and protect."[90] McNeil uses platinum to produce objects that will endure, images that will continue to speak in the future. Part of his *Global Climate Change* photo series, the image is a message to future generations. The "demented coal paradox" is the reality that the United States releases more coal emissions than any other country despite widespread awareness of the environmental impacts of burning coal. This inaction in the face of climate breakdown can only be read as demented, for when future generations ask why we kept burning fossil fuels, given our knowledge that it would lead to a "hellish existence for them," McNeil reflects that we can only say in response, "Yes, we knew."[91]

A latent theme running throughout this chapter is futurity: platinum printing was valuable because it would last; its permanence gave the prints a futurity. Anxiety over the impermanence of silver prints reflects the desire to indelibly fix an image, a desire seemingly attained with platinum. An often-repeated claim is that a platinum print will last one thousand years. What the world will look like in one thousand years is a more complex question. In an era shaped by the greenhouse gas effect, atmosphere and futurity are fundamentally linked. As literary scholar Eva Horn reminds us, air is the "infrastructure" of life; it forms the "*conditions of possibility.*"[92] A central tension emerges as a metal defined by its ability to last is routed into extractive processes foreclosing individual and collective futures. In the era of global climate change, the future will be shaped by atmosphere and atmospheric changes. Looking to the sky for prophecies has a long history: reading weather in clouds, omens in the heavens, one's destiny in the stars. Today, the monitoring of rising CO_2 levels makes a grim prognostication: the atmosphere cannot withstand the emissions produced by burning fossil fuels. The atmosphere cannot withstand the way of life fueled by extractive capitalism. As Nick Estes (Kul Wicasa, Lower Brule Sioux Tribe) concludes, "For the earth to live, capitalism must die."[93]

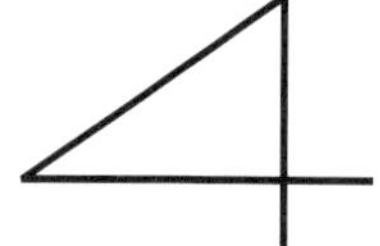

Iron and
Unstable Boundaries

LaToya Ruby Frazier, in "Intimate Debris" (2017)

CONSIDER THE SPIDERY TENDRILS of *Chordaria flagelliformis* superimposed over a flat, even background of Prussian blue (plate 17). The photographer and botanist Anna Atkins meticulously arranged the slender, elongated branches on the photographic paper, highlighting the formal structure of the plant. The emphasis on pattern and rhythm adds a sense of dynamic movement to the image, which is appropriate for a plant whose name, flagellum, comes from the Latin for "whip" or "scourge." The other part of the Latin title signals to the viewer the color of the algae, as *Chordaria* is a family of brown plants. In the photogram, however, the algae is desaturated, and the stark white of the contact print forms a striking contrast to the saturated blue background, which replicates the marine environment of the plant. What one loses in verisimilitude of color is compensated by the emphasis on pattern, which draws to mind the root structures of plants extending into the earth. The filamentary wisps of algae visually evoke mineral seams running underground and the veins in the human body. Once the algae was harvested and dried, Atkins placed the specimen on paper prepared with a mixture of iron salts. The print was then exposed to the sun. The print turned blue when washed, leaving a negative impression. Underneath the imprint of the plant, Atkins has written its Latin name, accented with tendrils of algae. The specimen is thus collected, catalogued, and identified. The print is one of nearly one thousand images produced for multiple albums that formed her self-published book series, *Photographs of British Algae: Cyanotype Impressions* (1843–53).

While Atkins was pressing algae under glass and exposing her prints to the sun in Kent's countryside, the world around her was undergoing tremendous change. Photography emerged in England in the transitional period between early industrialization and the Victorian era. Atkins's work is imbricated within this larger social context, though on the surface her photograms don't tell us much about the rise of industrial capitalism and the growth of empire. The prints as representations depict a world seemingly far removed from the changes brought by industrial growth; however, an ecocritical reading of Atkins's images shows that a focus on the prints' materiality nevertheless reveals that they would have been impossible without the advance of industry. Materially, the cyanotype is a siderotype, like platinum prints, an iron-based process that uses iron(II) oxalate as a light-sensitive material to make prints in Prussian blue/ferric hexacyanoferrate(II). Due to iron's low light sensitivity, iron-based processes require a secondary reaction to make a permanent print. In the cyanotype process, the image is fixed by a chemical reaction forming the pigment Prussian blue, giving cyanotypes their distinct blue hue.[1] The use of iron is an entry point to thinking through industrialization, as cheap and plentiful iron was the scaffolding for the emerging industrial economy.

Links between the material strength of iron and the soft power of Victorian values were forged in the cultural sphere. The architectural possibilities of iron were showcased in the Crystal Palace, the glass and cast-iron structure built to house the Great Exhibition of 1851 in London, which epitomized the scientific and industrial progress of the Victorian era. The exposition itself was also a tribute to the power of iron, as displays showcased the world-shifting applications of casting molten metal in locomotives, steam engines, and lampposts. However, the devastating impact on workers, ecosystems, and atmosphere caused by the extraction and refining of coal and iron was not part of the story told by the Great Exhibition's installations. Indeed, the prominence of iron in world's fairs would continue. As a contemporary wrote of the Exposition Universelle of 1889 in Paris, "We can say of this festivity that it has been celebrated, above all, to the glory of iron."[2]

While iron's primary role in cultural imaginaries is as an industrial metal, it is, as an element, essential to growth and metabolism in human bodies and the plant world. In humans, iron is central to the proteins hemoglobin, which carries oxygen from the lungs, and myoglobin, which provides oxygen to muscles. One human body contains enough iron to make a nail, reminding us of the metabolic permeability of the organic and the inorganic in human-nature interactions.[3] In plants, iron is essential for the synthesis of chlorophyll.[4] Ferric salts, or iron salts, were so widespread in plants that Sir John Herschel, inventor of the cyanotype, called them vegetable acids.[5]

This intimate connection between organic matter and the processed stuff of industry structured the Great Exhibition's central edifice. Visually, the Crystal Palace has an affinity with a greenhouse. Designed by the famed gardener Joseph Paxton, the structure was inspired by Paxton's Lily House, which was built to house the *Victoria amazonica* waterlily and whose structure reflected the architecture of the waterlilies themselves, a coming-together of natural and industrial forms. Paxton's Crystal Palace was designed to protect centuries-old trees in Hyde Park. Like the greenhouse, Atkins's prints also preserve (the traces of) organic life through the mediation of iron. In *The Arcades Project*, Walter Benjamin engages with A. G. Meyer's writings on the material properties of iron, particularly in his writing on the Crystal Palace itself. Meyer summarized the origins of the Crystal Palace, writing, "It was not a space-articulating architect who did the talking but a—gardener."[6] Meyer concludes that, as a result, the origin of iron-and-glass construction is the greenhouse. Benjamin reflected on the implications of this in the arcades in Paris, where the glass and iron structure designed to create an ideal environment for plant growth was applied to protect shoppers from the elements: "Curious that [the arcades] should be bound in its origin to the existence of plants."[7] Benjamin draws both a material and a conceptual link between the consumer worlds of the Paris arcades, the giant greenhouse of the Crystal Palace, and the role of iron itself as the material of industrial development. At present, another dimension of this connection is immediately apparent, as the greenhouse effect traps the sun's heat, causing the planet's mean temperatures to climb and drastically changing the climatic patterns that have sustained life across this planet. The consequences of industrial progress thus become differently tangible in this greenhouse: a suffocating covering that can destroy rather than nourish organic life.

Building from Benjamin's reflection on the link between culture, plants, and iron, I use iron to think through the material intimacies between the human and extrahuman worlds. While silver's extractive and photographic history is relatively straightforward, the material transformations of iron are harder to ground. Meyer writes on the architectural possibilities of iron:

> This material, in its first hundred years has already undergone essential transformations—cast iron, wrought iron, ingot iron—so that today the engineer has at his disposal a building material completely different than from some fifty years ago.... In the perspective of historical reflection, these are "ferments" of a disquieting instability. No other building material offers anything remotely similar. We stand here at the beginning of a development that is sure to proceed at a furious pace.... The ... conditions of the material ... are volatilized in "limitless possibilities."[8]

Meyer's reference to the limitless possibilities of iron echoes the belief in the potentially unfettered growth of industrial capitalism, driven by material possibility and human ingenuity. However, I want to draw attention to these "'ferments' of a disquieting instability" in iron's material transformations. While Meyer and Benjamin focus on the architectural possibilities of iron as a building material, its elemental form inhabits human bodies, plants, and photographic prints. Iron's prominence in

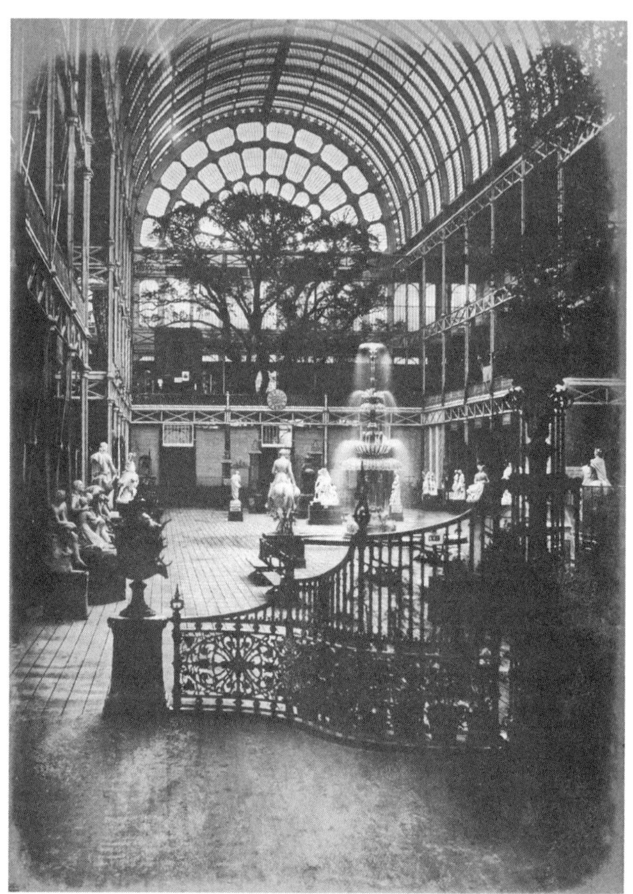

the plant, human, and industrial world highlights "a disquieting instability" between nature and culture. As a material, iron brings the unstable boundaries of nature and culture into view, inviting us to rethink the geological, botanical, and human together. By focusing on the themes of *growth* and *metabolism* introduced by iron as a mineral, I explore the history of iron in industrialization to consider how artists adapted visual vocabularies to respond to the Industrial Revolution's social, cultural, and ecological changes.

Taking iron as the starting point, this chapter turns to processes of industrialization and deindustrialization. I begin with a brief history of

4.1   William Henry Fox Talbot, *Interior of the Crystal Palace*, London, 1851. Mounted calotype.

iron as a material and its association with stability and industrial growth, which I read against its chemical instability. Iron's central role in industrialization situates us in Victorian England through a close study of Anna Atkins's botanical cyanotypes. I consider the material intimacies of making photograms before taking a broader view to situate the prints within social and historical memory. Representationally, Atkins's prints do not reveal very much about the seismic social and political upheavals of the period, but the materiality of the print registers industrial growth, linking cotton, paper mills, the railway, and plantation slavery. Atkins's precise, haptic prints give way to the applications of blueprint photography in the rapidly expanding steel industries of the American Northeast in the late nineteenth century, coinciding with the Second Industrial Revolution. I read cyanotypes of railroads against William Rau's albumen prints to show how blueprint photography promised, and indeed made materially possible, an industrial futurity. To bridge the industrial associations of cyanotypes with Atkins's material intimacies, I conclude with an analysis of LaToya Ruby Frazier's cyanotypes of deindustrialization in the US Rust Belt. Iron, in its movement between registers, destabilizes any neat binaries or bifurcations between nature, humans, and geographies. Instead, it forges a sturdy yet changeable web of relationships and connections.

Iron

Iron (Fe, atomic number 26) is the most common element on Earth by mass, but it is rare in the earth's crust in its metallic state. High-quality metallic iron is primarily found in meteorites. Prior to the Industrial Revolution, Britain produced brittle, low-quality iron. British iron production transformed over the course of the eighteenth century, growing from twelve thousand metric tons in 1700 to over two million metric tons by 1850. This explosion in production was the result of a series of innovations: in 1709, Abraham Darby powered a blast furnace with coke instead of charcoal, while Benjamin Huntsman invented the crucible method of steelmaking in the 1740s. Henry Cort patented the puddling and rolling technique in 1783–84, which enabled large-scale production by removing impurities from iron—a process initially developed by Black metallurgists in Jamaica.[9] Later, in the 1850s, Henry Bessemer developed a technique to mass-produce steel, an alloy of iron augmented with carbon. In its many forms, iron was an essential material in the industrial projects that formed the backbone of the expanding industrial and imperial economies, like steam engines, railway tracks, locomotives, and ship hulls.

The rapid pace of innovation brought massive social and economic transformations. While iron is defined by its strength and stability, in its raw form it has tendencies toward material instability. To achieve its famed stability, the material has to be extensively worked. In the symbolism of ancient Greece and Rome, the common periodization narrates a fall from grace moving from the Golden Age to the Iron Age. The mythology of the Golden Age promised a period when nature offers gifts without any effort. For Ovid, the Iron Age saw "men demarcate nations with boundaries; they learn the arts of navigation and mining; they are warlike, greedy, and impious."[10] In the Iron Age, important materials—like iron—needed to be mined rather than simply found.[11] According to the symbolist René Guénon, as the ages progressed away from the dawn of human society, they were characterized by increasing materialization.[12] The progression from the purest metal (gold) to the most malleable (iron) "implies involution."[13] In this framing, gold is a gift, but iron must be forged to perform its function. Gold is found in its natural form in the sand and streams of water; in contrast, iron always has to be mined and separated from its ore through high heat. It is perhaps for this reason that cultural imaginaries of iron center on the mill, where iron is processed and refined, rather than the mine. As Benjamin speculates, it may be "enthusiasm for machine technology and the faith in the superior durability of its materials that explains why the attribute 'iron' is used . . . whenever . . . power and necessity are supposed to be manifest. Iron are the laws of nature, and iron is the 'stride of the worker battalion'; [the] union of the German empire is supposedly made of iron, and so is . . . the chancellor himself."[14] Iron, in its worked form, stands in for human ingenuity, industrial power, and unbreakability.

The iron that provided the scaffolding of the Industrial Revolution was extracted from iron ore deposits deep in the earth or gathered from fallen meteorites. The iron that chemically interacted with light to produce the distinctive shade of blue in cyanotypes is in many ways removed from the extractive history of the iron industry. The prehistory of cyanotypes lies in an accidental ferment in an alchemist's lab in Berlin in the early sixteenth century. The generally accepted origin story begins with Johann Jacob Diesbach, a pigment maker who shared a workspace with Johann Konrad Dippel, an alchemist working on an elixir of life. While making the red pigment Florentine Lake, Diesbach bought potash from Dippel.[15] Dippel's potash was contaminated by an oil produced through the destructive distillation of bones.[16] As a result, nitrogen from the bones was added to Diesbach's pigment, which resulted in the unexpected shade of blue. The chemical reaction that produced the first Prussian blue pigment came from organic materials: blood, "hide, hair, feathers, horn, hooves or flesh," a decidedly *bodily* prehistory.[17] Any nitrogenous animal matter (protein) could produce the reaction. Dried ox blood was particularly popular.[18] Prussian blue is a deep and rich high-chroma pigment that quickly became in-demand: Pieter van der Werff's *The Entombment of Christ* (1709) is the oldest known painting that uses Prussian blue, in the sky and on Mary's mantle. By the 1730s, the manufacturing of the pigment was widespread, and the pigment went into factory production.[19]

The alchemist's lab played a significant role in many of the discoveries that led to photography; indeed, Dippel's oil was also used in Niépce's heliograph. The accidental cross-contamination of the pigment with nitrogen from bones created the first modern synthetic pigment, highlighting the unruly boundaries between chemical production and its organic "outside." A century after the discovery of Prussian blue, Sir John Herschel discovered that the chemical reaction could be replicated photographically. In 1842, Herschel observed the effect of light on potassium ferricyanide and ammonium ferric citrate. He called his invention the cyanotype, after the distinctive shade of blue that surfaced on the print. Cyanotypes, as we saw with platinum prints, are siderotypes, or iron-based prints, which require a secondary reaction to make a permanent impression. In platinum prints, for instance, platinum salts must be reduced to the metal. In the cyanotype process, the image is fixed by forming the pigment Prussian blue.[20] The formula for cyanotypes has changed very little from Herschel's original formulation. Typically, paper is coated with a mixture of iron salt (typically ferric ammonium citrate) and potassium ferricyanide.[21] An object or negative is placed on the top of the paper and held under glass. After several minutes of exposure to sunlight, the object is removed, and the paper is rinsed. The chemical interaction of light with the mixture leaves the parts exposed to light in Prussian blue (ferric ferrocyanide). The subject is silhouetted in white—what Carol Armstrong describes as evoking a cameo.[22] The most striking aspect of cyanotypes is their color, but they have likewise had the benefit of being affordable and relatively simple to produce.

Herschel's discovery of a silverless photography came three years after Daguerre and Herschel's friend William Henry Fox Talbot announced their success in setting photographs in silver. Chemist Mike Ware concludes that the motivations driving Talbot's and Herschel's inventions were different:

> while Talbot was creating several thousand silver images and assembling them into a creative oeuvre that is now recognised as the richest vision in early photographic art, his fellow-scientist Herschel, rather than making new pictures, was experimenting widely with the processes of photography, for he was motivated by a preoccupation with the medium rather than the message. As one of the leading physical scientists of his day, Herschel was driven by a desire to understand photochemical phenomena, and to enlist them as tools for probing the electromagnetic spectrum outside the narrow optical limits imposed by human vision.[23]

While Herschel was more interested in the process than in the images themselves, his personal network spearheaded some of cyanotypes' most iconic applications. He was friends with John George Children, a scientist who served as vice president of the Botanical Society of London, who then introduced his daughter Anna Atkins (born Anna Children) to the cyanotype process. In 1843, Atkins began working on a series of photobooks, *British Algae: Cyanotype Impressions*, applying Herschel's formula.

Anna Atkins's Material Intimacies

Despite the striking formal beauty of Atkins's prints, the cyanotype process has generally been relegated to the realms of amateur production and female domestic craft, marked as distinct from serious photography. As a process, cyanotypes are affordable and simple, which made them suitable for Atkins's ambitious undertaking, but the process was considered less sophisticated than printing in silver or platinum. Photographer Peter Henry Emerson, a member of the Council of the Photographic Society, reflected that "no one but a vandal would print a landscape in red or in cyanotype."[24] R. Child Bayley used the biological metaphor of evolution, writing that the "simple blueprint . . . survives [just] as the Darwinians tell us some of the lower forms of life survive from the extreme simplicity of its structure."[25] However, the ease of exposing and washing the prints belies the painstaking labor of making botanical reproductions, reflected in Atkins's meticulous prints and inscribed text. *British Algae* is considered the first book to use photographic illustration, predating Talbot's *The Pencil of Nature* (1843) by eight months. Atkins hand-lettered the captions before making the cyanotypes, integrating text and image, and she managed all aspects of the book's production.

The gendered framing of Atkins's work as amateur—though amateurs drove nineteenth-century photography—reflects the devaluing of female artistic and scientific production. Victorian culture framed collecting botanical specimens as an appropriate and polite amusement for women rather than as botanical science, which was the realm of men.[26] Atkins structured her project to follow William Henry Harvey's algae taxonomy, revealing the scientific ambitions of the project as a comprehensive catalogue. Her cyanotypes are situated within the history of Victorian botanical science and carry on an evolution in botanical illustration that began in the 1840s. The rise of mechanically reproduced botanical illustrations—photography and intaglio prints—marked a distinct shift from the dried specimens, drawings, and prints-after-drawings that were the most common form of botanical illustration up until that point.[27] Atkins's photograms followed Talbot's experiments with cameraless botanical photographs. Talbot had approached the eminent botanist William Hooker about collaborating on a book of native plants illustrated by photographs. Hooker refused, suggesting that while the images were pretty, they lacked the detail to make them useful to scientists.[28] This notion of scientific truth is rooted in visual observation, the artist's own and then the scientist's, in which a drawing mediated by a skilled artist could conceivably show more critical detail than a contact photograph. Atkins makes a different truth claim: that direct contact with the plant itself is a form of accuracy and truth.[29] Contact print photographs gained their authority from their status as images made by the direct hand of nature without human intervention—an unmediated impression. As Atkins explained, the difficulty of making accurate drawings of small specimens led her to "obtain impressions of the plants themselves."[30] This is not to suggest that there was no artistic intervention. In their unnatural coloration, Atkins's images deny photography's claims to represent the world truthfully. We cannot imagine that the image is an unmediated impression of reality. Mike Ware suggests that the "cyanotype was seen as both unnatural and anti-natural, and so of little use to the photographic 'artist,'" and presumably, of even less use to the scientist.[31] But the imprint of the plant registers its trace. Each specimen stands in for the species, but due to the nature of the photogram process, each print ultimately directs attention to its singularity.[32]

Atkins's photograms carry the physical trace of the object. Iron-based processes have low light sensitivity and can only make contact prints and photograms.[33] Visually, Atkins's cyanotypes anticipate the illuminated inversions of the X-ray and 1920s modernist experiments with cameraless photographs, or photograms, by practitioners like László Moholy-Nagy and Man Ray. An intimacy of touch is required to make contact prints of plants. The sharpness and softness of each print reflects the physical process of flattening and pressing the specimen, driven by the hand rather than the eye. The tactile dimension of making grounds the alchemy between light and minerals in the solidity of the object world.[34] The haptic facet of making in Atkins's practice is redoubled, as the prints are formed into books meant to be touched. The choice to gift the books as unbound volumes positions the project as relational, for this process integrates its mutability and potential evolution, rather than asserting the book as a fixed site of knowledge to be disseminated.[35]

Upon first glance, Atkins's taxonomical project fits neatly into an Enlightenment norm of observation, classification, and scientific distance. But Atkins's print shows unruly specificity rather than standardized types. The cameraless photographs are unique, singular, handmade images, not the anonymous, objective products of mechanical reproduction. While harvesting and cataloguing are small-scale practices of extraction, Atkins's cyanotypes introduce a material intimacy that blurs boundaries between the artist and the material. The result are prints that direct our attention to the natural world through an invitation to close looking.

What happens when we expand the frame of our analysis from Atkins's domestic encounter with algae, light-sensitive paper, and iron salts to take a broader view? Or following John Berger, what new insights are cultivated when we situate the photograph within social and political memory?[36] Another cyanotype by Atkins, *Gleichenia immersa, Jamaica*, is found in the 1853 album *Cyanotypes of British and Foreign Ferns* (plate 18). The tapered, curvilinear forms of the forking fern cut diagonally across the page. Toward the end of *British Algae* and in her later project *Cyanotypes of British and Foreign Ferns*, Atkins collaborated with her childhood friend Anne Dixon, who shared her interest in botany. Unlike the meticulous attempt to catalogue all British algae in *British Algae*, the volume on ferns did not endeavor to be encyclopedic. Atkins and Dixon included specimens gathered abroad, such as the *Gleichenia immersa*, or umbrella fern, from Jamaica. Formally, the project also makes a break from symmetry as a visual device, which was deployed to reflect the scientific ambitions of *British Algae*. Here, symmetry is abandoned for the impression of movement.

In *Gleichenia immersa, Jamaica*, we see the other side of Atkins's domestic practice, as the project reveals the reach of British imperialism. Collectors travelled the world, gathering exotic specimens and species, fusing scientific inquiry with imperial ambitions. This fern specimen underwent a journey from the colonies to the heart of the empire, where it was codified into knowledge and, in this case, aesthetics. Here, the fern is an index of the transnational flow of materials that characterized the Victorian period. Atkins's artfully arranged ferns seem far removed from the rise of industrial capitalism, the growth of empire, or the constitutive role slavery played in funding industrial growth. This is another moment where materials can show us something that representation cannot, for the materials used in the image-object reflect an ecosystem transitioning to an industrial economy.

England's rapidly industrializing manufacturing centers were plagued by child labor, the despoliation of the environment, and the alienation of the factory system.[37] Industrial growth was also linked to the United States' plantation economy. Britain's textile mills, a vital pillar of the industrial economy, relied on cotton produced in the antebellum South: before the American Civil War, the South accounted for 77 percent of the eight hundred million pounds of cotton used in Great Britain.[38] Raw cotton was transported to Britain where it was woven into cloth—occasionally in the form of photographic paper—which constituted the largest British export in the nineteenth century. How does the image-object embody these broader histories?

As literary scholar Ann Garascia writes, "The particulate matter of [Atkins's] images narrates an unseen environmental history of the area's shift from an agrarian to a more industrial landscape."[39] Some of the materials used to produce the photographs were local: the water came from the well on the family's Halstead property; the potassium ferricyanide was generated by a battery built by Atkins's father, John Children; and the image was exposed on J. Whatman Turkey Mill writing paper. Many early photographers favored Turkey Mill paper. It was produced in Kent as the region became a center for paper production within the industrializing economy. Before steam power was applied to papermaking, it was a small-scale, rural industry limited in scope by the productivity of the waterwheel.[40] The scale of production and increasing mechanization at Turkey Mill shifted the craft of paper to a high-output process standardized across its many steps.[41] In processes like cyanotypes or Talbot's calotype process, the fibers of the paper itself hold the chemicals—unlike glass, metal, or film, where the emulsion is on the surface. Because of this, the paper was an essential part of the photochemical processes, and any impurities or chemical treatments affected the outcome of the print. Whatman's woven paper was strong and smooth, making it well-suited for photography. Soaked in a gelatin bath of hooves and bones, it was robust and appropriately absorbent. Talbot recommended the paper to photographers.

The paper that forms the basis of the photographic surface introduces a secondary question about the transnational flow of materials. In the nineteenth century, paper was made from rags. Rags produced in textile factories were gathered by ragpickers and sold to papermaking mills, where they were disassembled and processed into paper. Art historian Katherine Mintie writes that in the case of albumen paper, the raw materials came from two mills: Blanchet Frères et Kleber Company (Rives paper) in Grenoble, France, and Steinbach and Company (Saxe paper) in Malmédy, in present-day Belgium.[42] In the factories, primarily female workers separated and beat the eggs that gave the prints their name before rolling the sheets of paper. The paper was a mixture of 85 percent linen from Ireland and 15 percent cotton from the United States. Like albumen prints, Whatman's paper was made from a mixture of rags, including linen, hemp, and cotton. It is reasonable to conclude that some of the cotton used in Atkins's prints came from plantations in the antebellum South, which would be the case for most paper prints in this period given the predominance of American cotton on the British market.[43] A focus on paper illuminates the largely unrecognized role of enslaved peoples, the working class, and women in the early history of photography and foregrounds the international dimensions of its material production.[44]

Yet the links between Atkins's prints and ties to slavery run deeper. Despite the abolition of the British slave trade in 1807 and of British colonial slavery with the Slavery Abolition Act of 1833, economic ties mediating ongoing slavery remained central to Britain's economy in the aftermath of abolition within British territories. This included direct slave ownership as well as financial and commercial ties with the plantation system in the Americas, establishing deep patterns of investment and returns that withstood the formal banning of the practice.[45] *Capitalism and Slavery*, a classic study by Eric Williams, a historian and later the first prime minister of Trinidad and Tobago, demonstrated how slavery was central to financing the Industrial Revolution in England, as plantation owners, shipbuilders, and merchants who were implicated in the slave trade used their accumulated wealth to establish centers of finance and heavy industry while expanding the reach of capitalism globally.[46] By the 1760s, British slave ships transported a majority of the eighty thousand Africans who were forcibly taken to the Americas annually. The Caribbean islands, particularly sugar colonies, became central to the British Empire, and the profits extracted from chattel slavery provided substantial seed capital for the Industrial Revolution.[47] Over 3,500 trading voyages originating in Britain have been identified between 1776 and 1807, primarily to the West Indies. John Atkins—Anna Atkins's father-in-law—was the lead investor of two voyages.[48]

Anna Atkins was born in 1799 to a wealthy family living in the countryside of Kent. In 1825, she married John Pelly Atkins, a railway promoter and the co-owner with his father of plantations in the Caribbean. The Halstead Place estate in Kent where Anna Atkins made her prints was purchased by her father-in-law, John Atkins, in the early 1820s.

John Atkins was an English politician who made his fortune as a merchant in the Caribbean and is identified by Catherine Hall's Centre for the Study of the Legacies of British Slavery as a significant enslaver.[49] John Atkins co-owned an estimated seven plantations with John Pelly Atkins under the firm J&A Atkins, which became John Atkins & Son in 1820. The plantations primarily grew coffee, but others supplied sugar, rum, and logwood. In 1825, the year Anna married John, one coffee plantation, Jamaica Port Royal, held 105 enslaved people, while the Dublin Castle Plantation, which also grew coffee, held 175.[50] When John Atkins died in 1838, he left the majority of his estate, including the Halstead Place property and his mercantile concerns and investments, to Pelly Atkins.[51]

Pelly Atkins's inheritance also included the six-figure sum paid to John Atkins by the British government as compensation for the loss of his enslaved workforce, following the Slavery Abolition Act of 1833. The act freed eight hundred thousand Africans enslaved by British plantation owners, though formerly enslaved people had to provide an additional forty-five hours of unpaid labor per week for four years following their freedom. Another clause in the act financially compensated Britain's 46,000 slave owners for their loss of property. Paid for by British taxpayers, it made up 40 percent of the total government expenditure for 1834 and was the largest financial bailout in British history until the 2009 bailout of banks following the 2008 financial crisis: roughly £16 billion to £17 billion in today's currency. John Atkins and John Pelly Atkins filed twelve claims on abolition bailout funds. This taxpayer payout funded the building of mines, factories, and railways. According to geographer Kathryn Yusoff, "As a ledger, the financial benefits of ending slavery reshaped the world to provide the material preconditions for the Industrial Revolution and the metamorphosis of capitalist forms."[52] It also reinvested money in mining, funding both iron production in Britain and additional investment in mining in Potosí.[53] The Industrial Revolution and the changes enabled by the steam engine were underpinned by empire and slavery, by the entangled geographies and histories of extraction and forced labor. Economic growth was—and is—predicated on exploiting the global majority, particularly the extraction of labor from Black and Brown workers. Meanwhile, Anna Atkins began her photographic project in 1843, gathering plants and exposing images on a property purchased by wealth extracted from the labor of enslaved people in the Caribbean. The imperial flow of materials is itself

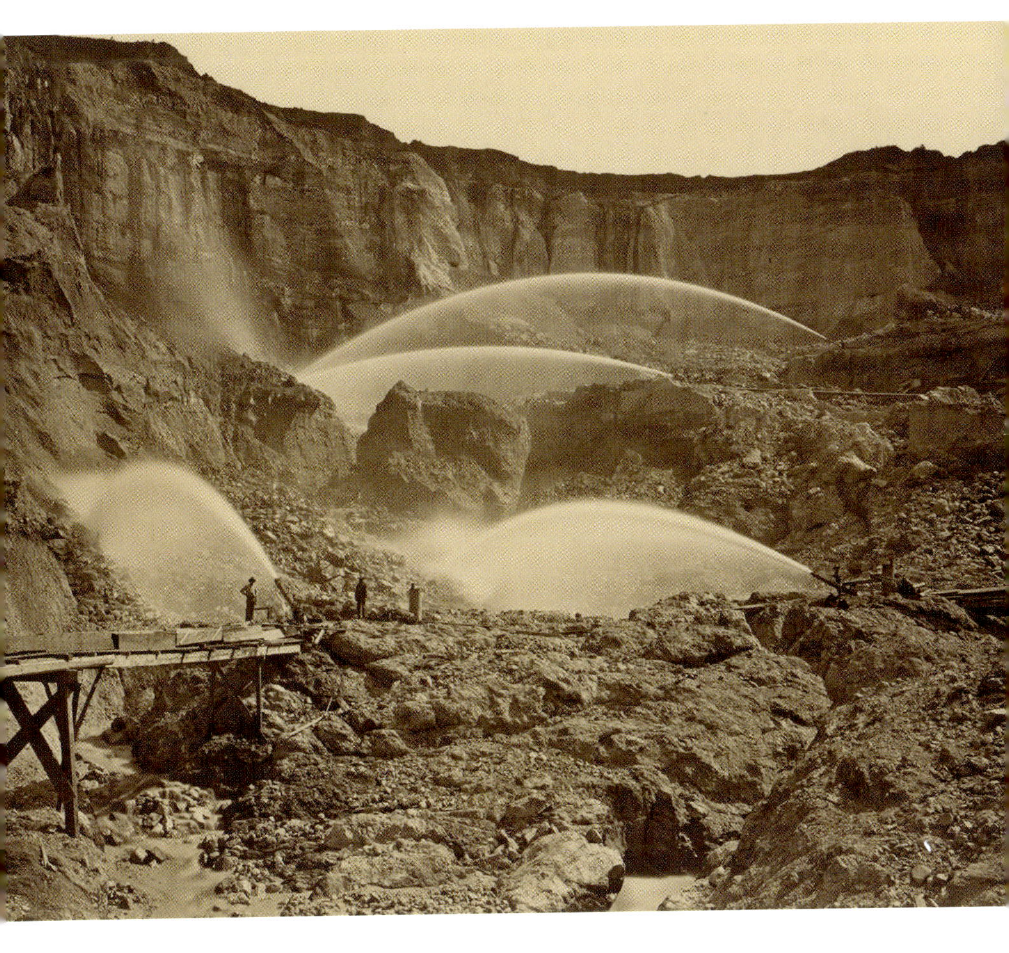

1 Carleton Watkins, *Malakoff Diggins*, Nevada County, California, 1871. Albumen print, 15⅘ × 20⅗ in. Courtesy of the Bancroft Library, University of California, Berkeley.

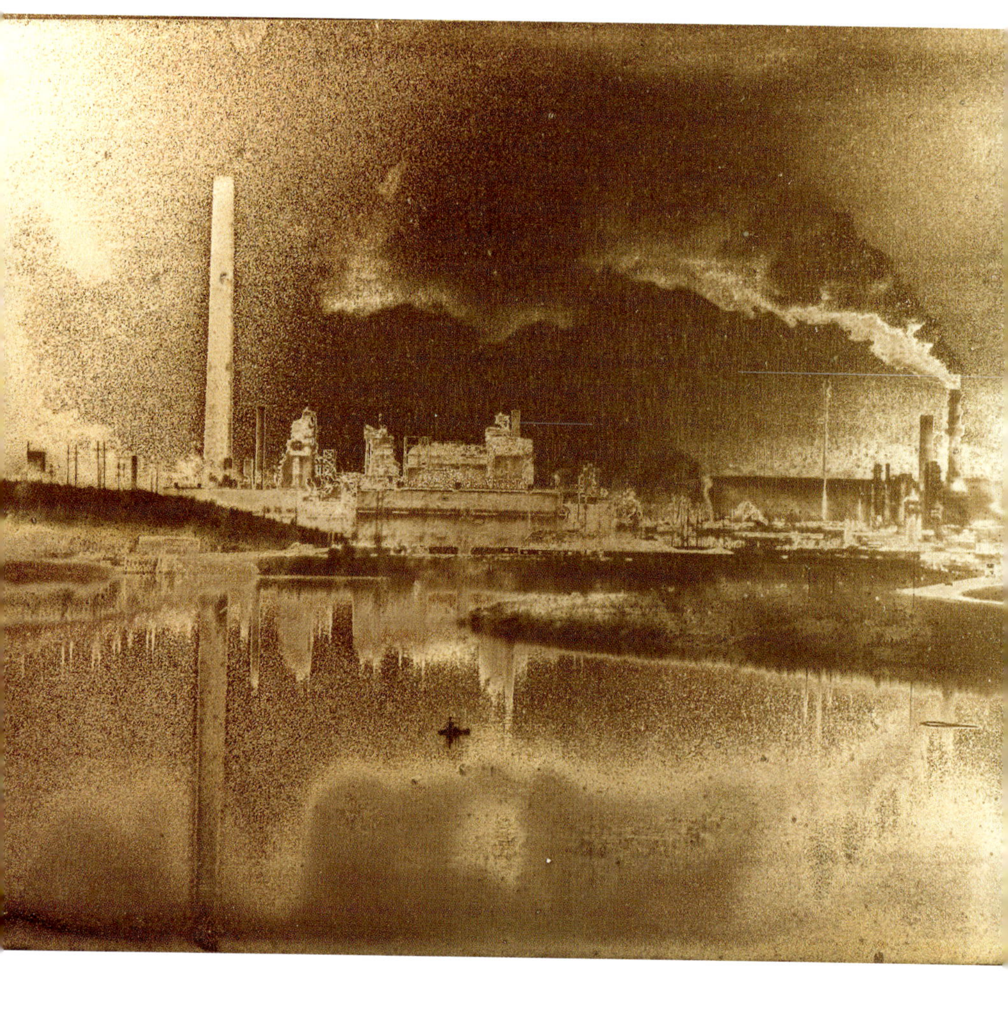

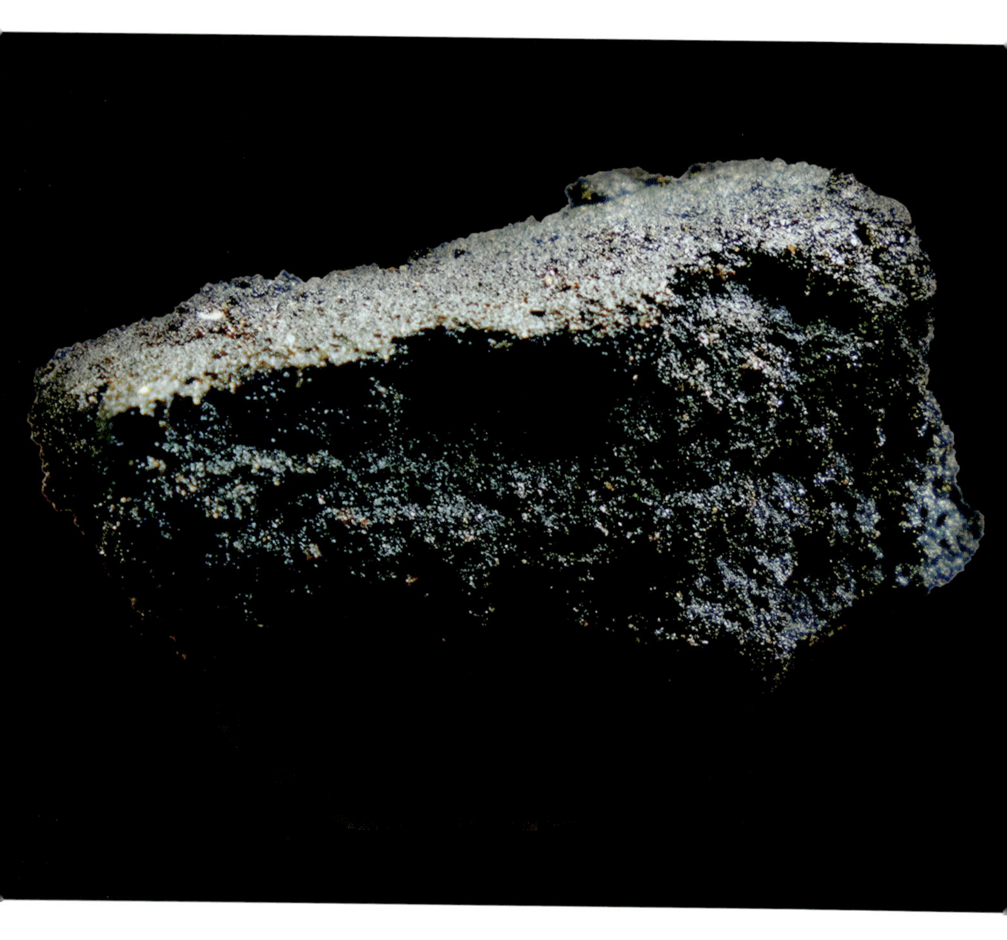

  Ts̱ēmā Igharas, *Emergence: Bitumen*, 2018. Back-lit digital print.

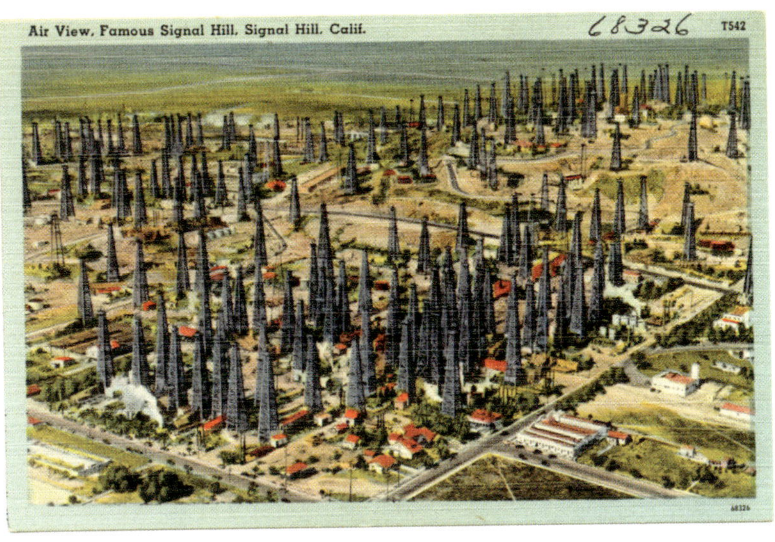

5   Postcard of Signal Hill oil field, Long Beach, California, 1937.

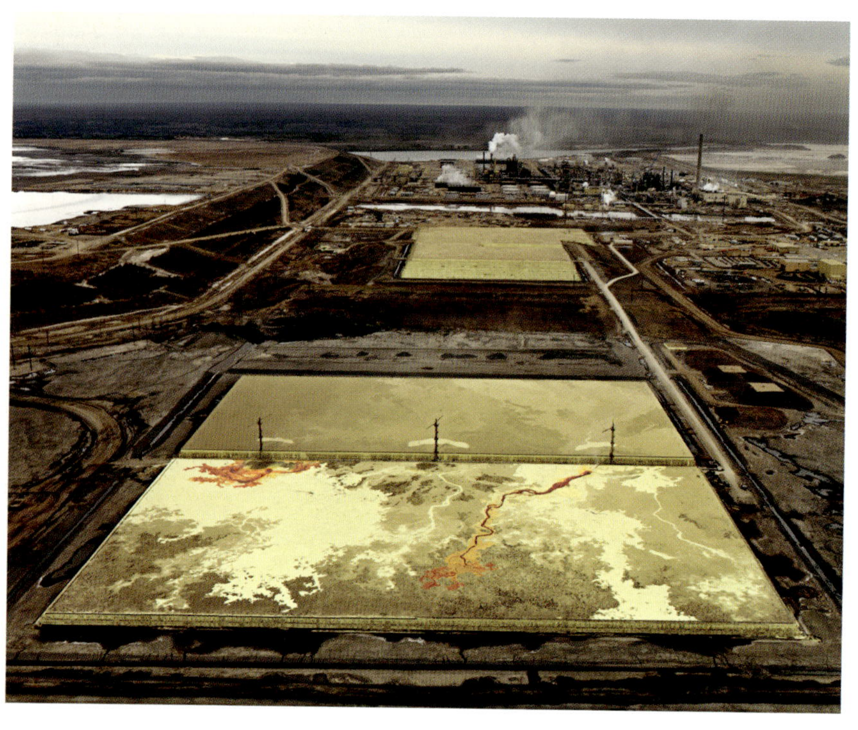

6 Edward Burtynsky, *Alberta Oil Sands #6, Fort McMurray, Alberta, Canada, 2007*. Inkjet print, 48 × 64 in. © Edward Burtynsky, courtesy of the Nicholas Metivier Gallery, Toronto.

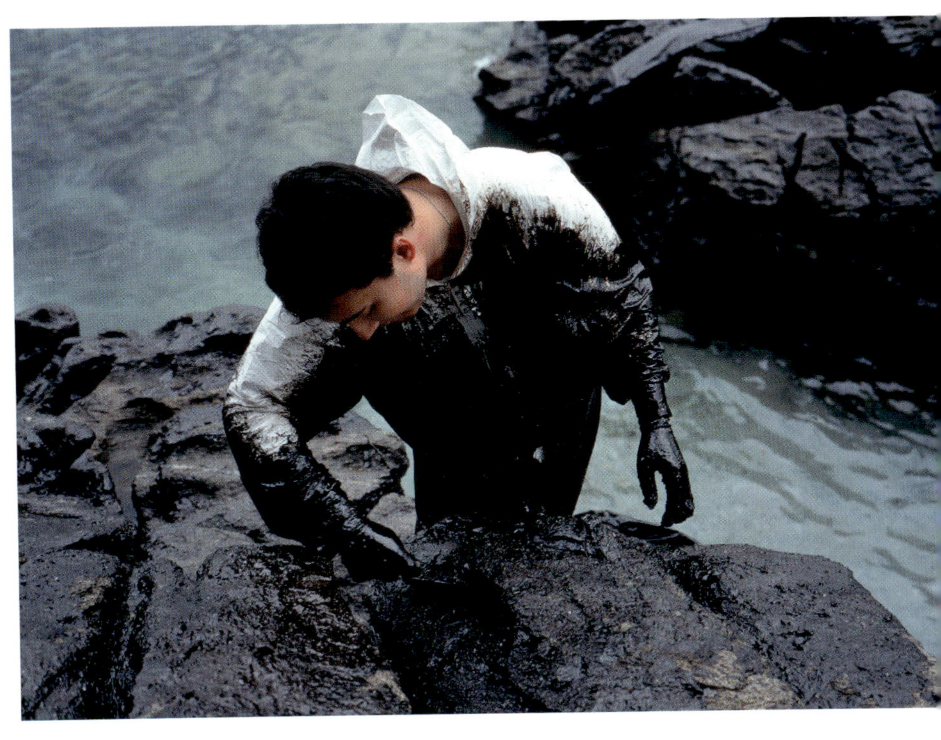

7   Allan Sekula, *Volunteer on the Edge*, Cíes Islands,
December 20, 2002. From the series *Black Tide/
Marea Negra*, 2002/2003, 29⅝ × 42¼ × 2½ in.
Courtesy of the Allan Sekula Studio.

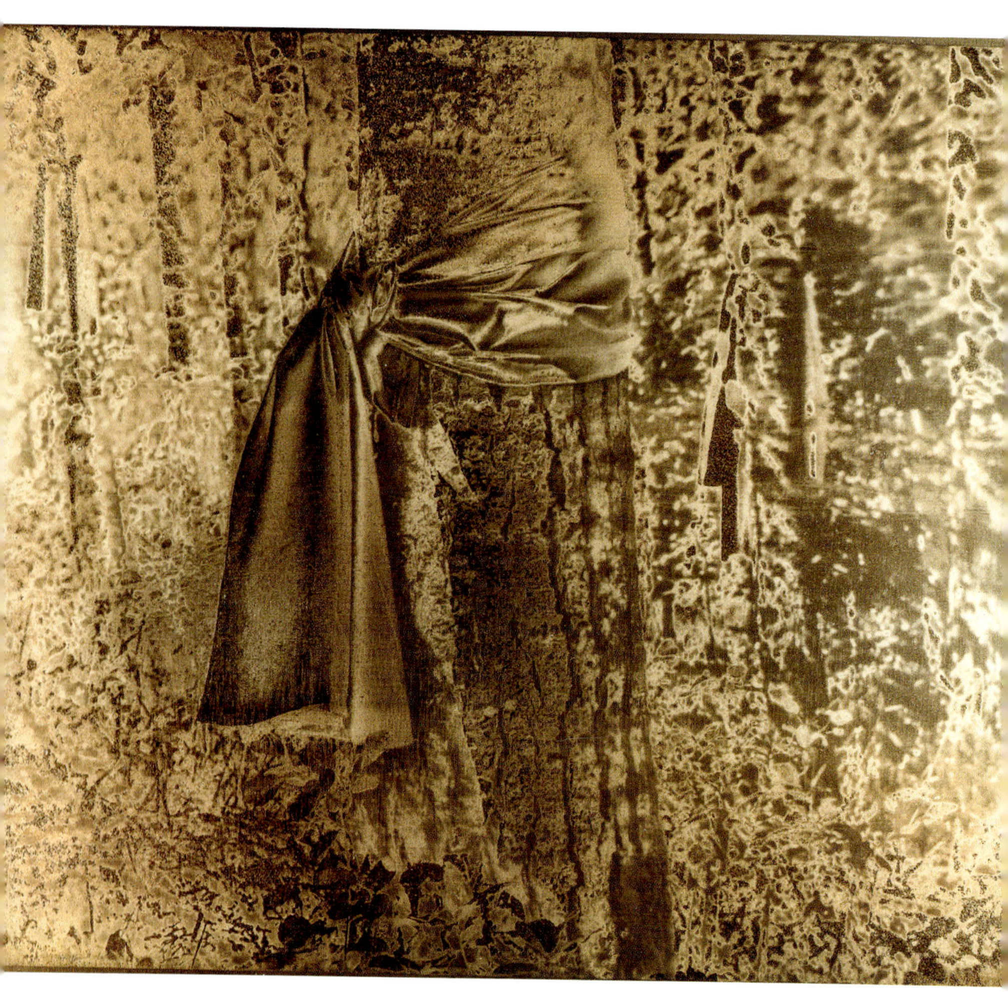

8

Warren Cariou, *Prayer Tree*, 2017. Petrograph on aluminum, 8 × 10 in. Courtesy of the artist.

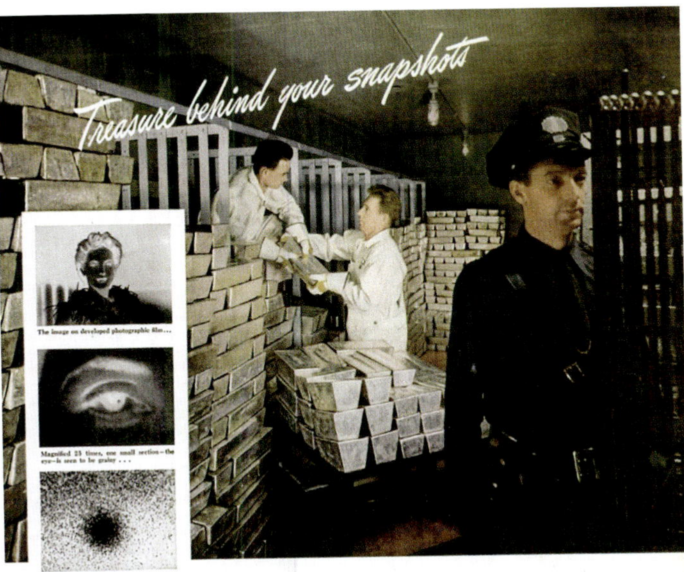

9   "It's pure silver that 'gets the picture' on Verichrome and other Kodak Films." Advertisement, reproduced in *Life*, January 15, 1945.

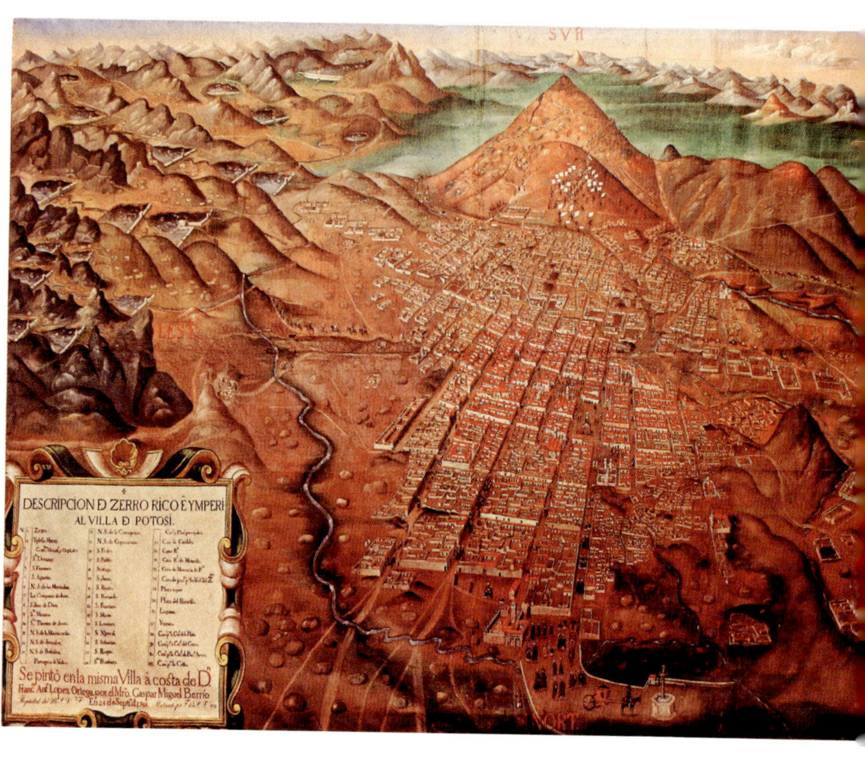

10 Gaspar Miguel de Berrio, *Cerro Rico and the Imperial Municipality of Potosí*, 1758. Oil on canvas, 103 × 71 in.

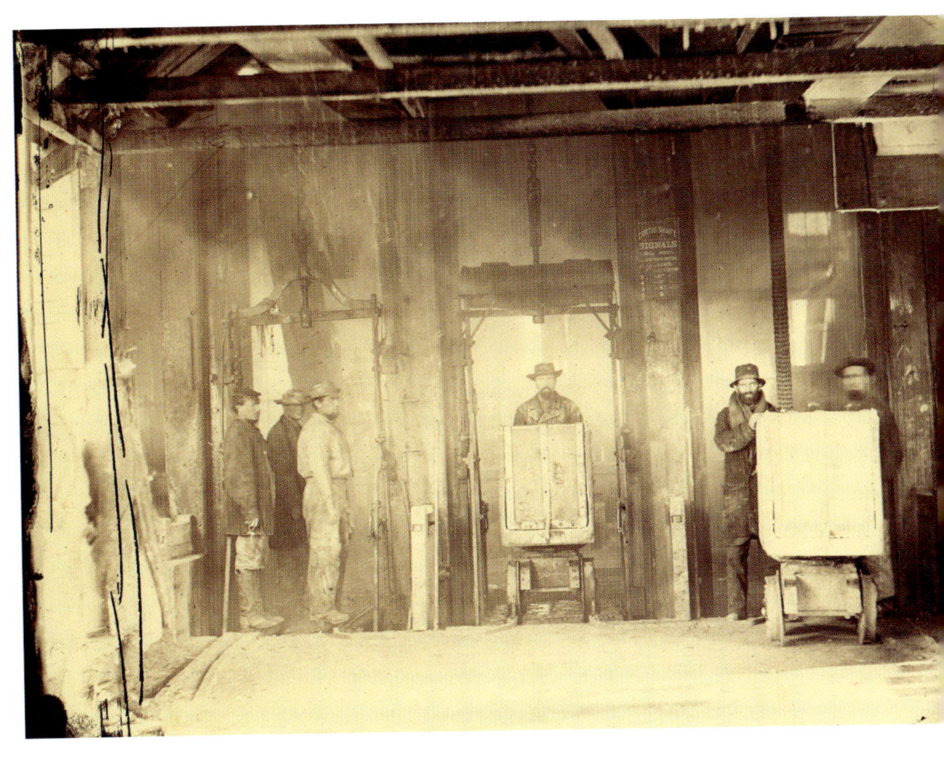

11   Timothy H. O'Sullivan, *Savage Mine, Curtis Shaft (Nevada)*, 1867–68. Albumen print, 7½ × 9⅞ in.

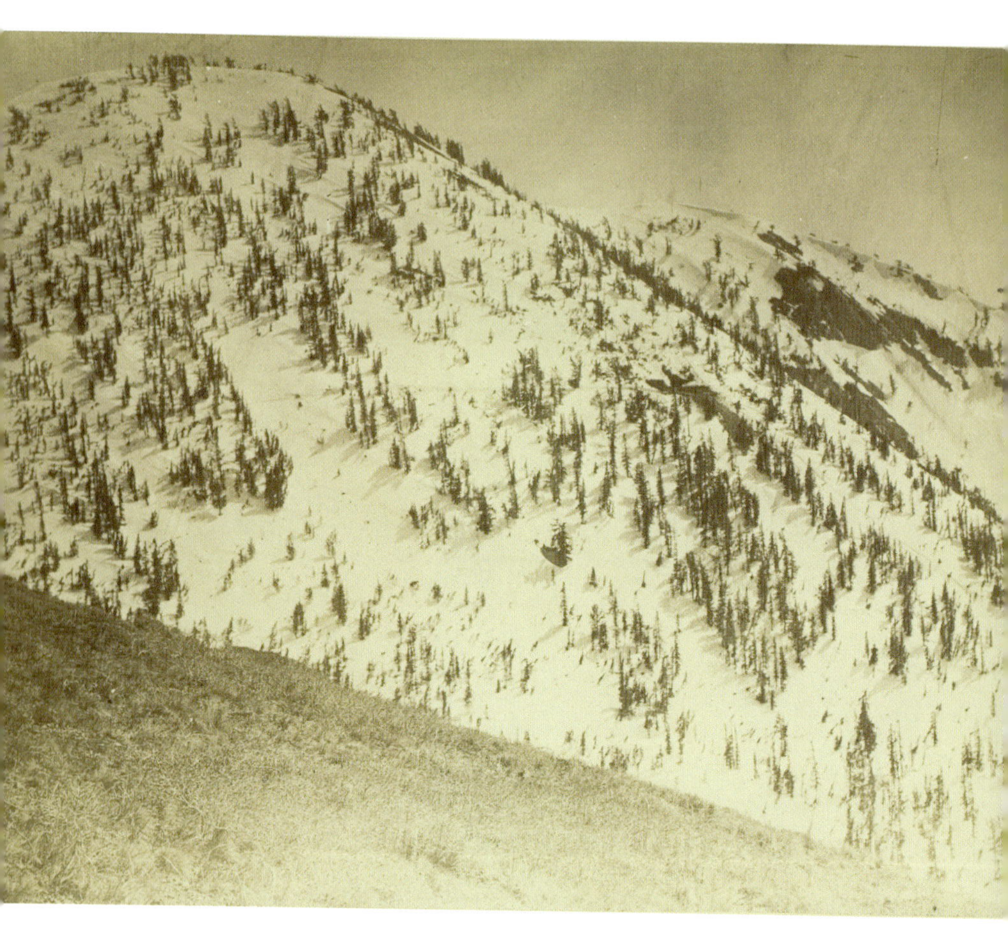

12   Timothy H. O'Sullivan, *Snow Peaks, Bull Run Mining District, Nevada*, 1871. Albumen print, $7^{15}/_{16} \times 10^{13}/_{16}$ in.

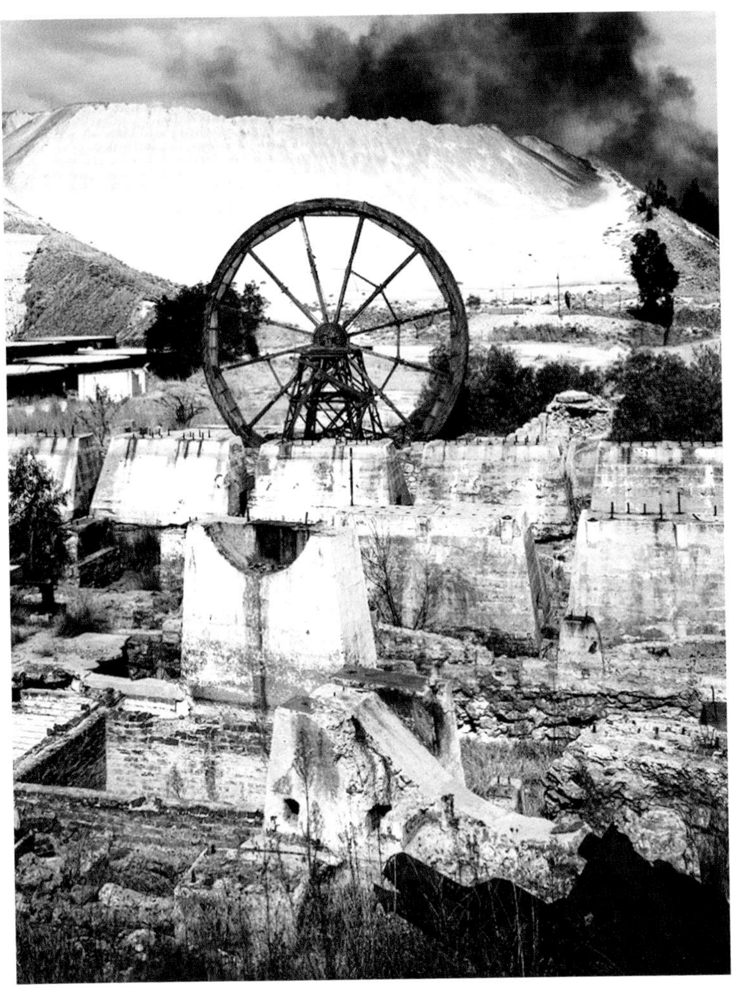

13 David Goldblatt, *Old Mill Foundations, Tailing Wheel and Sand Dump, Witwatersrand Deep Gold Mine, Germiston, August 1966*. Platinum print on Arches Platine sheets, 310 gm, 22 × 30 in. Courtesy of the Goldblatt Legacy Trust and Goodman Gallery.

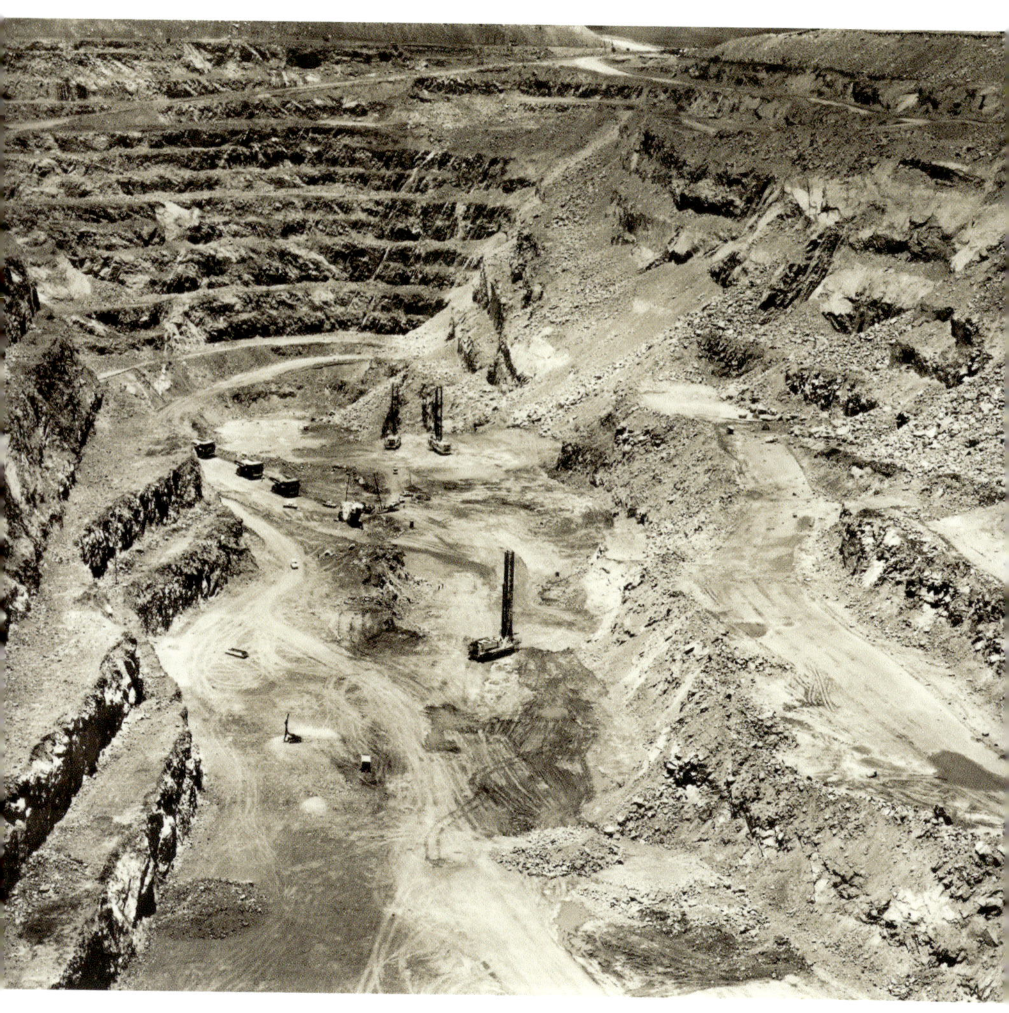

14 Simon Starling, *One Ton, II: 5 Handmade Platinum/ Palladium Prints of the Anglo American Platinum Corporation Mine at Potgietersrus, South Africa, Produced Using as Many Platinum Group Metal Salts as Can Be Derived from One Ton of Ore*, 2005. Set of five platinum/palladium prints framed in acrylic boxes, each 65 × 85 cm. Courtesy of the artist and the Modern Institute/Toby Webster Ltd., Glasgow.

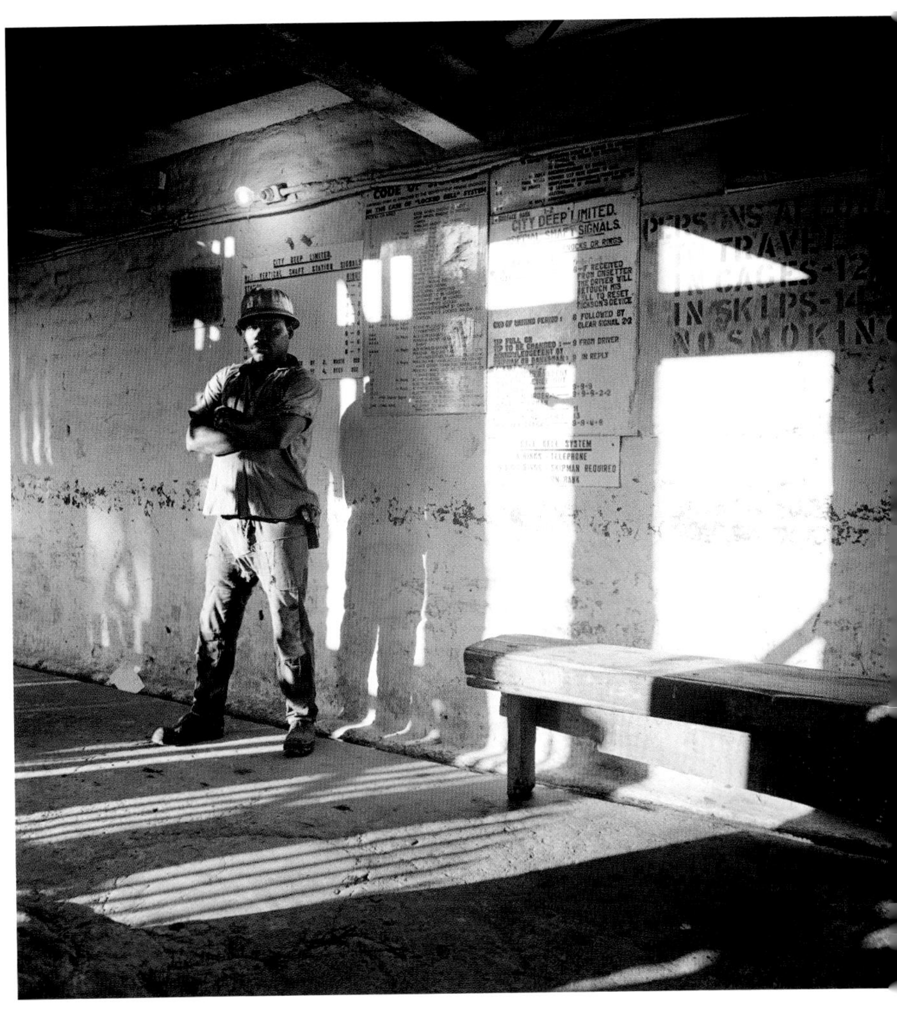

15 David Goldblatt, *A Miner Waits on the Bank to Go Underground, City Deep Gold Mine, 1966*. Platinum print on Arches Platine sheets, 310 gm, 14.5 × 14.5 in. Courtesy of the Goldblatt Legacy Trust and Goodman Gallery.

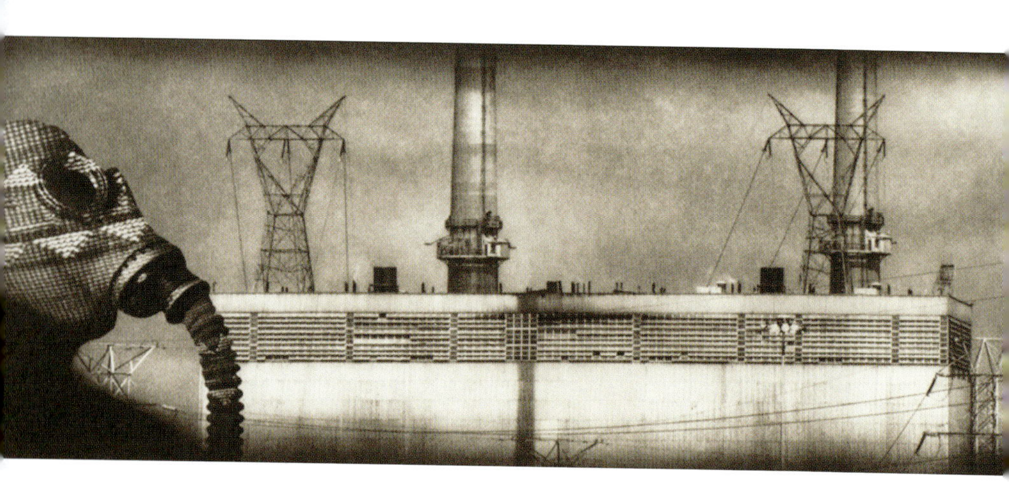

16 Larry McNeil, *Demented Coal Paradox*, 2013. Platinum print.

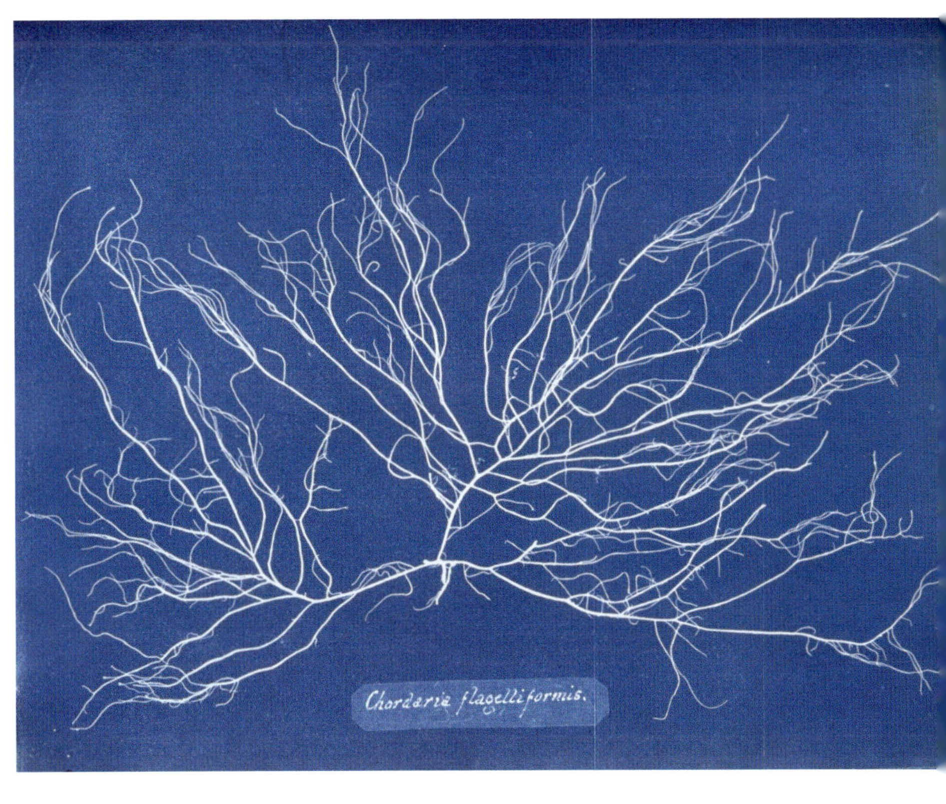

17   Anna Atkins, *Chordaria flagelliformis*, 1844. Cyanotype, 10 × 8 in.

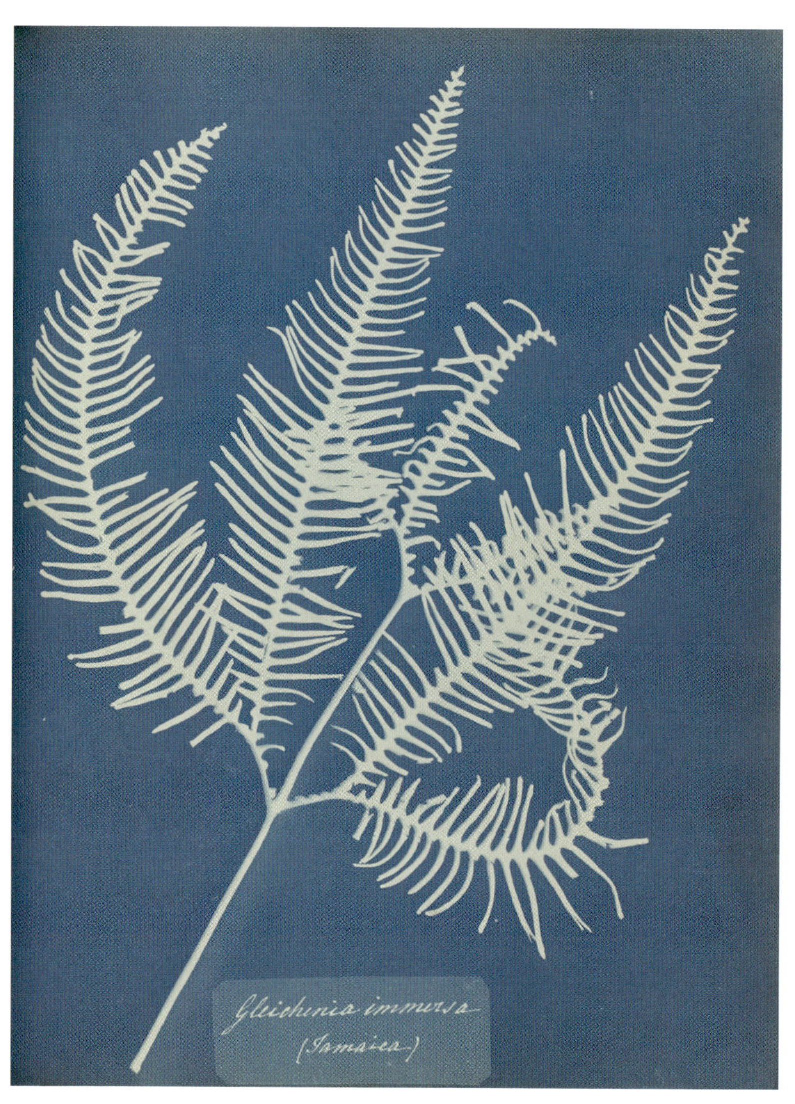
Gleichinia immersa
(Jamaica)

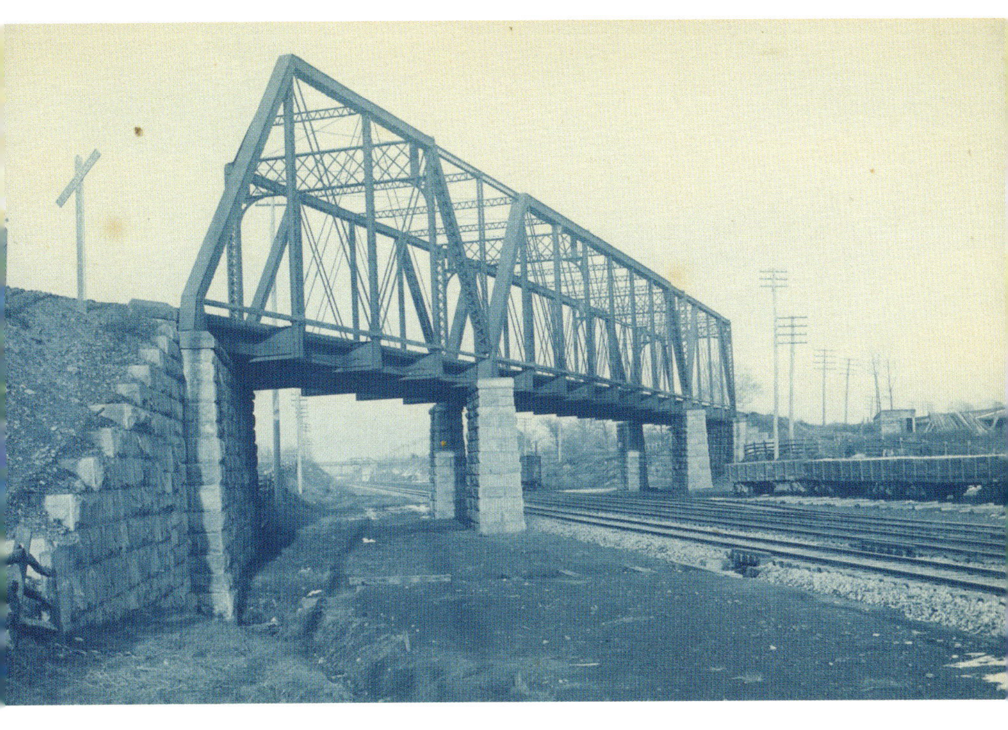

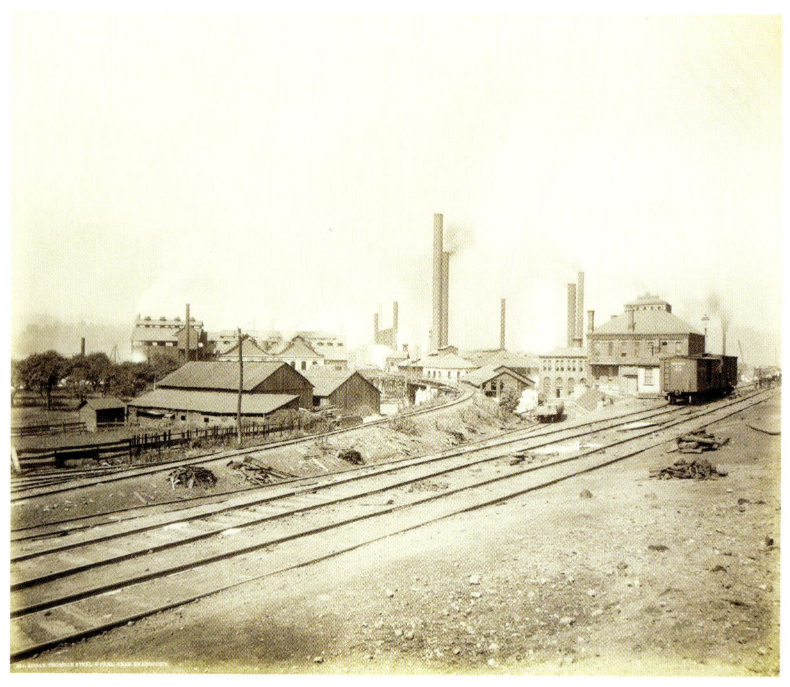

20 William H. Rau, *The Edgar Thomson Steel Works in Braddock, PA*, 1891. Albumen print, 17⅛ × 20⅝ in. The Library Company of Philadelphia. Courtesy of American Premier Underwriters, Inc.

21 LaToya Ruby Frazier, *If Everybody's Work Is Equally Important? (II)*, 2017. Four cyanotypes, 28 × 22 in. each. Courtesy of the artist and the Gladstone Gallery.

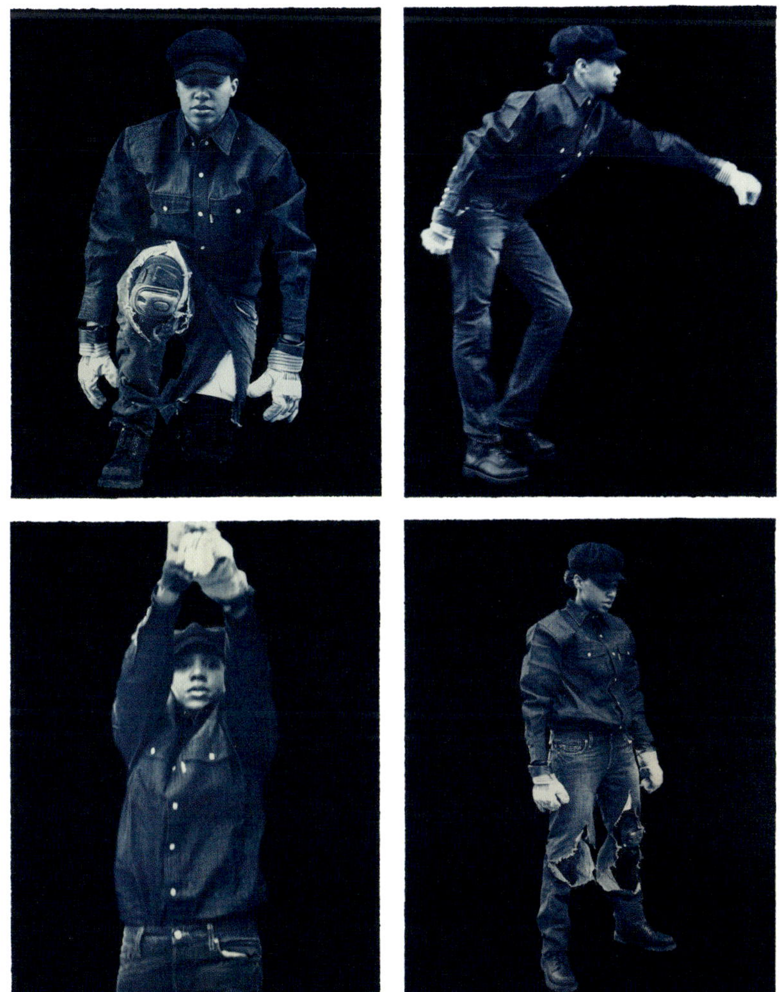

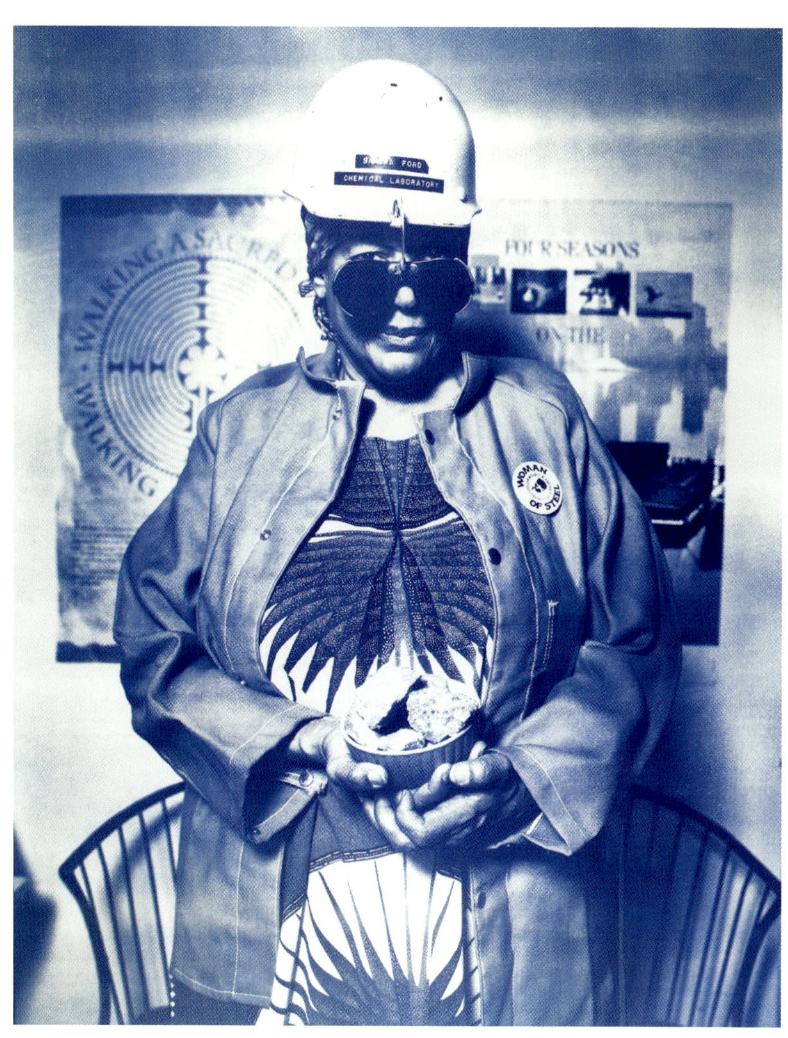

22 LaToya Ruby Frazier, Sandra Gould Ford wearing a work jacket and hard hat holding pink granite in her meditation room in Homewood, Pennsylvania, 2017. Cyanotype print, 24 × 19 in. From *On the Making of "Steel Genesis," Sandra Gould Ford*, a collection by LaToya Ruby Frazier, 2017. Courtesy of the artist and the Gladstone Gallery.

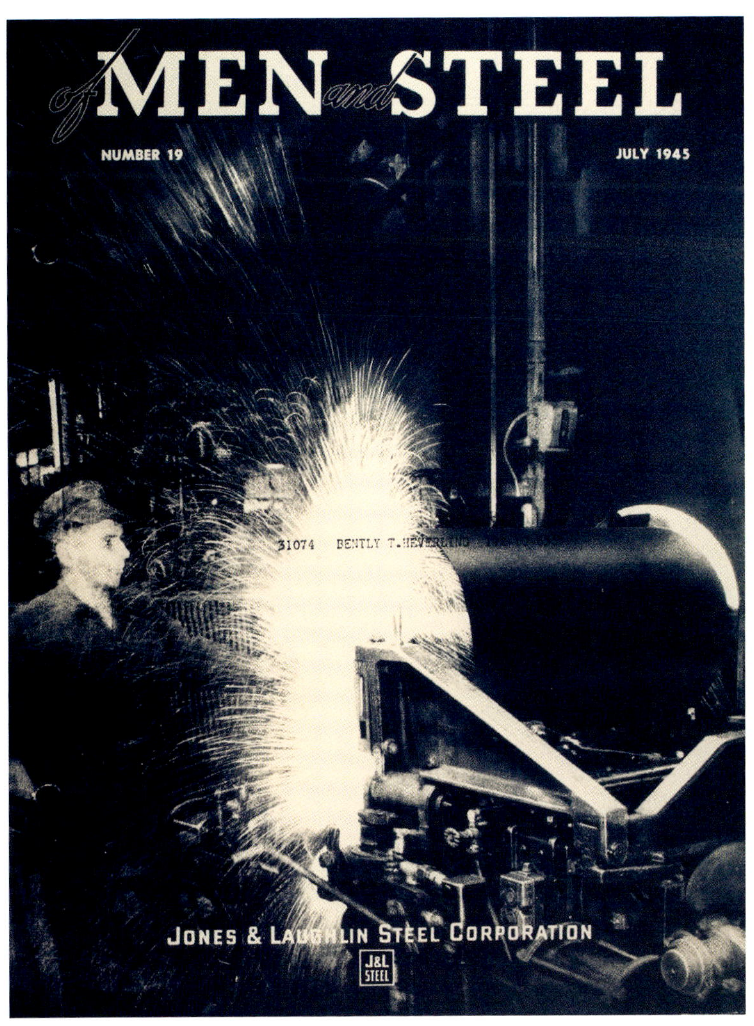

23   LaToya Ruby Frazier, *Of Men and Steel*, Jones & Laughlin Steel Corporation employee shareholder magazine, no. 19, July 1945. Cyanotype print, 11 × 8.5 in. From *On the Making of "Steel Genesis," Sandra Gould Ford*, a collection by LaToya Ruby Frazier, 2017. Courtesy of the artist and the Gladstone Gallery.

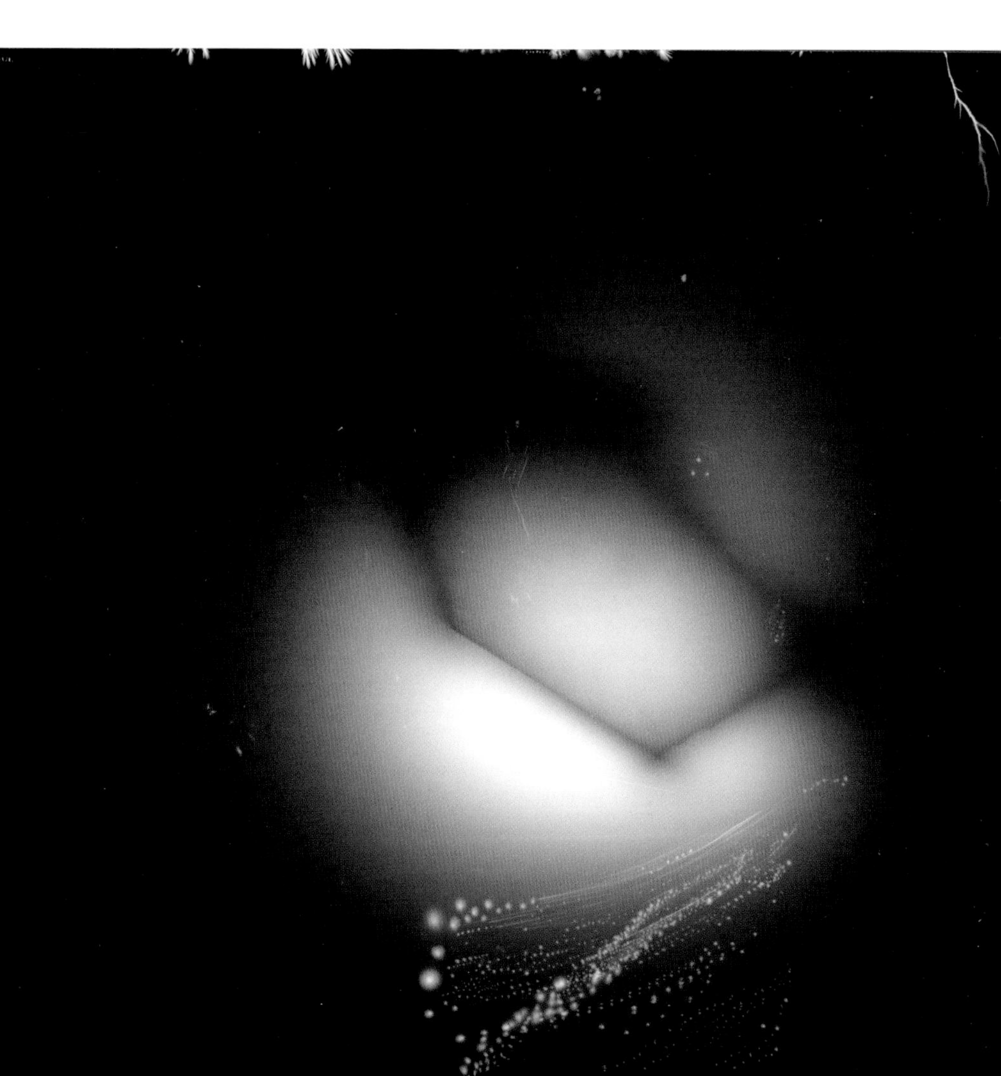

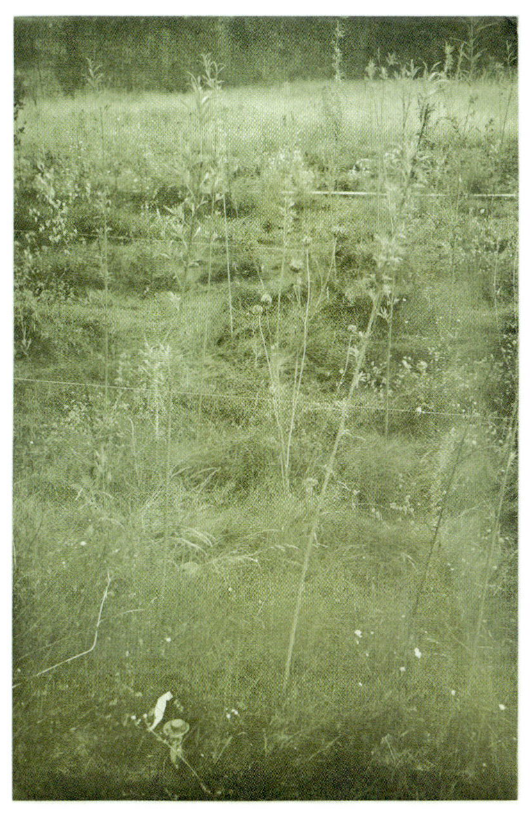

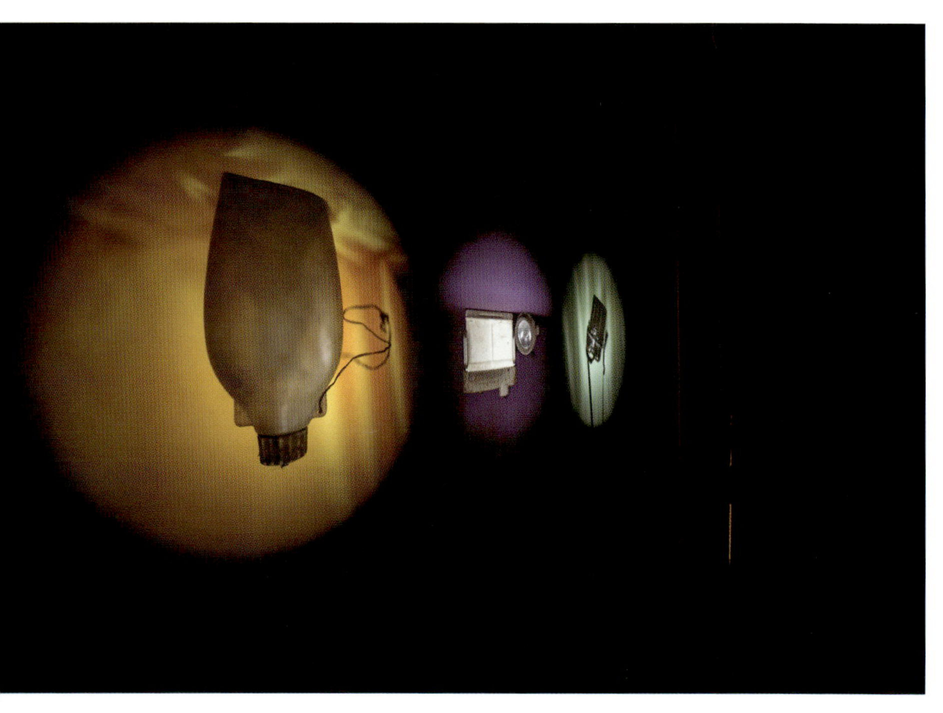

26 Susanne Kriemann, *Pechblende (Chapter 1)*, from the exhibition *In the Belly of the Whale*, 2016. Witte de With Center for Contemporary Art, Rotterdam, the Netherlands.

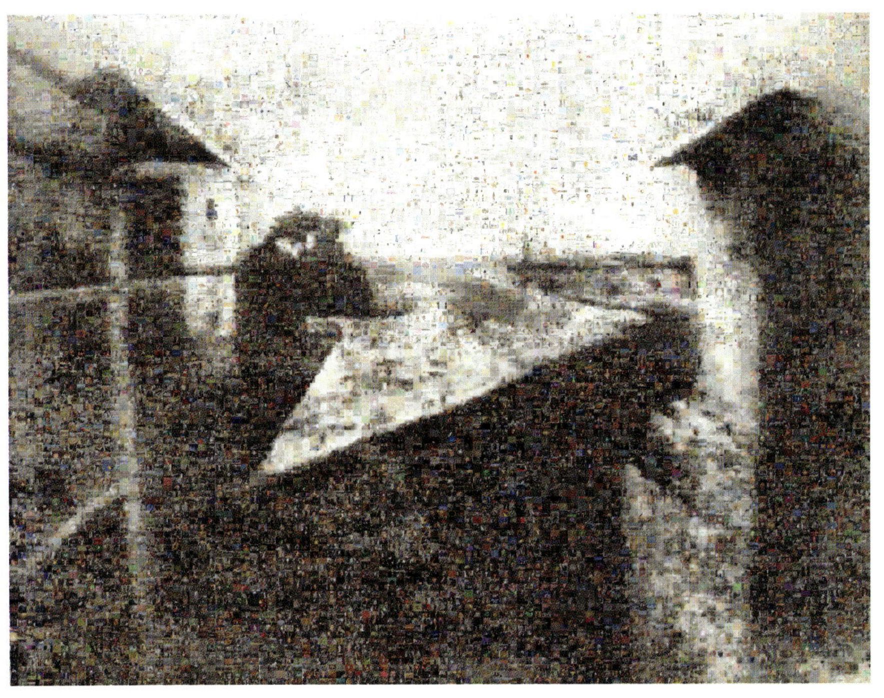

Joan Fontcuberta, *Googlegram: Niépce*, 2005. 47¼ × 63 in. The first photograph in history, taken by Nicéphore Niépce in Gras, France, 1826. The photograph has been refashioned using photo-mosaic freeware, linked to Google's Image Search function. The final result is a composite of ten thousand images available on the Internet that corresponded to the words *photo* and *foto* as search criteria. Courtesy of the artist.

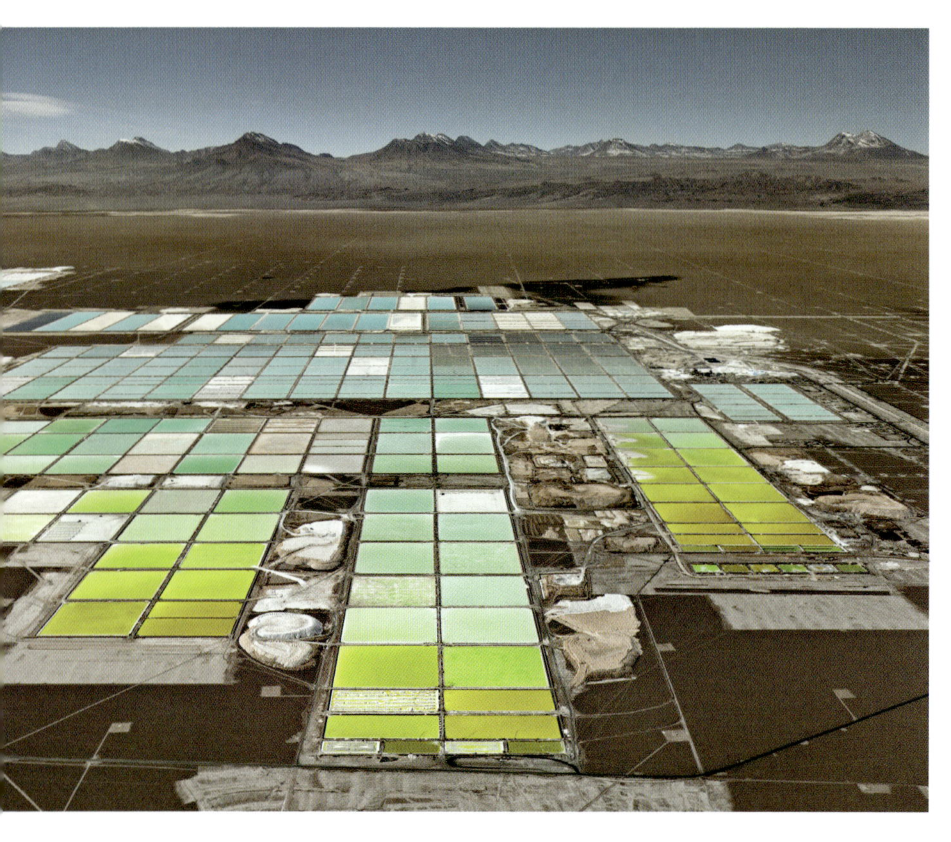

28 Edward Burtynsky, *Lithium Mines #1*, Salt Flats, Atacama Desert, Chile, 2017, 48 × 64 in. Inkjet print. © Edward Burtynsky, courtesy of the Nicholas Metivier Gallery, Toronto.

29 Trevor Paglen, *NSA-Tapped Undersea Cables, North Pacific Ocean*, 2016. C-print, 48 × 72 in. Courtesy of the artist and Altman Siegel, San Francisco.

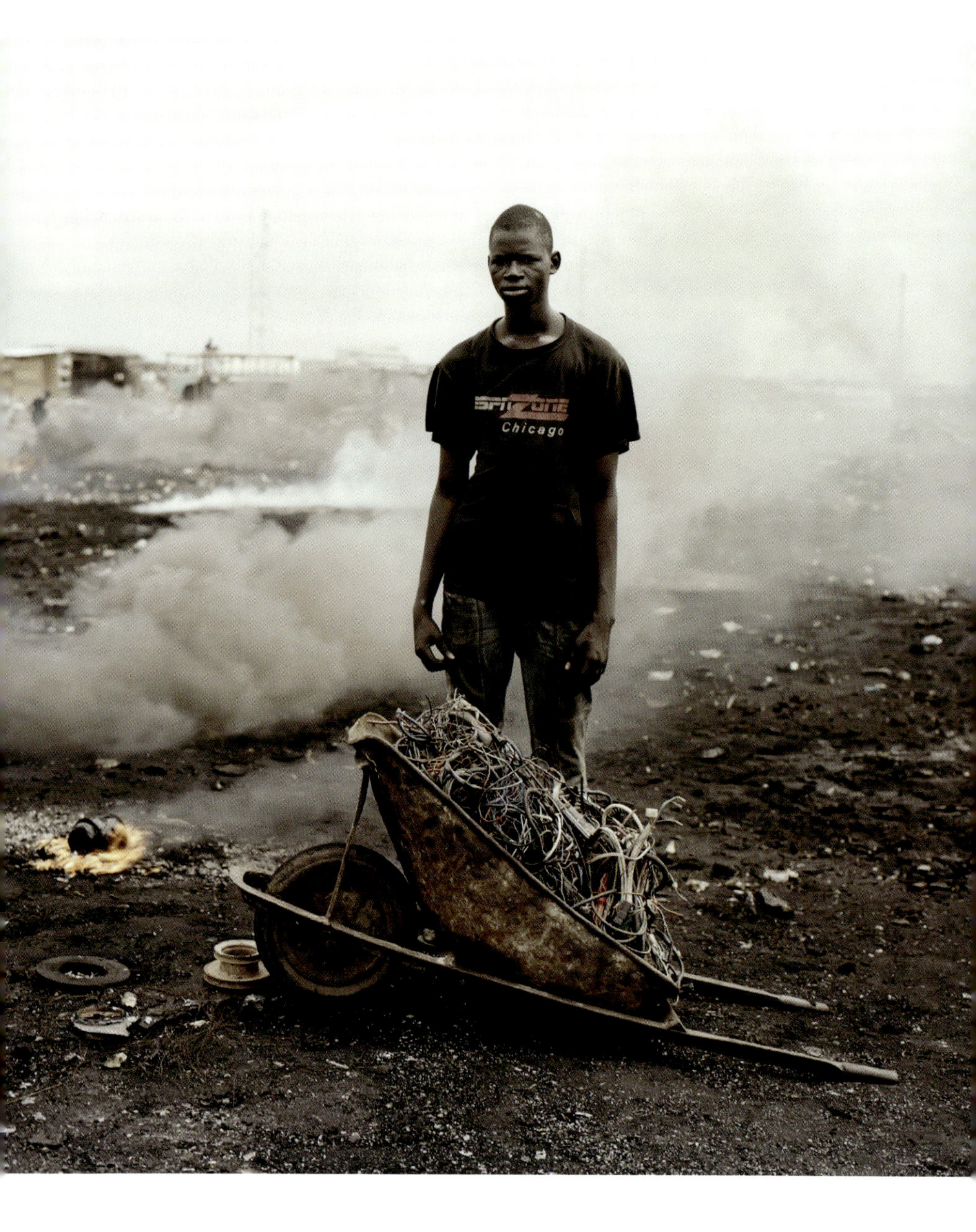

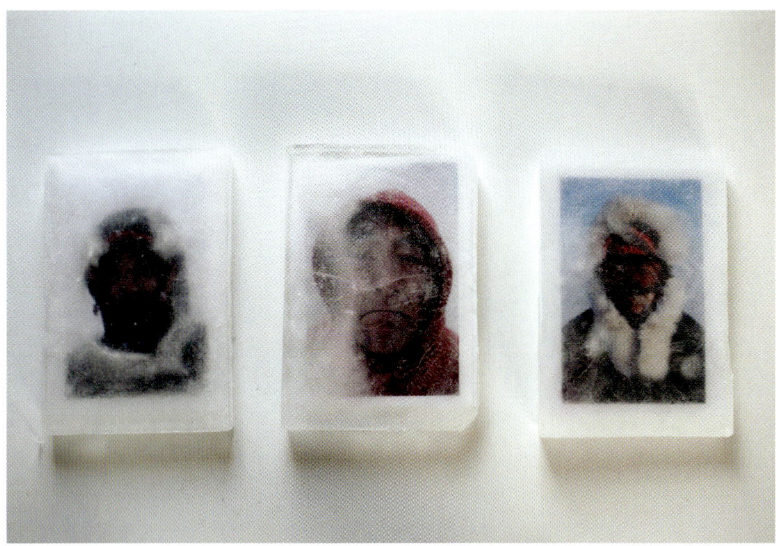

30 Pieter Hugo, *Al Hasan, Agbogbloshie Market, Accra, Ghana*, 2009. Digital c-print, 38½ × 38½ in. © Pieter Hugo. Courtesy of the Yossi Milo Gallery, New York.

31 Louie Palu, *Arctic Passage*, 2019. Ice block installation project. Large- and small-format prints frozen in ice blocks that melt as a single-day-or-less installation, first premiered at SXSW in 2019 at the Harry Ransom Center at the University of Texas at Austin. © Louie Palu.

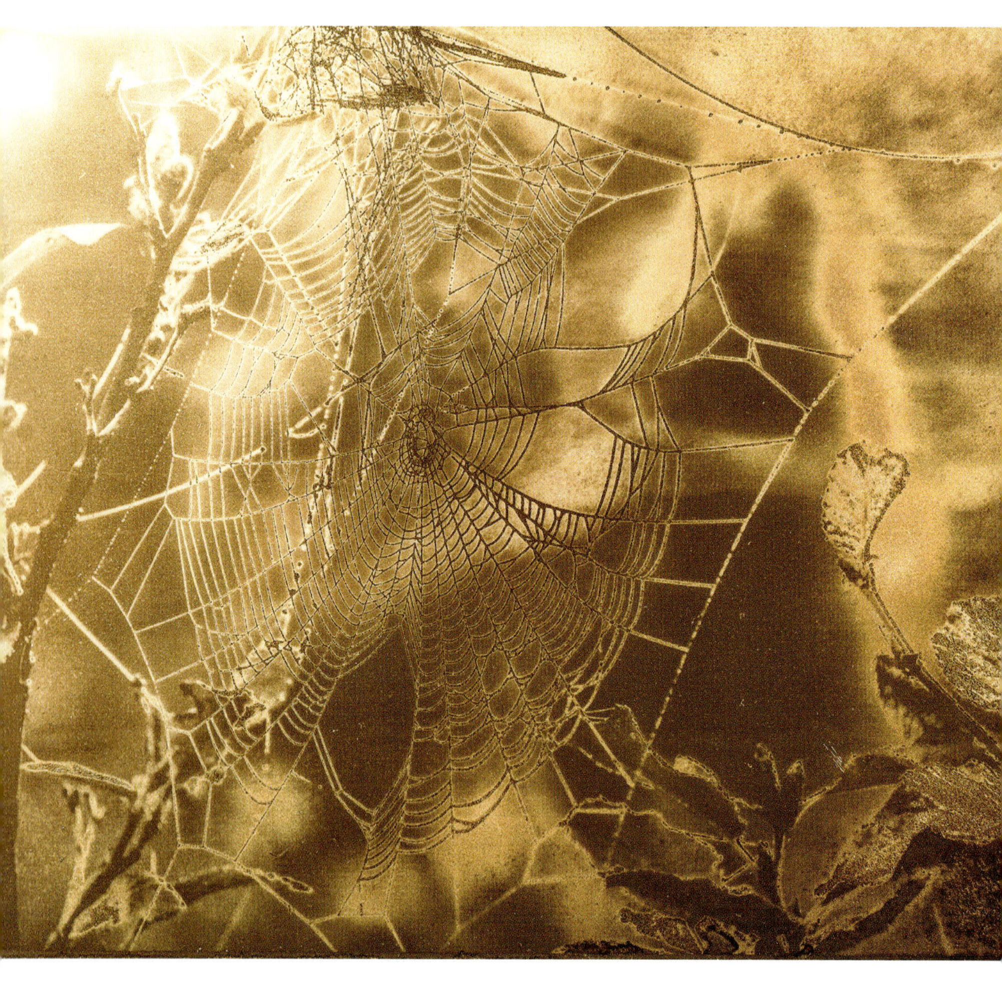

32   Warren Cariou, *Boreal Web*, 2017. Petrograph on aluminum, 8 × 10 in. Courtesy of the artist.

materially reflected in Atkins's prints: some of the Jamaican specimens Atkins used came from her husband's plantations.[54]

The use of Whatman's paper also reflects the transformation of Kent, known as the Garden of England, into an industrial region. Here, again, Atkins's biography complicates a sympathetic twenty-first-century reading that links her careful, striking prints of plants to ecological concerns. By the 1850s, Pelly Atkins had turned his attention to railway promotion with the Mid-Kent and London South-Western Junction Railways, registered with a capital of £1 million.[55] In advertisements, the wealth produced by paper factories was emphasized as enticement to invest in the region's infrastructure. The iron tracks laid over wooden planks that cut through the countryside form one of the more tangible manifestations of the industrial conquest of the landscape—the annihilation of space by time.[56] Here, we see a concrete case study for Williams's argument that the wealth produced by slavery enabled regional industrialization in the metropole. I highlight these links between Atkins's biography and the materials used in her images to note the impossibility of separating photography from the world within which it is enmeshed. Photography emerges from complex chains of extraction, production, refining, and labor, all deeply embedded in interlocking and often life-extinguishing social relations of production and social reproduction across geographies. It also demonstrates the impossibility of separating the past from the present, for the world as we know it today was in many ways forged in England's industrial furnaces and exported worldwide by force, persuasion, and competition. The bucolic contemplation of Atkins's cyanotypes may register industrialization subtly, but their contribution would ultimately enable the medium to find its primary application as a tool in mass production itself.

Blueprint Photography and the Second Industrial Revolution

Outside of Atkins's practice, cyanotypes are perhaps most known as blueprint reprography. The cyanotype process was commercially produced as Ferro-Prussiate or blue print paper. Mass-produced cyanotype paper was introduced in 1872 by Marion & Co., the company that popularized the carte de visite. In 1876, a commercial blueprint machine, manufactured by a Swiss company, debuted at the Philadelphia Centennial Exposition.[57]

Due to its low cost and speed, the blueprint machine was used to reproduce drawings and plans and was quickly applied in industrial contexts. The paper became popular with railway companies, engineers, architects, shipbuilders, and governments—anyone needing to rapidly reproduce a plan for the reproduction of something else. Architectural blueprints became commercially popular during the 1890s Klondike Gold Rush in Canada's Yukon Territory as a cheap map-making method.[58] Again, we see how the needs of extraction exert a powerful pull on the spread of new visual forms.

An 1891 album of cyanotypes documents bridges, culverts, and buildings spanning the Philadelphia Division of the Baltimore and Ohio (B&O) Railroad. The photographs, 129 in total, were taken by a group of B&O Railroad employees surveying the line in March 1891. The group inspected and photographed seventy-eight different bridges and culverts and thirty-seven of the nearly seventy stations along the line. In many of the photographs, the angular lines and modernist forms of bridges dominate the surrounding landscape. For example, a through-truss bridge at South 57th Street spans the Philadelphia, Wilmington, and Baltimore railroad tracks (plate 19). Thick beams of steel form crisscrossing patterns in cool shades of blue, framed by the triangular forms of the truss bridge. The bridge is superimposed over the flat, light-blue sky, desaturated in the print. The angle of the photograph creates a bisecting linearity between the bridge and the railroad track, which meet in the middle and extend into the distance. The bridges and railroad tracks were forged in iron and steel, drawing a link between the iron basis of the medium, its applications in industrial processes, and the material foundations of the subject matter. In the images, industrial growth cuts through the landscape. Unlike Atkins's botanical prints, the cyanotypes are not imprints of the object but rather a contact-print reproduction of a negative. Here, the haptic, handmade dimension of Atkins's botanical prints gives way to the speed and precision of industrial mass production, applied to celebrate the tensile power of steel.

By the 1890s, Pennsylvania was a major industrial center, as the United States had become a significant steel producer in the second half of the nineteenth century. Andrew Carnegie brought the Bessemer process—the first process to mass-produce steel—to the United States, where he founded the Carnegie Steel Company. By 1899, Carnegie's factories alone produced about half as much steel as Britain. The labor conditions in Carnegie's mills were backbreaking, compounded by heat from the furnaces, while wages were low, accidents were common, and labor unrest was boiling over. In 1901, J. P. Morgan formed the United States Steel Corporation by financing a merger between Carnegie Steel, Federal Steel, and National Steel. The Pittsburgh-based company was the first in history to be capitalized at over a billion dollars. At the time, it was the world's largest corporation and the biggest steel producer. In 1873, the United States produced 220,000 tons of steel, a number that would rise to 11.4 million tons by 1900, more than England and Germany combined. American steel built bridges, roads, and skyscrapers; it laid round after round of railroad track across the country.

The development of the steel industry had significant environmental impacts on Pennsylvania. In 1868, a writer for the *Atlantic Monthly* described Pittsburgh as "hell with the lid taken off," as smoke from blast furnaces polluted the air with soot and clay.[59] The language echoes descriptions of industrializing England. Writing on the iron works in Shropshire, Arthur Young described how the natural scenery was "too beautiful to be much in unison with that variety of horrors art has spread to the bottom. The noise of the forges, mills, etc., with all their vast machinery, the flames bursting from the furnaces with the burning of the coal and the smoak [*sic*] of the lime kilns, are altogether sublime."[60] The "smoak" was horribly sublime, but sublime nonetheless.[61] Young locates a rupture between the natural world and the "variety of horrors" within the ironworks. For Young, as for many observers of industrialization, industrial production provoked an ambiguous response: awe shadowed by disgust, fear tinted with admiration.

Photography was used to document and celebrate industrial growth. In a series of photographs commissioned for the Pennsylvania Railroad (PRR), the Philadelphia-based photographer William H. Rau photographed the Edgar Thomson Steel Works in Braddock, Pennsylvania, the site of Carnegie's first steel mill in 1891 (plate 20). The mill was built between the PRR and the B&O lines to entice both railroads to reduce their rates. In the photograph, railway tracks run alongside the complex, as smoke from the stacks pours into the sky. The scene is industrial, devoid of human figures. Commissioned by the advertising department of the PRR, Rau took 463 photographs, primarily mammoth-print albumen photographs and large scale panoramas, while travelling in a passenger car equipped as a photographic studio. The photographs were intended for display at the 1893 World's Columbian Exposition in Chicago, where the PRR had its own building. Rau summarized:

> Railroad companies now use photography very extensively in advertising their routes. The object is to make public the beautiful scenery along the line of their roads, and by that means to tempt patronage from the travelling public. This is by no means a bad plan, and it is a great accommodation to the traveller. . . . The shrewd railroad men know this, and they employ no servant whose labors repay them so handsomely as the photographer.[62]

Reflecting the commercial origins of the commission, Rau photographed the railway merging into the natural landscape, a harmonious vision of economic growth and nature coexisting. The images do not show the upheaval caused by the railway: the displacement of Indigenous people, the forced labor of Chinese workers, burgeoning labor unrest, or train accidents. Instead, they render it beautiful, making industry *natural*. Within the context of the larger commission, the steel refinery becomes another feature of the landscape, a peaceful attraction.

In the late nineteenth century, the railway became an important theme in photography, particularly in the industrial centers of America's Northeast. Photography and the railroad both emerged in the 1830s and fully developed in the second half of the nineteenth century. In 1830, there were twenty-three miles of railroad in America; by 1860, there were around 27,000 miles. The B&O was the first railway to use photography for promotional purposes in 1858, but railway photography did not become a prominent subject until the 1890s. As historian John Stilgoe writes, in 1891—when both the cyanotypes and Rau's albumen prints were taken—the railroad was wholly of the present and, without historical precedent directing the viewer on how to read the images, it is "the lack of context that glistens everywhere in the images."[63] The technological sophistication and aesthetic power of Rau's photographs reflect an artist grappling with representing new visual forms, for there were few visual precedents for representing railroads at that time. By finding formalist geometry in the tracks and mills, Rau invented "a genuinely new way to see the world."[64]

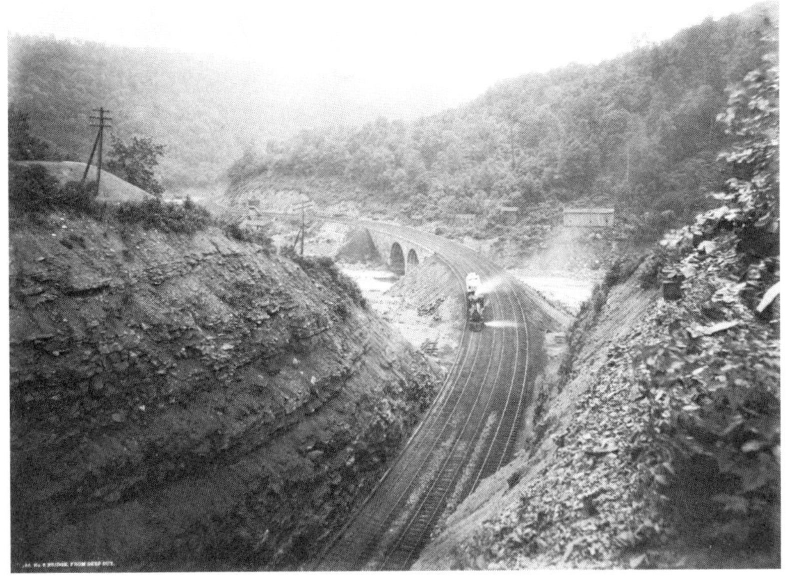

An illustrative comparison can be made between the industrial application of the B&O cyanotypes and Rau's railroad photographs. In Rau's photographs, nature often dwarfs or swallows the cool steel rails cutting through the landscape. In the cyanotype images, the emphasis is on the power and solidity of the steel construction. Here, photography was applied to document industrial power, not normalizing industrial development through aesthetic harmony. Rau's photographs seduce with technological sublimity woven through the natural landscape, pulling the viewer into the future while promising that nothing will be lost. In contrast, the cyanotypes show industry dramatically conquering nature. In part, the difference is tonal. The cyanotype prints are sterile, unnatural, and render the natural landscape through the detached view of the builder's blueprint. This is an industrial gaze, with no public to entice and becalm. This mode, however, would nevertheless find its own audience.

4.2 William H. Rau, *No. 6 Bridge from Deep Cut, Pittsburgh Division*, 1891. Albumen print, 17⅛ × 20⅝ in. The Library Company of Philadelphia. Courtesy of American Premier Underwriters, Inc.

Over the course of the twentieth century, the industrial sublime in photography developed into a significant genre. The industrial sublime describes an aesthetic emphasis on the structure and form of machines, factories, and other sites of production, reflecting a fascination with the scale of industry.[65] Some of the most famous photographs from this period center on steel, such as Lewis Hine's 1930–31 photographs of the Empire State Building, which was built using sixty thousand tons of steel supplied by US Steel. The glorification of the worker and industry in early twentieth-century photography celebrated the progress of industrial capitalism and the wealth produced by fixed capital investment. Particularly in the post–World War II period, many members of the working class, particularly those racialized as white, moved into a sort of middle class. Read through the lens of the present, the industrial landscape—and industrial photography—often evokes nostalgia. For example, when a set of William H. Rau albumen railway photographs of Altoona, Pennsylvania, were found in the mid-1980s, the town fundraised to buy them for the Altoona Area Public Library.[66] Altoona was founded by the PRR to connect Philadelphia and Pittsburgh, and it became an important node as steel production consolidated in Pittsburgh following the American Civil War. The decline of the railway in the second half of the twentieth century strongly impacted the town, reflecting how the fortunes of some communities were linked to industry. According to labor studies scholars Jefferson Cowie and Joseph Heathcott:

> What may be most troubling about these ruined industrial landscapes is not that they refer to some once stable era, but rather that they remind us of the ephemeral quality of the world we take for granted. If Karl Marx was right in saying "all that is solid melts into air," then the industrial culture forged in the furnace of fixed capital investment was itself a temporary condition. What millions of working men and women might have experienced as solid, dependable, decently-waged work really only lasted for a brief moment in the history of capitalism.[67]

Today, Altoona, like many former Steel Belt towns, is economically depressed. Rau's photographs were celebrated in this context of slow economic decline. Photo historian Mary Panzer uses a geological metaphor, describing Rau's work depicting Altoona as a "fossil record," as "Altoona's most valuable legacy was its past as Rau had preserved it."[68] Materially, Rau's prints evoked nostalgia even at the moment of their creation. Rau's choice to use albumen prints reflects a conscious linking of his photographs to the earlier era of Western expedition photography, whose distinctive purple-brown hues had largely been displaced by the more modern silver gelatin print by the time he took the photographs. The choice of an older process was another way of naturalizing the railroad's incursion into the landscape.[69] Through the already-nostalgic albumen print, Rau's photographs looked backward, even at the height of industrial growth for the region. In contrast, the cyanotype prints looked only to the future. As a material form, the blueprint anticipates its own futurity, always gesturing past its own objectness toward something yet to come, a replicable and predictable future premised on growth.

All That Is Iron Corrodes into Rust: Postindustrial Decline

Rau's photographs contain a history that the town of Altoona, Pennsylvania, sought to recover—a way of life that, before it turned to rust, was fixed in the photographic print. Rau's optimistic renderings of steel merging into the wooded Allegheny Mountains—nature and industry in balance—is undone over a century later by contemporary artist LaToya Ruby Frazier's frank look at the outcome of these processes. Frazier is a native of Braddock, Pennsylvania, which, like Altoona, was a twentieth-century crucible of the steel industry and the site of Carnegie's first steel mill, the Edgar Thomson Steel Works. Braddock is located in what we would now call the Rust Belt, the former industrial cities of America's Northeast: Altoona, Pittsburgh, Detroit, Gary, Youngstown. At the peak of industrial production in the postwar period, these centers of heavy industry in the Great Lakes and Midwest were celebrated as the Steel Belt. The industrial sector's decline, automation, and consolidation in the last quarter of the twentieth century were accompanied by regional economic decline, population loss, and urban decay. The Rust Belt is a colloquial nickname that uses a material metaphor to describe the corrosion that happens to iron as it is exposed to the elements. The former industrial centers have metaphorically turned to rust. Braddock shares with Altoona the experience of postindustrial decline driven by job loss, fiscal crisis, and the "organized abandonment" of the neoliberal period, captured so vividly in Frazier's work.[70]

A series of four portraits capture the artist in motion, superimposed over a flat, dark-blue background (plate 21). The portraits are documentation from a 2011 performance piece titled *If Everybody's Work Is Equally Important?*, held outside the Levi's store in SoHo, New York. In Frazier's performance piece, she rubbed her body on the sidewalk, wearing out a pair of jeans. In the portraits, the artist is dressed in head-to-toe denim, wearing work boots and heavy gloves. The most obvious thing to say about the images is that they are *very* blue. The Prussian blue coloration doubles the symbolic reference to blue-collar work evoked in the head-to-toe denim uniform. Frazier's cyanotype prints make a clear link between materiality and meaning, using iron as a light-sensitive material to create prints illuminating histories of steel—an iron alloy—and its communities.

Frazier's self-portraits responded to a 2010 advertising campaign by Levi Strauss called "Go Forth," that featured Braddock and highlighted a line of Levi's workwear, celebrating the pioneering spirit of America. Levi's cast residents from Braddock to star in the ads under slogans, including "Ready to work," "Everybody's work is equally important," and "We are all workers." In "Ready to Work," a voice-over narrates, "Maybe the world breaks on purpose so we can have work to do" while shots aestheticizing the gritty realism of industrial zones appear on screen. Visually, the romanticized decay of the Levi's ad reflects the larger cultural fascination with "ruin porn," aestheticized images of urban and industrial sites in states of disrepair, often being at least partially reclaimed by nature. Ruins, like photographs, are traces of the past and, as such, appear to have a privileged relationship to the real.[71] These images promise a working-class authenticity to people who feel increasingly alienated from the real economic base of their world: an authenticity that can be bought with blue jeans. The ad concludes by suggesting that Braddock is another frontier to be opened up.[72] The frontier reference renders the violent history of white settlers "conquering" the West as the undercurrent of the Levi's campaign, which evokes terra nullius and the doctrine of discovery, legal theories that provided a justification for the seizure of Indigenous territory in the Americas, here expanded to invite gentrification in working-class communities of color. Frazier rejects Levi's romanticization of Braddock as a postindustrial frontier full of opportunity for (presumably wealthy) people to "seize the land and build their lives."

While ruin porn is the dominant way mass culture imagines de/industrial landscapes, Frazier uses self-portraiture to place her body in front of the viewer, insisting on the ongoing histories that animate these postindustrial spaces. Frazier has a personal connection to Braddock's history of industrial promise, decline, and eventual abandonment. One of the workers in Andrew Carnegie's steel mills was Frazier's grandfather, one of the few African American workers employed in the mill. Artistically, Frazier is situated within another lineage of social documentary photographers who, to borrow Gordon Parks's famous expression, use the camera as a weapon to reveal systemic injustice. However, her practice has an important distinction from early social documentary projects. Walker Evans, Lewis Hine, Dorothea Lange, and many others were avid observers of the social suffering brought on by economic crisis, most famously documented under the aegis of the Farm Security Administration, a New Deal agency created in 1937 with a photography division that told a story of poverty during the Great Depression. Frazier, however, uses the camera to tell her own story. In the cyanotype portraits, Frazier uses her body to ground the vast abstract forces that make up deindustrialization and its effects, challenging the romanticization of decay presented in the Levi's ad. While photography struggles show us attritional violence in action, Frazier's practice makes visible the larger systems and structures that cause this violence, by directing our attention to the interlocking structures that made regional economic collapse happen and spread its costs unevenly among the population. In the process, the structural socioeconomic choices that build environmental sacrifice zones are laid bare.

The formal structure of Frazier's self-portraits alludes to Pennsylvania's industrial history. Visually, the portraits borrow a vocabulary from mid-century steel industry advertisements. Frazier uses repetition to insist on presence. Deindustrialization decenters the worker, physically leaving them behind, and is often accompanied by automation, a literal disembodiment of the worker. In the portraits, Frazier places her body in front of the viewer to contest the erasure of Black working-class women in industrial histories. The figure centered in histories of deindustrialization is typically the white male manual worker, making the working class erroneously synonymous with whiteness.[73] Deindustrialization has in fact had an outsized impact on racialized communities. Historian Thomas Sugrue traces the impact of deindustrialization in Detroit, Michigan, showing how racial discrimination magnified the effects of deindustrialization on Black communities.[74] Existing labor market segmentation and discrimination meant that for Black workers in Detroit lower levels of seniority, as well as their concentration in lower-paid, low-skilled positions, made them more vulnerable to industrial decline. Despite their apparent neutrality, last-hired, first-fired layoff structures would redouble the effects of decades of racist occupational segmentation. Sugrue's study brings to the fore the interrelations between deindustrialization, racism, and white-flight suburbanization. The region's industrial history was never entirely white.

Within her larger body of work, Frazier intervenes in the imaginary of whiteness in the Rust Belt by centering three generations of women in her family to show a history of Black women in Braddock that spans the rise of the local steel industry and its ensuing collapse. Her first major series, *The Notion of Family*, was a fourteen-year project documenting Frazier, her mother, and her grandmother, charting the rise and fall of local industrial agglomerations through a multigenerational experience of industrial labor. Frazier's work undertakes a durational analysis of the aftermath of both industrial activity and the loss of work. Here, a comparison to Allan Sekula is illuminating. Sekula spatially explores the outsourcing of labor in *Fish Story*, visually linking the deindustrializing Global North with labor in the Global South. He traces transnational shipping flows to insist on the ongoing materiality of production that relies on labor. Frazier, in contrast, explores the consequences of these shifts through the lens of temporality, as *The Notion of Family* locates connections through time, concretized in the bodies of her mother and grandmother.

Frazier's embodied portraits materialize the porous barriers between bodies, industry, and elements, drawing attention to the unstable boundaries between the body and the landscape. Iron, as we have seen, is a mineral necessary for human health, but refining iron in industrial processes introduces industrial toxins and mineral imbalances into bodies of workers and those who live nearby. The three women have dealt with the health consequences of such pollutants: Frazier suffers from lupus, her mother has cancer and an undiagnosed neurological disorder, and her grandmother died of pancreatic cancer and diabetes in 2009.[75] In *Self-Portrait, Lupus Attack* (2005) Frazier stares directly into the camera in a simply framed shot. Another portrait in the series shows a scar from her mother's spinal surgery. Frazier says, "I believe that the history of a place is written on the body of its inhabitants and their environment. Often in my photographs, whether it's a landscape of a house or an aerial view of railroads or a steel mill, I see the landscape as a portrait, a portrait of the body." She then summarizes, "I carry Braddock in my body. I have so much metal in my blood from living in that town. My body is that town."[76] Through this focus on the body, Frazier puts the Black working-class experience at the center of histories of deindustrialization and questions of environmental justice. Frazier's work is a powerful testimony to how the legacies of environmental toxicity and systemic racism directly impact the health of people living in environmental sacrifice zones like Braddock and their continuing determination to make life differently.

The shaping role of industry and economy is inescapable in Frazier's work, affecting employment, unemployment, health, social relations, and politics. The complex role that industrial growth played in working-class lives echoes the often-contradictory role that extraction plays more broadly. It is this tension that Frazier subtly explores. Her practice is not simply a demand for visibility, or for the right to see and be seen on her own terms, but rather, as Clorinde Peters writes, she makes visible the "processes and effects of disposability itself," highlighting the structural socioeconomic choices that made Braddock an environmental sacrifice zone.[77] Frazier's practice shows these choices are precisely that: the decision to abandon entire communities.

In 2017, Frazier took a cyanotype portrait of Sandra Gould Ford wearing a hard hat and protective eyewear, holding samples of metal ores (plate 22). The portrait is from the series *Steel Genesis*, which focuses on Gould Ford, a working-class Black woman who worked as a clerk and secretary at Jones & Laughlin (J&L) Steel Corporation in Pittsburgh until the factory closed in 1985. In an echo of Anna Atkins's collaboration with Anne Dixon on *Cyanotypes of British and Foreign Ferns*, *Steel Genesis* is a collaborative project between Frazier and Gould Ford. *Steel Genesis* includes Frazier's portraits alongside photographs that Gould Ford took while working at J&L and documents from the firm that Gould Ford archived. In *Steel Genesis*, the documents were transformed by Frazier into cyanotype prints on a manual platemaker at Pittsburgh Filmmakers. The documents include records of workplace accidents, insurance claim settlements, and workplace safety concerns alongside company promotional photography. For example, a blueprint labeled "Fatal Accidents for South Side Works" documents nine fatal accidents in 1912. Above the list of accidents are covers of J&L's in-house magazine. The magazines showcase a portrait of a worker calmly performing his task (plate 23). Through this juxtaposition, Frazier highlights the gap between the idealized vision of corporate communications and the realities of work while contesting the erasure of women's labor in industry, exemplified by the naming of the magazine: *Of Men and Steel*. The archival documents and Gould Ford's photographs frame the portraits of Gould Ford that form the heart of the series. In these keystones of the series, we see Gould Ford in a variety of contexts: quilting, in Homewood's Carnegie Library, surrounded by plants. The constitutive role of the steel industry in Gould Ford's life is clear, but in Frazier's portraits, her subjectivity is not limited to her labor or her connection to the industry.

Steel Genesis also includes black-and-white photographs taken by Gould Ford that capture the day-to-day realities in industrial landscapes alongside explanatory captions. The photographs contribute to and nuance an archive of steel by photographers like W. Eugene Smith, who celebrated the towering monuments of industry. Gould Ford's photographs include the hallmarks of steel production—coke ovens, barges, and smelting furnaces. However, in contrast to the modernist forms of the industrial sublime, which forms a visual language based on the dynamism and power of industry, Gould Ford's black-and-white photographs are prosaic scenes. Gould Ford's *Coke Oven Peaches* (1988) is still, quotidian. She tells us that the peaches grow near the toxic Hazelwood coke ovens. In a short essay accompanying the image, Gould Ford reflects on the health benefits of peaches, which contain potassium, fluoride, vitamin C, antioxidants, and iron, which, as she notes, are necessary for red blood cell formation. In the shadow of the coke ovens, however, the organic growth of the fruit is poisoned by industrial growth. The emphasis on plants harkens back to Atkins's cyanotypes, which embodied but did not directly register industrial production. Here, the deformations of industry to nature and human bodies are the subject. Gould Ford explains that coke ovens "burn away coal's hydrogen, methane, nitrogen, carbon monoxide, carbon dioxide, ethane, propane, tar vapors, naphthalene vapor, ammonia gas, hydrogen sulfide gas and hydrogen cyanide gas," as well as remove oxygen from iron ore, making it metal. In 1911, Jeremiah Jenks reported that "the smoke and gas from some ovens destroy all vegetation around the small mining communities." The health benefits peaches could offer are overwhelmed by the carcinogens that saturate the landscape—hyperconcentrations of elements necessary for life can be lethal. The carcinogens introduce a third kind of growth, the mutating, out-of-control growth of cancer cells so often found in the regional aftermath of industry. The peaches do not show this history, though they have absorbed it. Gould Ford concludes that the peaches had "the audacity to send out roots that claimed sustenance. Amidst toxic and brutal conditions, peaches grew and delivered sweetness, beauty and healing."[78]

Gould Ford thus posits that beauty and a will to live and provide exists amid the wreckage of industry. Similarly, the inclusion of genesis in the title positions steelmaking as a metaphor for people forging their lives and identities, what Gould Ford describes as a process of "constantly coming into being, constantly evolving." Gould Ford uses the steel mill as a metaphor for becoming:

> Steel is a manmade item; steel is essentially iron, which occurs in nature, but steel is superior to iron because of what people do to the iron. Mettle is the courage and willingness to stick with something, to endure, to strive—that's another human quality. So, I see the two as similar, and that is really what I intertwine as I carry this work forward. The fact that we can make ourselves into steel.[79]

In *Steel Genesis*, Gould Ford is both symbolic and specific, standing in for a larger economic system while resisting the move to abstraction. Portraiture can function as a space of self-construction by allowing the subject to present as they want to be seen. Frazier's portraits show Gould Ford in a multifaceted way: they acknowledge her status as a worker without

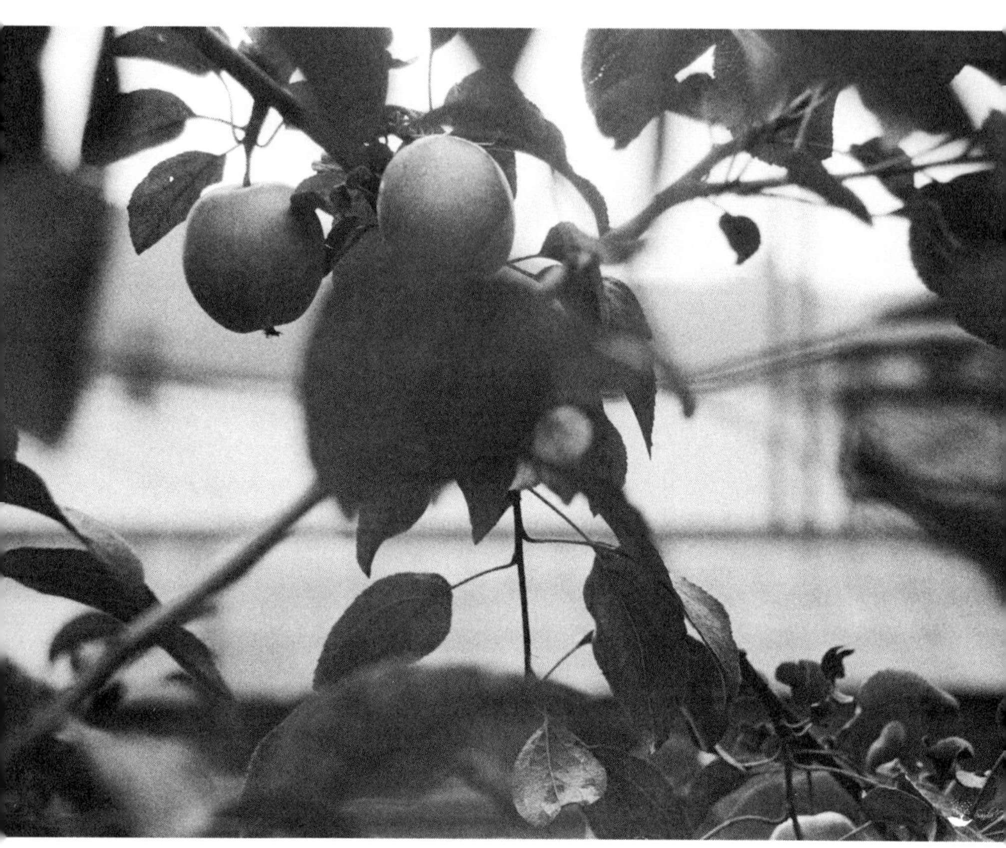

reducing her identity to it. Here, we see a person in charge of their own life and destiny without eliding the complex role of industrial racial capitalism in shaping and constraining that life. Frazier's work contests the erasure of blue-collar workers in Pennsylvania, but she does not idealize industrial work either. Rather, the photographs probe the larger working of neoliberal capitalism, organized abandonment, and worldmaking in the wreckage. Frazier's cyanotypes are not photograms of tangible objects, but through this symbolic connection between the materiality of the image and its subject matter, the images ground us in the solidity of the world. In the process, she shows us something about how our world was—and continues to be—made.

4.3   Sandra Gould Ford, peaches growing near the toxic Hazelwood coke ovens, 1989. Silver gelatin print. © Sandra Gould Ford, SteelGenesis.com.

Blue is a color rarely found in nature, though it colors the sea and the sky, giving it a kind of vertical symbolism.[80] According to the symbologist J. E. Cirlot, blue has contradictory meanings: purity and clarity, as well as loss and sadness, "to have the blues."[81] White is associated with illumination and ascension but in the cyanotype, also absence, as white forms a negative impression. Following Cirlot, it could be argued that within the cyanotype, the combined visual effect of white superimposed over blue is that of absence and loss. The color compounds Roland Barthes's observation that photography has an implicit connection to death, loss, and absence. Linking the color blue to memory, Carol Mavor describes Atkins's algae as "fragments of short stories carried by the sea," suggesting that the algae might speak in the present through the memory trace.[82] I want to suggest that Atkins's white impressions of algae are a kind of haunting. The plant begins to die once it is harvested. In this sense, the image becomes a funeral shroud. Read through the lens of the present, Atkins's stark-white algae draw a visual association with coral bleaching. Algae has an endo-symbiotic relationship with coral: algae live inside coral tissue and, in turn, provide up to 90 percent of the energy required for coral to live. As climate change causes temperatures to rise, coral expel the algae that live in their tissue. In a process called coral bleaching, the coral turns white and begins to starve. Coral provides coverage for many marine species, and coral collapse reduces overall fish biodiversity while making the coastline more vulnerable to damage by heavy waves, flooding, and erosion.

The ghostly forms of the plant cyanotypes—*there-then*—are a type of haunting that can speak in the present, *here-now*. As climate crisis has materially demonstrated, the traces of the past never entirely disappear, and the future is always prefigured in the present. As the editors of *Arts of Living on a Damaged Planet* write, "The winds of the Anthropocene carry ghosts—the vestiges and signs of the past ways of life still charged in the present."[83] The future called into being in the Victorian period required a "willingness to participate in great projects of destruction while ignoring extinction as collateral damage. . . . The terrain carved out by this future is suffused with bad death ghosts."[84] The ghosts of industrial development thus form a type of haunting. Atkins's photograms prefigure the mass extinctions of the twentieth and twenty-first centuries, an extinction whose seeds were planted in the socioenvironmental transformations of the Victorian period. They remind us of what has been lost; through a presentist lens, they are an invitation to mourn.

While Atkins's work engages with loss on the level of visual form, Frazier's subject matter engages directly with what has been lost or, perhaps more accurately, expropriated. Her photographs meditate on the complex ways that working-class communities remember industrial work and make sense of its dissipation. The closing of steel mills had detrimental effects on communities, but when they operated and provided employment, they also had destructive effects, polluting air, soil, and waterways. These legacies of industrial production resurface as industrial disease in bodies and polluted landscapes. Frazier's camera shows the hollowness of the slogans touted by Levi's, confronting the viewer with the human and environmental legacies of industrial work, as she shows focused attention to communities that experienced grueling work in mills, factories, and mines. Her work turns us toward the environmental injustice of these kinds of production; the disorganizing ravages of globalization of trade and competition, automation, and the closing of local steel mills; the mushrooming of poverty and the punitive attempts to create unjust stability out of capitalist disorder, such as the war on drugs—other hauntings that continue to shape the present and whose material impacts function to foreclose more just possibilities for the future. Eve Tuck and C. Ree write that "haunting is the cost of subjugation. It is the price paid for violence, for genocide."[85] While Tuck and Ree specifically reference settler colonialism—a project of empire—these hauntings are inescapable throughout worlds defined by racial capitalism—the subjugation of nature, the privatization of land in the enclosure, the discipline of the factory. These are debts that society has inherited, hauntings that make themselves known in species extinction, in rising temperature, in melting ice caps, in climate migration.

Iron as a material is rich with narratives of development, trade, the factory, empire-building, migration, and conflict. Through its material trace in the photograph, it forms a visual archive showing the collective choices made during industrialization. Iron is the material emblematic of progress as measured by economic growth. Central to the growth of human and plant metabolisms, the very collective human ability to concentrate and augment iron's properties likewise created concentrations of it and other minerals that are toxic to those same beings. The byproducts of mining and refining iron cause carcinogenic growths, mutations that destroy life. The costs of valuing economic growth over bioflourishing worlds are painfully clear. As Frazier shows us, industrial progress on these terms has had significant and lasting consequences for the working-class communities who live and work in the shadow of industry. And yet there is a potential distinction between the brutal organized abandonment of deindustrialization and more ethical models of planning to provide for large-scale human needs without mass sacrifice, a shift that would require a fundamentally different system of values.[86] Materially, the cyanotype poetically troubles the boundaries between the *geos* and *bios*, between the mineral, plant, and human worlds. If capitalist manners of viewing the world reinforce dualisms such as nature and culture, these messy mixings—of ecosystems, of human and nonhuman activity, of nature and culture—can point to new ways of living and relating. In both a tangible and symbolic way, it insists we take seriously the profound entanglement between the human, botanical, geological, and industrial dimensions of our existence.

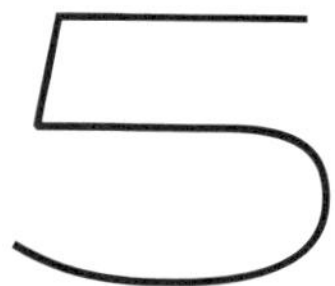

Uranium and Photography beyond Vision

> The uranium stockpiles, their ash-grey emissions, the thin,
> hair-like covering of birches at the foot of this mountain be-
> longed to the horizon of my childhood as for others maybe
> the Alps do, or the eves of a neighboring row of houses. When
> I draped myself across the gate that opened back out onto
> the fields, their horizon reached all the way into my dreams.
>
> Lutz Seiler, "The Territory of Tiredness" (2016)

A SERIES OF EIGHT-BY-TEN-INCH PHOTOGRAPHS captures a ghostly white light emerging from a black void. The prints are hung between sheets of Plexiglas, which evoke microscope slides from the laboratory and the lantern-slide plates once used in scientific lectures. The images are cameraless exposures, called autoradiographs, exposed by atomic light (plate 24). The autoradiographs were included in the installation

Pechblende (Prologue) (2016), staged by Berlin-based artist Susanne Kriemann, which forms an archive that explores the material histories and visual possibilities of uranium. Indeed, the history of radiation and the medium of photography have unexpected connections.

Experiments with uranium were common in early photography. Uranyl nitrate was used as a photosensitive salt and as a toner to tint and shade silver prints. In 1895, Wilhelm Röntgen discovered X-rays, opening up a new realm of perception that called into question the boundaries of visibility and invisibility. A year later, Henri Becquerel observed radioactivity—the spontaneous emission of radiation—after leaving pitchblende on a photographic plate. The pitchblende essentially took a photograph of itself, thus establishing that uranium emitted radiation without an external energy source like the sun. Röntgen's and Becquerel's discoveries centered on photographic plates, for radiation exceeded human vision but could nevertheless be captured photographically.

Using Becquerel's process, Kriemann stages her archive around these indexical but abstract images. The exposure time shapes the visual form, ranging from three days to 112 days. Installed behind Kriemann's autoradiographs are X-ray photographs and archival autoradiographs from *Operation Crossroads: The Official Pictorial Record*, a series of nuclear tests conducted by the United States military at Bikini Atoll in the Marshall Islands in 1946. One such archival image shows a fish cut in half, splayed on a photographic plate, ironically tagged "hot supper." Like Kriemann's prints, the image is a cameraless photograph exposed by radioactive soft tissue.[1] The radioactive algae eaten by the fish exposed the image, the hot—or radioactive—supper. The fish's body was imprinted with the violence of the blast, a delayed violence that then became imprinted onto the photographic surface.

The cameraless photographs make visible a type of disaster that often eludes vision. As radioactivity enters ecosystems, it spreads through food chains, waterways, and the atmosphere. While narratives of nuclear catastrophe typically focus on the potential of a singular catastrophic event, atomic events leave behind a legacy of slow violence. Visually, the impacts of uranium are most associated with the mushroom cloud produced by the atomic bomb, but as artist Susan Schuppli writes, "The mushroom cloud separates the visual field from the material field it documents, [but] the radiological contact print is immanent to, and continuous with, the event."[2] Atomic fallout continues beyond the initial disaster, as radiation

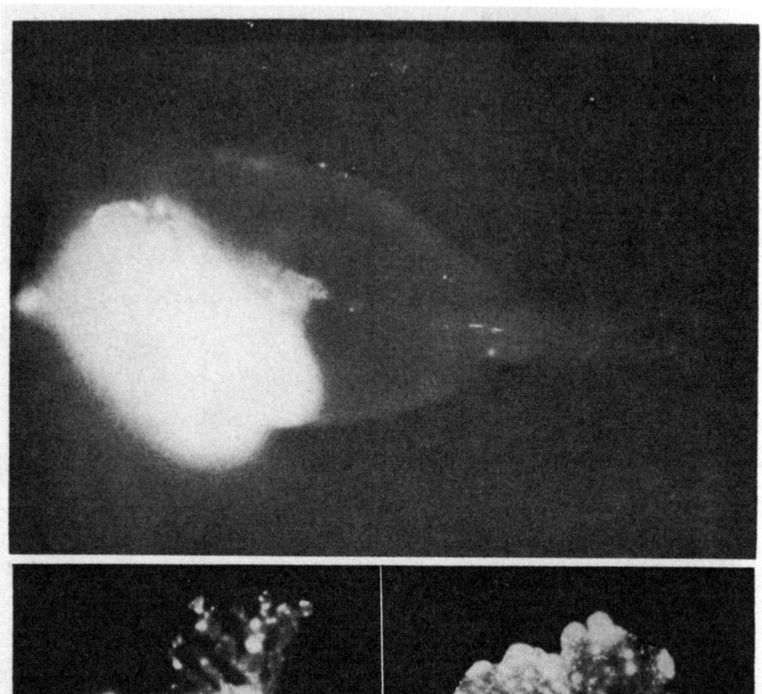

INSIDE THE FISH, A "HOT" SUPPER. UPPER. This puffy surgeon fish has gorged itself heavily on radioactive algae, the common seaweed of the Bikini Lagoon. The plant itself had previously ab-, sorbed radioactive fission products deposited in the water by the atomic bomb. Radioactivity of the plants will last a long time, is not affected by passing into the fish's stomach. Previous similar dinners already digested and distributed about his body are revealed by the radioactivity of regions in the neighborhood of nose and eyes. The fish was not killed by the radioactivity, it being apparently true that the more elemental the form of life, the less it is affected by radiation. LOWER. Radioactive algae. These "radioautographs" were made by placing the subjects on a photographic film overnight. They suggest Becquerel's 1896 discovery of radioactivity when a key left near some uranium-bearing pitchblende ore similarly "took its own picture" on a piece of film.

5.1   "Inside the fish, a 'hot' supper," 1946. From *Operation Crossroads: The Official Pictorial Record* (Washington, DC: Joint Task Force One, 1946).

exposure is a cumulative affliction based on the interplay of time and intensity. In human bodies, the immediate impacts of exposure include burns, hair and nail loss, and nausea while long-term exposure damages genomes, disrupts metabolic processes, and causes carcinogenic mutations.

Kriemann's archive stages a journey that guides the viewer through multiple sites of interaction in the life cycle of pitchblende, tracing a reverse journey back to the mine. Her project centers on uranium mining regions in the former German Democratic Republic (GDR), where the mining company SDAG Wismut mined pitchblende (*Pechblende* in German). Pitchblende, a type of uraninite, has a radiation level four times that of uranium.[3] SDAG Wismut extracted more than two hundred thousand metric tons of uranium oxide—yellowcake—from pitchblende between 1947 and 1990.[4] The uranium-rich mineral was primarily used as fissionable material for Soviet atom bombs. The poet Lutz Seiler, whose father was a uranium miner at SDAG Wismut, describes his hometown as "tired villages—what a beautiful way to describe the effect of low doses of constant radiation," where "the space we knew—which right before our eyes had been expelled or pumped away into a tailing—had become a landscape of slagheaps." The landscape of Seiler's childhood was "excavated away, buried alive."[5] Seiler's evocative description of the landscape being excavated and buried alive indicts Wismut while provoking the reader to reckon with the impact of toxic mining landscapes on communities and workers.

Mining caused "locally extreme devastation": the groundwater and soil in the region are laced with arsenic, uranium, and radium while wastepiles emit radon gas.[6] Remediation projects are underway—involving levelling slag heaps, covering them with soil, and regrassing them; filling or flooding mine workings, shafts, and tunnels; and decontaminating and reclaiming waste ponds—but scientists estimate that environmental damage will continue to affect the region for hundreds of years. While radioactivity is often narrated in graphs and charts, the autoradiograph is the physical trace of a violence that eludes human vision. Graphs are a representation of radioactivity, mediated by humans. In these images, the radioactivity reveals *itself*. With both uranium mining and the atomic bomb, the release of invisible radioactive isotypes into the surrounding ecosystems contaminated water, soil, and air, entering bodies and landscapes. Radiation moves through temporal registers and space in often unpredictable ways. The considerable damage to both environments and human bodies caused by uranium extraction continues to affect the present and shape the future, though these costs also often exceed the visible. On the photographic surface, these traces of slow violence become tangible in our vision.

Experiments on the mineral pitchblende were at the heart of many significant atomic discoveries. *Pitch* refers to the black color of the rock while *blende* comes from the German *blenden*—to blind, deceive, or make unable to see—as the density suggested a metal, but before the discovery of uranium, it was considered worthless or an impostor mineral.[7] Literature scholar Birgit Dahlke connects *blende* to the camera shutter, noting that *Verblendung* ("blindness, infatuation") and *Ausblendung* ("fade out, suppression") are also associated with the term. Writing about the poetry of Seiler, Dahlke suggests that he "enriches the terminology of mining with the dimension of camera technology but also of psychological memory and thus poetics."[8] Building from this insight, I turn to the considerable costs of uranium extraction on workers and landscape to tease out what histories of labor and displacement have been faded out and suppressed in the visual culture of uranium.

Examining uranium allows us to explore the limits and possibilities of vision and visibility in the context of forms of environmental damage often outside the realm of the visible. Uranium is significantly less central to the production of photography than the other materials explored in this book, but the material possibilities of the mineral can reveal the integral connections between geology and vision that have been driving our investigation—in particular, by directing our attention to what photography theorist Shawn Michelle Smith calls "the edge of sight."[9] The edge of sight could also describe extraction, systems that are omnipresent but for so many hard to "see" and integrate into our conception of our social lives. For this reason, uranium will serve as our guide to consider one of the most pressing questions animating the intersection of the visual and the environment: the phenomenon Rob Nixon calls "slow violence."[10] The protracted temporality of environmental damage "occurs gradually and out of sight, a violence of delayed destruction that is dispersed across time and space, an attritional violence that is typically not viewed as violence at all."[11] Critically, Nixon writes, slow violence resists representation as it develops unevenly and moves between temporal scales and geographic regions. Slow violence eludes human vision and the quick flash of photographic capture, revealing the limitations of photography in making sense of environmental transformations.

Radiation is a material phenomenon that cannot be seen, smelled, touched, or tasted, waves of energy that present us with various paradoxes regarding the limits of representation in understanding both matter and environmental crisis. However, while the slow violence of uranium exposure eludes vision, the strategic use of radiation also extends human vision. Atomic light's materiality pushes the boundaries of the visible and the invisible, as most tangibly shown in the X-ray. The perceptual possibilities introduced by radiation bring to mind Walter Benjamin's notion of "nature that speaks to the camera rather than to the eye," a phenomenon he termed the "optical unconscious."[12] Despite its inherent invisibility, the culture surrounding radiation is surprisingly visual: the spectacle of the mushroom cloud defines the visual culture of the Atomic Age. The applications of radiation in military technology layer on complex questions of visibility and invisibility as the secrecy in developing the bomb made some places of extraction and refining socially invisible for reasons of national security. This also reflects which lives, landscapes, and histories are within the configuration of the visible that defines what is sayable and seeable and what, in turn, is considered insignificant or unworthy of entering into public consciousness. Radiation challenges vision and provides an entry point to consider how toxic histories might be productively reworked.

So far, I have primarily focused on histories of extraction, but uranium's history introduces a different set of questions. The emphasis in this chapter remains on mined materials, but uranium's histories are not primarily told through this lens. Accordingly, this chapter toggles between the material realities of extraction and the more spectacular imaginaries of atomic culture. Historian Gabrielle Hecht argues that "our fetishes have kept us close to bombs and reactors but far from other places," like mines and laboratories where the nuclear emerges.[13] In this way, it is similar to oil, which emphasizes energy over extraction, raising larger questions about what is seen and unseen. I return to the question of visibility explored in chapter 1 and focus on the invisible to probe the material limits of seeing in the context of slow violence.

I begin by outlining uranium's unusual materiality and its use as a colorant and in photography. A number of critical discoveries in atomic culture centered on photography, and I analyze early photographic experiments by Niépce de Saint-Victor, Wilhelm Röntgen, and Henri Becquerel to reveal how scientific discoveries centered on uranium challenge existing conceptions of vision. From there, I turn to the visual culture of the atomic age and photography of the bomb to consider what photography reveals and what it suppresses. Then I return to the question of uranium's materiality to think through how the atomic bomb challenges the notion of the photographic, as the light and heat of the bomb turned entire cities into photographic surfaces. Finally, I conclude by returning to Susanne Kriemann's exploration of uranium mining in the GDR, which brings together plants, photography, and phytoremediation to consider slow healing as an antidote to slow violence. Throughout, I attempt to sense the limits and possibilities of vision and visibility in the context of both slow and spectacular violence.

Uranium

Approximately 6.6 billion years ago, uranium (U, atomic number 92) was formed in a supernova, a catastrophic explosion whereby a star ejects most of its mass. Uranium is naturally radioactive, as the nucleus of the element is unstable. As the element seeks stability, it is in a continuous state of decay as the element ejects nucleons and other particles in an attempt to find balance. The slow processes of radioactive decay are the main source of heat inside the earth, generating continental drift and convection.[14] Radium and polonium are the daughters of uranium, created from the decay of the uranium atom, which eventually turns into lead at the end of the radioactive decay chain. The timescale of the half-life (the time required for a quantity to reduce to half of its initial radioactivity) of uranium is around 4.5 billion years, roughly the same as the age of Earth.

Uranium was formally discovered in 1789 by the German chemist Martin Klaproth in the mineral pitchblende. Klaproth named this unstable mineral after the planet Uranus, which had just been discovered by William Herschel in 1781. The ringed planet spins on its side, a fitting namesake for an unstable element in constant decay. Uranium is rare and often found in combination with other elements, including black oxide, silica, oxide of lead, oxide of iron, lime, and phosphorus.[15] Uranium is rarely sufficiently concentrated in ore bodies for extraction to be economically viable, but it is commonly mined as pitchblende. As pitchblende is often found alongside silver, the pitchblende used in Klaproth's experiments came from a silver mine in Johanngeorgenstadt, in Saxony's Ore Mountains.[16] The earliest reference to pitchblende dates back to 1565,[17] while early eighteenth-century sources reference its presence in the Erzgebirge Mountains in Bohemia (Czechia).[18] By 1790, a year after Klaproth's discovery, mining for the newly valuable pitchblende had begun in Joachimsthal, which had historically mined silver for currency.[19] The roots of atomic culture proper—culminating with the atomic bomb—date back to an 1841 experiment by the French chemist Eugène-Melchior Péligot, who isolated uranium metal by heating uranium tetrachloride with potassium. The purification of uranium ores forms uranium oxides, or yellowcake.

Centuries before it could be deliberately produced, uranium oxide was used as a coloring agent for ceramic glazes and in glass as early as 79 CE.[20] Pitchblende was primarily used to produce olive-green, gray, and ochre pigments. As stated by art historian Eva Wilson:

> Painters in the seventeenth and eighteenth centuries used concoctions of oxidized uraninite mixed with fresh-egg tempera or oil made from local walnut crops to produce paintings that are radioactively contaminated: look at them and they will look back at you, not by means of the refracted sunlight glistening off their varnish in all the colors of the visible spectral range, but by the stealthy emanation of their own energies, radiating not just to your eyes, but through your clothes, your hair—and all the cells of your body.[21]

This uncanny phenomenon—"look at them and they will look back at you" or, for our purposes, the rock taking a photograph of itself—reveals uranium's power to trouble static ideas of vision.

With the dawn of photography, several early practitioners experimented with uranium salts, and uranium was used as a toner in liquid film-developing baths. An 1858 edition of *Photographic Notes: Journal of the Birmingham Photographic Society* stated, "Uranium is not a costly metal. It is obtained from a mineral termed Pechblende and from varieties of uranitic mica found at Callington, in Cornwall."[22] The assessment of uranium as cheap would change drastically over the course of the twentieth century as uranium, radium, and polonium found previously unimaginable uses in nuclear projects. Radium would peak at a price of $125,000 per gram, making it the most expensive material in history.[23] Uranium, however, became decisively linked to photography at the end of the nineteenth century. Following Röntgen's and Becquerel's discoveries, Marie and Pierre Curie discovered polonium and radium in pitchblende in 1898, which emit stronger radiation than uranium, thus cracking open the fields of radiotherapy and nuclear medicine.[24] The immediate applications of X-ray technology were in medicine. It was quickly established that radiation could cure tumors, but untargeted exposure could also cause illnesses. Many innovators in the field of radiation died of diseases caused by radiation exposure. Marie Curie died of aplastic anemia, Anna Bertha Röntgen died of sarcoma, and Wilhelm Röntgen died of carcinoma of the intestine. Henri Becquerel died of unknown causes in 1908 at fifty-five, but his skin was severely burned from handling radioactive materials. High levels of industrial disease characterize every step of uranium's production chain, as uranium rays and particles enter the body and mutate its internal structure.

In 1935, Arthur Jeffrey Dempster identified Uranium-235, the only naturally occurring isotope capable of sustaining a nuclear fission reaction. In 1938, the German chemists Otto Hahn and Fritz Strassmann discovered the possibility of nuclear fission, which was then theoretically developed by Lise Meitner and Otto Frisch, laying the groundwork for an atomic bomb. This initiated a flurry of research and a series of discoveries across leading industrial states, culminating in the Manhattan Project, formally founded in 1942. With the development of the atomic bomb, the slow violence of radiation was augmented by the spectacular violence of the bomb.

Visibility, Atomic Light, and Material Traces

To demonstrate the significance of the visual in building public knowledge around the nuclear, I turn to a speculative image, a photograph that no longer exists. A photograph that, when it was first displayed, was illegible within existing conceptions of vision.

In 1857, Niépce de Saint-Victor—a cousin of Nicéphore Niépce—reported that uranium salts left in the dark made marks on photographic plates. The process was described as "uranium paper" or a "uranium copying process," and Niépce showed pictures made with this process at the Third Exhibition of the Société Française de Photographie in Paris in 1859.[25] Niépce recognized that neither conventional phosphorescence nor fluorescence was causing the exposure, as the salts exposed photographic plates *without* sunlight. Niépce demonstrated that "a drawing traced on a piece of carton with a solution of uranium nitrate … whether or not exposed before to light, and applied on a piece of sensitive paper prepared using silver chloride will print its image."[26] The uranium salt essentially took a photograph of itself, thus establishing that uranium emitted radiation without an external energy source such as the sun, resulting in a spontaneous photograph. Niépce concluded that uranium salts emitted a form of energy that exceeded human vision but could be made visible photographically.

This was the first documented experiment to capture the material traces of radioactivity, the process later named by Becquerel and used by Susanne Kriemann in her artistic practice. The phenomenon Niépce described is radioactivity—the spontaneous emission of radiation—though it would not be named and understood until nearly forty years after Niépce's initial findings. He announced his experiments in photographic journals, and other photographers replicated his experiments, including William Crookes, the inventor of the Crookes tube—an experimental electric tube central to the discovery of the X-ray—and an editor of the *Photographic News* and the *Journal of the Photographic Society*. Crookes wrote:

> Niépce de St. Victor discovered that uranium salts possessed the property of storing up light and giving it out in the dark, and in 1858 I took what was perhaps the first radium photograph in this country, by writing with a solution of uranium nitrate on a card, isolating it, and then putting it face to face in the dark with a sheet of photographic paper; the image of the writing was reproduced on the paper.[27]

However, writing in response to Niépce's experiments, the editor of *Photographic Notes: Journal of the Birmingham Photographic Society* concluded:

"We see nothing in any of these experiments to lead us to believe in any 'new action of light,'" and he instead determined that the creation of an image by exposure to light is simply "common photography."[28] Failing to recognize that the light that Niépce had identified was atomic light, the editor suggested printing with uranium might be of use to people taking portraits to be colored by an artist but that carbon printing was a much more exciting area of research.[29]

Like many early photographs, Niépce's photograph no longer exists and was not reproduced in any journals. I begin with this image for two reasons. First, it highlights photography's central role in scientific discoveries related to uranium. Second, it is an image that did not enter the realm of the visible. I suggest that this is, in part, a question of legibility. Studies of uranium salts on photographic plates date back to the 1789 discovery of uranium by Martin Heinrich Klaproth. There are many examples of scientists encountering uranium exposure in experiments before Röntgen's discovery, as people noticed "strange shadowy pictures" surfacing on photographic plates.[30] William Crookes, experimenting with what would become the Crookes tube, complained about fogged and blackened photographic plates to his supplier, not realizing he was seeing the visual artifacts of radioactivity.[31]

While Niépce's experiments forged an early link between uranium and photography, the German physicist Wilhelm Röntgen's discovery of X-rays decisively fused photography to experiments with uranium. Consider a photograph taken by Röntgen on December 22, 1895, titled "Hand with Rings." Today, the image is immediately recognizable as an X-ray, a ghostly illumination that reveals the interior of the body. The X-ray photograph was a formalized experiment that emerged from an earlier accidental discovery. While conducting experiments with cathode rays in November 1895, Röntgen placed a Crookes tube on a book with a key inside. Underneath the book was a photographic plate. When he later used the plate in his camera, he discovered the outline of a key. He concluded that the glass tube emitted a light that penetrated the book's pages. Röntgen called these rays X-rays, *X* indicating an unknown quantity in mathematics. X-rays have a shorter wavelength and a higher frequency than visible light. Because of this, the rays can penetrate and illuminate solid matter but cannot be seen by the human eye. Photography is the medium through which these rays become visible.

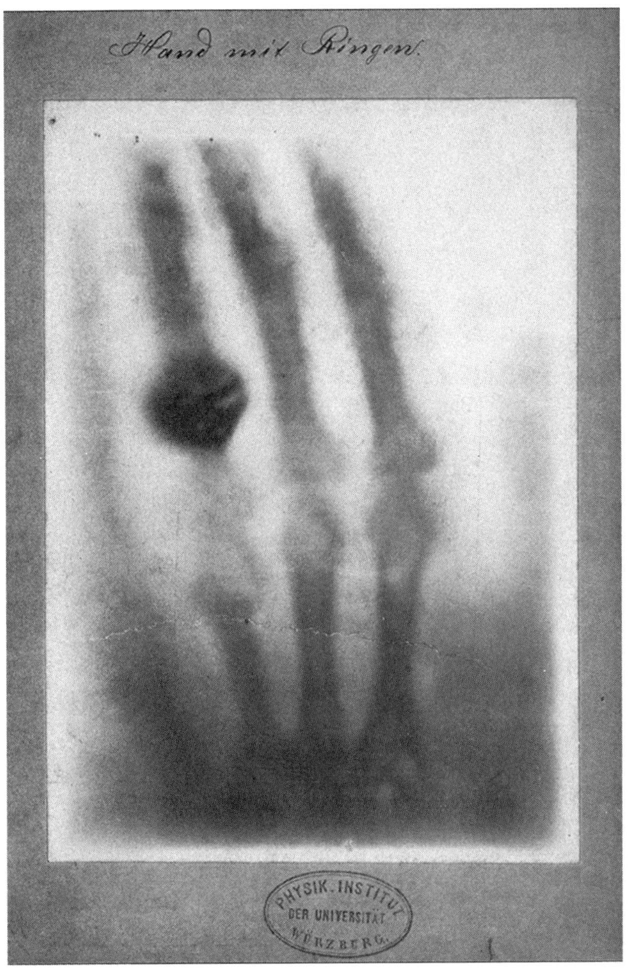

Röntgen's photograph of his wife's hand was an attempt to re-create the experiment in more controlled conditions. After an exposure time of fifteen minutes, the impression was recorded on the photographic plate. The photograph showed that the invisible rays could pass through human tissue.[32] In the photograph, the bones in the hand and the ring she wore are visible, while the flesh and other soft tissue surrounding the hard forms are invisible. As film scholar Akira Mizuta Lippit describes, "A living image of death and the deathly image of life are intertwined in the X-ray."[33] This destabilizing shift is noted in Anna Bertha Ludwig's remark "I have seen

5.2   Wilhelm Röntgen, *Hand with Ring*, 1895. X-ray.

my own death" upon seeing the X-ray of her hand. Bertha alludes to the instability between visibility and invisibility, the interior and the exterior, but also to the instability between temporalities, feeling she *sees* her future. This evokes Karen Barad's reflection on how radiation's "temporalities are specifically entangled and threaded through one another" as radioactive decay "elongates, disperses, and exponentially frays time's coherence."[34] Radiation also frays the coherence of vision.

Röntgen widely circulated reproductions of his visually striking X-ray photograph, which showed that mysterious rays could extend human vision beyond the limits of solid matter.[35] When the discovery was announced in the Vienna newspaper *Die Presse* in January 1896, the author wrote that it sounded like "a fairytale or like a daring April Fools' joke" but would surely be "epoch-making" research.[36] The article emphasizes the "fantastic futuristic speculations in the style of Jules Verne."[37] X-rays had wide-ranging implications, for they showed the inadequacy of human vision and initiated a series of experiments on uranium and visual perception.[38]

Shortly after Röntgen's discovery, Henri Becquerel announced that he had discovered "radiation produced by phosphorescence" after leaving pitchblende on a photographic plate. Becquerel was shown an X-ray by Röntgen in January of 1896 and intuited that there might be a connection between radiation and uranium compounds, which generate phosphorescence. Becquerel's father, Edmond, experimented with uranium in addition to researching the effects of light, which would prove to be key for the research of Becquerel the younger. Edmond Becquerel was a founder of the Société Française de Photographie and was one of the first people to photograph the spectrum of sunlight.[39] Both Henri and Edmond recognized that the visual was an important scientific tool, and they played an essential role in adapting photography to science.[40] Henri Becquerel's initial experiments involved placing uranium on a photosensitive plate made from gelatin and silver bromide, which he wrapped in thick black paper. The wrapped package was exposed in sunlight for several hours. When the plate was subsequently developed, a shadow of the phosphorescent substance was visible. It was a later discovery that would have more impact, however. When a series of cloudy days interrupted an experiment, he left his sample of uranium on a photographic plate in a drawer.[41] When he developed the plate, an image of the uranium—much clearer than in his earlier experiment—was visible.

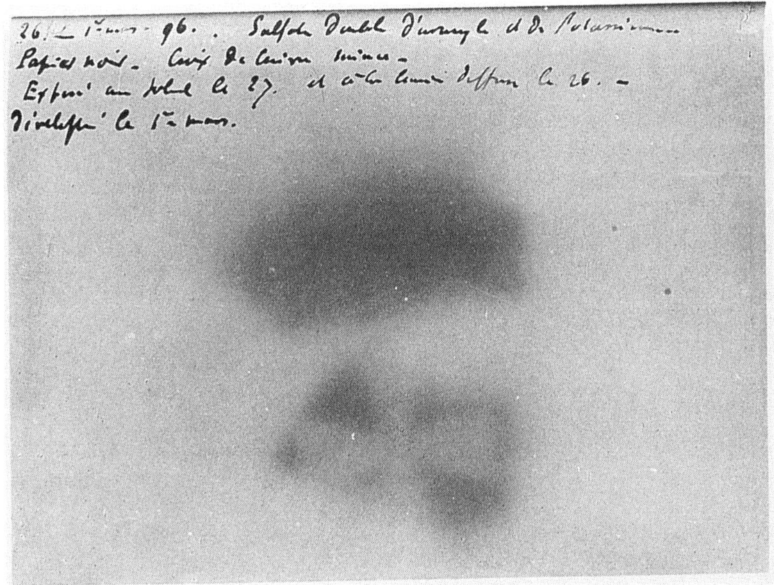

While Röntgen's X-rays added the energy of the Crookes tube, Becquerel's photographs did not use an external source of energy. To summarize, Röntgen's X-rays allowed people to photograph the interior of objects, but Becquerel made pictures of the *rays themselves* by capturing the material trace of energy emitted by uranium salts. The resulting image, *Becquerel's Plate* (1896), is more abstract than the X-ray. While visually quite different, both photographs are created through the interplay of radiation with photographic plates, extending the possibilities of human vision. While X-rays became commonly known as photographic documents, for Röntgen, photography had only been "the means to the end."[42] In contrast, historian of science Kelley Wilder draws attention to Becquerel's innovative use of photography, writing:

Photography was not just about generating visual material from a new and invisible source or radiation, or about recording what he saw in permanent or semipermanent form. It was about how he gave radioactivity a materiality where it had none, and how he chose

5.3   Henri Becquerel, pitchblende exposed on photographic plates, 1896.

to describe its physical characteristics in pictorial rather than numerical form. He went beyond using photography as an instrument merely to detect the presence or absence of radioactive emissions.[43]

Becquerel initiated a novel way of seeing by extending photographic vision, giving "radioactivity a materiality."[44] Many of his trials occurred on the level of the photographic medium, as Becquerel experimented with different photographic plates (though he primarily used Lumière Bleue plates), emulsions, exposure times, quantities of photographic plates, and varying levels of contact with uranium salts.[45]

Following Becquerel's announcement, a debate began as to who actually discovered radioactivity, as the process and results of Niépce's earlier experiments were quite similar to Becquerel's. In an echo of Nicéphore Niépce's contributions to photographic history being largely obscured by his collaborator Daguerre, who renamed the process after himself, Becquerel received the Nobel Prize and is widely credited with the discovery of radioactivity. While Niépce distributed his results and exhibited the images, the material traces of uranium salts were not as distinct as Röntgen's X-ray image. After Röntgen's strikingly visual experiments clearly proved to all viewers that there were forms of light that were invisible to the human eye, Becquerel's subsequent findings were more legible. Further, Becquerel preserved his discovery by reproducing his experiment as a photolithograph to function as evidence of his findings.[46] He consciously enlisted the evidentiary potential of the visual. In contrast, Niépce's images were not preserved.[47] Becquerel's experiments entered into the realm of the visual, and thus the scientific record, in a more legible way than Niépce's.

As we can see, the foundational experiments in the discovery of radioactivity have direct ties to photography. Photography becomes the medium through which the complex material transformations enacted by uranium and its daughters become visible—in fact at all detectable. As *Photographic News* summarized, "The human eye, indeed, was a photographic plate prepared for the reception of certain rays, and unaffected by other rays."[48] Comparing the human eye to the photographic plate responds to the discourse in early photography about the role of the camera in extending human vision.[49] Uranium played a crucial role in uncoupling the visual from the visible-to-the-human-eye. The rays of atomic light are outside of human perception, but photography registers its material traces. In doing so, things that are otherwise invisible become visible. Over the course of the twentieth century, these ties between photography and atomic culture became even stronger—and more planetarily decisive.

The Seeable and the Unseeable

Early experiments with radiation brought it into the realm of vision. However, it was a realm defined by the uncanny aesthetics of atomic light. Over the course of the twentieth century, more traditional photographs were used to make atomic culture legible. The Atomic Era officially began, in most histories, on July 16, 1945, when the US military detonated the first bomb—codenamed Trinity—on the traditional territory of the Pueblo, Apache, and Navajo Nations in the New Mexico desert. At this epoch-defining moment, the people organizing to drop the bomb made sure to document it for posterity. Fifty different cameras took both still and motion picture images, including pinhole cameras to document gamma rays, spectrograph cameras to record the wavelengths of light, high-speed Fastax cameras to document the explosion at ten thousand frames per second, and a 35-millimeter Perfex 44, which captured the detonation in color. The prominence of photography at the inception of the Atomic Era speaks to the centrality of visualization to the construction of the collective memory of atomic culture.[50] Anthropologist Joseph Masco concludes that this created a "dual form of exposure: simultaneously embodied and photochemical."[51] Many of the photographers employed in documenting the explosion suffered radiation exposure, and military doctors tested the impact of the bomb on their vision—and overall health. The commitment of the US military to making and circulating images of the bomb tells us something significant about how atomic culture would come to operate. A photographer was also present when the US military dropped the bomb on Hiroshima. On August 6, 1945, the B-29 bomber *Enola Gay* released "Little Boy," a 9,700-pound uranium bomb on Hiroshima. On the *Necessary Evil*, one of the two planes accompanying the *Enola Gay*, Russell Gackenbach snapped two photographs from a height of 31,000 feet and a distance of sixteen miles from Hiroshima as the bomb exploded. Gackenbach, a navigator and photographer, was unaware that the plane was

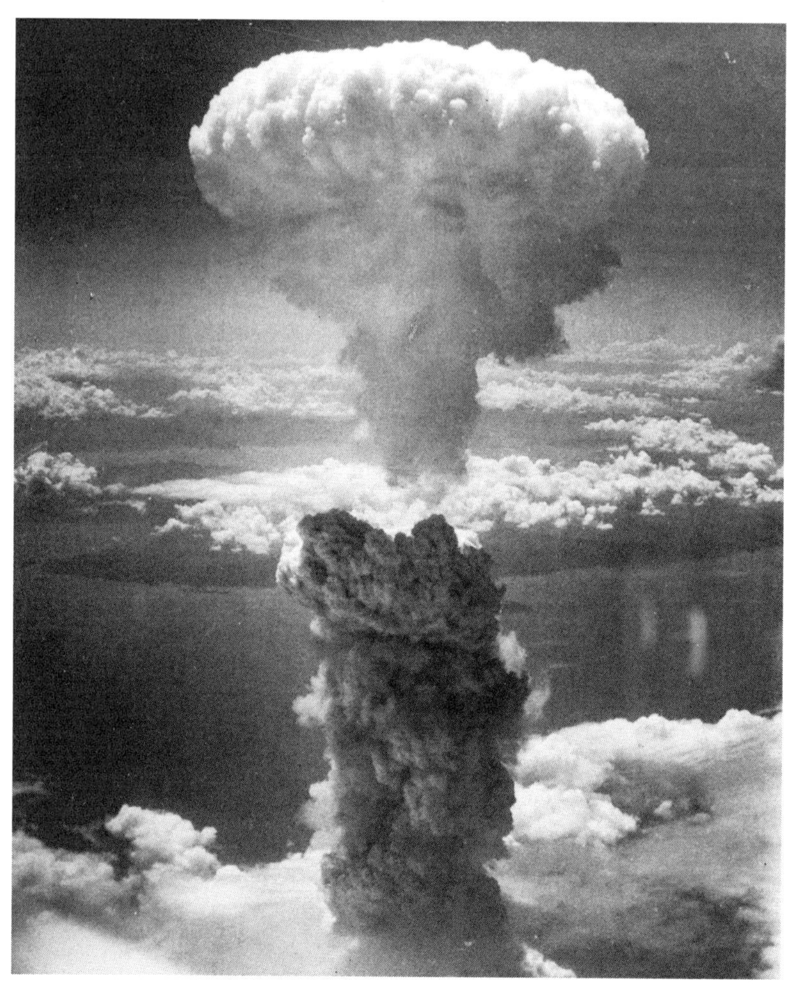

dropping an atomic bomb. Three days later, on August 9, Charles Levy captured the mushroom cloud over Nagasaki with his personal camera. Levy was aboard the B-29 aircraft *The Great Artiste*, an observation plane that accompanied the strike plane *Bockscar* to document the blast. Levy's widely circulated photograph shows the mushroom cloud as an imposing swirl of light and debris. The cloud is suspended in the sky, appearing like an act of God, disconnecting the spectacular impression of power from the destruction it wrought on the ground.

5.4   Charles Levy, *Mushroom Cloud over Nagasaki*, after atomic bombing on August 9, 1945.

Cameras were also present at Operation Crossroads, a series of nuclear detonations between 1946 and 1958 at Bikini Atoll in the Marshall Islands.[52] Eighteen tons of camera equipment documented detonations at seven test sites on the reef, in the air, and underwater.[53] More than one million still images were generated, though the images selected for public release hid any concerns about the effect of radiation.[54] The landscape of Bikini Atoll features prominently in the photographs. Art historian Peter B. Hales has referred to the aestheticization of the bomb as the "atomic sublime." Hales argues that the visual language of the sublime recontextualizes the bomb as a man-made marvel of nature and enacts the "mythic embedding of the Atomic Bomb in the grandeur of nature, as the manifestation of God's will."[55] The idyllic island paradise visually neutralized the destructive effects of the atomic bomb, rendering it purely aesthetic.[56] Photographs were carefully selected for circulation and played a central role in constructing a public memory of nuclear energy. For instance, while autoradiographs that documented radiation exposure, including the image Kriemann reproduced of the radioactive fish, are included in the official pictorial record of Operation Crossroads, they were not widely circulated.

In one photograph of Operation Greenhouse, a 1951 atomic detonation in the Pacific, observers sit in beach chairs. As Kyo Maclear writes, watching a nuclear test "as if watching some 3D 1950s movie" enacts a casualness that reveals complicity: "The very passivity inherent to the image is its message, its aggression."[57] The Atomic Age developed a militarized gaze, making viewers spectators to horror, neutralized through aesthetics and repetition—photography's ability to reveal and obscure normalized atomic culture. Art historian John O'Brian concludes that photographs by "some strange calculus" both "alert viewers to nuclear threat [and] numb them to its dangers."[58] As we have seen, photography does not have one function in the context of atomic culture—or extraction more broadly—but, rather, occupies a contradictory space of revelation and concealment, complicity and critique.

Visually, atomic culture is linked to the visual signifier of the mushroom cloud rather than to the violence of the bomb. The mushroom cloud directs attention to the atmosphere, which draws the gaze away from the reality of destruction on the ground. As with the perspectival shifts explored in previous chapters, I now turn to what was hidden by the spectacular visuality of the mushroom cloud.

Yoshito Matsushige, who worked at the newspaper *Chugoku Shimbun*, took the only known photographs from August 6, 1945, in Hiroshima. Over ten hours, he took seven exposures, of which five survived. A widely reproduced image shows a crowd of people at Miyuki Bridge, two miles from ground zero. Shot from behind, their clothing is muddy and torn as they stare at the fire and smoke. On the left side of the image, the figures sit, knees pressed against their chests, while a police officer and school children stand in the center. In another haunting photograph, a man stands in the rubble of a destroyed cityscape, the skeletal outline of a building and trees in the background against a flat, gray sky. Making the photographs was complicated as Matsushige's darkroom was destroyed in the blast. He washed his film in a creek and hung it out to dry on the branch of a tree. If the flash of the bomb made the event itself photographic, the darkroom also left its stable, contained confines. The material traces of the bomb registered in these photographs on the chemical level, introduced through the radioactive residue in the water.

No photographs exist from August 9, 1945, in Nagasaki, but on August 10, a photographer for the Japanese News and Information Bureau, Yosuke Yamahata, documented the aftermath of the bombing. Yamahata took one hundred exposures. Some of the images were printed as early as August 21 in the newspaper *Mainichi Shimbun*. However, the arrival of American forces led to the censorship of the images, and Yamahata hid the negatives. Images of the destruction on the ground—resulting in the deaths of an estimated 66,000 at Hiroshima (from a population of 255,000) and 39,000 at Nagasaki (from a population of 195,000)—were suppressed under censorship codes enforced by the US occupation of Japan between 1945 and 1952.[59] These images were marked as unseeable, making the devastating costs of the bomb unsayable in domestic Japanese and international discourse. The images were not widely circulated until the occupation of Japan by American forces ended.

Images of atrocity were not only censored by the US regime to prevent outrage among the occupied population but likewise to prevent a shift in public opinion surrounding Cold War nuclear policies more broadly. The censorship of the images suggests that the violence of the bombings of Hiroshima and Nagasaki could potentially challenge moral arguments about the necessity of the bomb. This is not to suggest that visibility is in and of itself political or can move people to action in any straightforward sense. As Susan Sontag famously argued, the numbing function of photographs raises serious questions about the political potential of photographs of atrocity. In her later writing, however, Sontag allowed more possibility for the potential of photographs to critique the power structures responsible for mass suffering and to move people to action.[60] As John Berger wrote in 1981, there is a need "to reinsert those events of 6 August 1945 into living consciousness," reflecting, "At how many meetings during the first nuclear disarmament movement had I and others not recalled the meaning of that bomb?"[61] The shocking violence of the bomb was obscured through "a systematic, slow and thorough process of suppression and elimination."[62] In 1952, *Life* published previously censored photographs of the aftermath of Hiroshima by Matsushige and Yamahata in an article titled "Atom Blasts through Eyes of Victims." Following the end of occupation-era censorship codes, a new visual culture emerged in the United States that focused on the actual and potential horror of the bomb, what Hales termed the "atomic gothic."[63] The Castle Bravo test accident in 1954 and the resulting nuclear fallout led to a panic that nuclear war—let alone a series of nuclear accidents—could instantly destroy civilization. A paranoid visuality would come to characterize the nightmare subconscious of the Atomic Age.

Depictions of the aftermath of the bombings are not the only images we do not see. As the heavily documented nuclear tests were very intentionally and selectively presented to the public, the area beneath the bomb, where atomic tests forcibly displaced people indigenous to the territory, is not visible within the photographic frame.[64] The Trinity bombings in New Mexico took place on the territory of several Pueblo communities, including the nearby San Ildefonso Pueblo community. During Operation Crossroads in 1946, the inhabitants of Bikini Atoll in the Marshall Islands were relocated, as the islands and lagoons became military sacrifice zones. A photograph in *Operation Crossroads: The Official Pictorial Record* of the Marshall Islanders leaving was titled *Moving Day, Bikini to Rongerik*. A group of people walk across the sand, carrying thatching to rebuild the church and community house on Rongerik. The human figures are shadowy blurs,

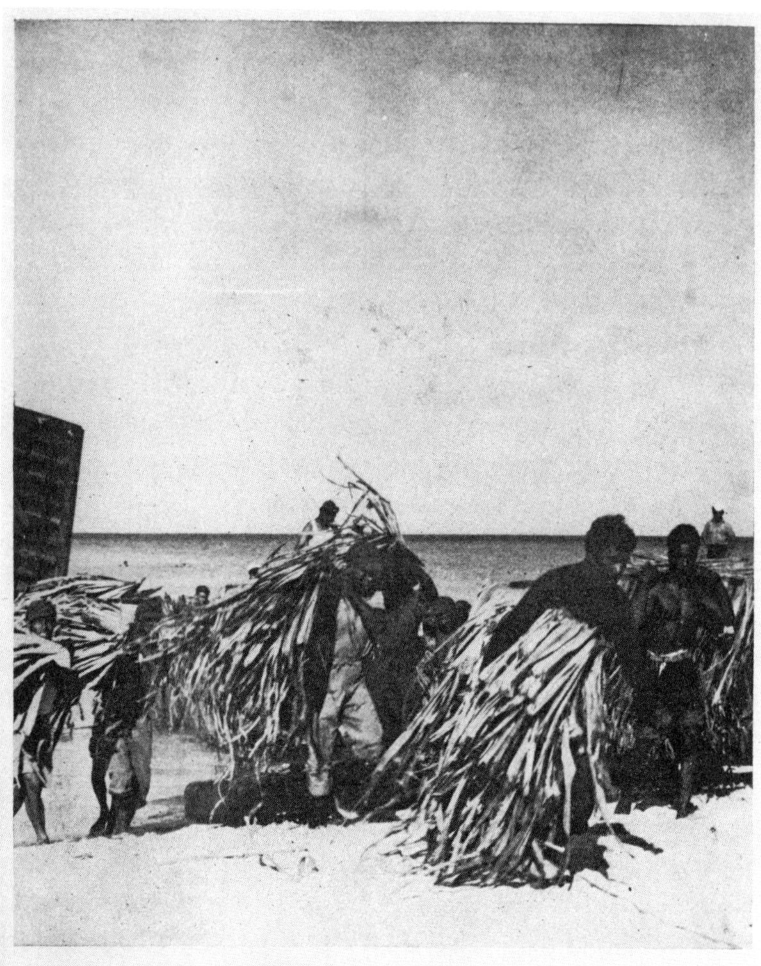

MOVING DAY, BIKINI TO RONGERIK. On March 7, 1946, the population of Bikini was moved to Rongerik in LST 1108, a total of 161 persons making the trip. Rongerik had been the first choice of nine of the eleven family heads, called alaps, as the new home for the evacuees. The island is roughly triangular in shape with good topsoil and relatively heavy growth of coconuts, pandanus, breadfruit, and arrowroot. Beaching facilities were good. The Bikini church and community house were dismantled and transported to Rongerik. Pandanus thatching for the new village on Rongerik was prefabricated. OPPOSITE. Seabees and Marshallese at work installing temporary canvas roofs on the new houses. Later, thatch replaced the canvas.

5.5   *Moving Day, Bikini to Rongerik*, 1946. From *Operation Crossroads: The Official Pictorial Record* (Washington, DC: Joint Task Force One, 1946).

their faces invisible, dwarfed by the vegetation they carry. The emphasis is not on the humans who are being displaced but on a speculative, future-oriented vision of the rebuilding of the community. The bomb embodied progress, and the photograph confirms that the resettlement it necessitated was in the service of progress. The displacement of a community and their spatial marginalization are mirrored in the photograph's framing wherein the political and aesthetic work of the photograph operate from the same principle of removal.

Once resettled on Rongerik Atoll, the islanders began to starve, as the island could not produce enough food to sustain the population. However, they could not return to the heavily contaminated Bikini Atoll, so they were relocated to Kwajalein Atoll and eventually Kili Island, an island one-sixth the size of their home island, less suitable for sustaining a large community. After the Bravo hydrogen bomb—a bomb one thousand times more powerful than the bombs dropped on Hiroshima and Nagasaki—was dropped on March 1, 1954, ash fell from the sky onto islanders on Rongelap Atoll, 125 miles from Bikini Atoll. The fallout had severe health consequences. In the 1950s, the Atomic Energy Commission launched a medical research program, Project 4.1, that studied the effects of radiation exposure on people in the Marshall Islands without consent from the patients. The study included experimental surgery and injections of chromium-51, radioactive iodine, iron, zinc, and carbon-14.[65] In the 1970s, islanders were permitted to return to Bikini Atoll. However, cesium-137 was found in their bodies in high doses due to its prevalence in the regional food chain, and they had to move once again. On top of this direct environmental damage, the Marshall Islands face further threats from the industrial-capitalist world the US military secures globally, as they are at high risk of vanishing altogether due to climate-change-induced sea-level rise.

As Marshall Islanders experienced exile, the iconic status of the bomb was celebrated in the press. In 1946, *Newsweek* described the "Atomic Bomb: Greatest Show on Earth" while tourists flocked to the Yucca Flats test site in Nevada.[66] In a *Business Week* article from 1952, the reporter stated:

If there had been a city instead of sagebrush below that cloud, you realize later, half a million people might be dead or dying at this instant. But that's not what you think at the time. You think fleetingly, "is that all?" And then you suck in your breath and hold it because—welling up from the center of the cloud and spilling in easy, leisurely magnificence down the side comes a wave of glowing pink foam, one of the most strangely beautiful of human creations.[67]

The "strangely beautiful" bomb separated its sublime spectacle from its destructive potential, the visual separated from material consequences. Turning the mushroom cloud into a tourist spectacle obscured other durational disasters. In her study of photographs of workers exposed to radiation, art historian Julia Bryan-Wilson notes the limitations of photography to document the "slow blooming invisible damage that accompanies these blasts and shoots itself into our atmosphere and into bodies, damage that can take years, even decades to reveal itself."[68] Photography cannot, with any firm causality, connect long-term damage to any particular exposure, as the effects of radiation exposure can take so long to manifest in symptoms. The material consequences were felt acutely by workers who extracted the uranium that built the bomb and worked in the factories where the uranium was refined. People who live downwind or down-water from mining or production facilities are also exposed to industrial contaminants. For example, the Hanford plutonium plants near Richland, Washington, emitted at least two hundred million curies of radioactivity— twice that emitted at Chernobyl—leaving hundreds of square miles uninhabitable due to contaminated water, air, and soil.[69] Chernobyl was spectacular, immediate. But the "ordinary disasters"—both natural and economic—that result in the slow poisoning of people and communities are often forgotten in public memory when the burden of risk falls on low-income communities.[70] There are numerous sites worldwide where communities are experiencing displacement, high rates of disease, and environmental damage—histories that rarely come into official discourses of atomic culture.

The Crisis of Vision and Photography at the Limit

Photography's central role in the discovery of radioactivity—through the transformation of its invisible chemical operations into visible effects— could make these occlusions seem strange on first consideration. But as we have seen throughout our inquiry, every aspect of photography's regimes of visualization has involved foregrounding some phenomena at the expense of others. Atomic culture is no different: just as uranium's production and use has manufactured a racial-spatial geography of radioactivity's presence and absence, its central visual symbol of the bomb and its spectacular cloud hides as much as it shows, distracting us from the material realities of its process of enabling new forms of vision. Röntgen's X-ray photograph initiated a new era of vision that questioned the boundaries of visibility and invisibility, surface and depth. Film scholar Akira Mizuta Lippit writes:

> The new rays facilitated the realization of certain drives intrinsic to photography. The photographic project had always involved more than the mere duplication of nature or the accurate representation of the visible world. Within the depths of what one might call the ideology of photography was a desire to make the invisible visible, but also to engender a view of something that had no empirical precedent. . . . Tearing through the opaque materiality of bodies, X-rays transformed photography from an exercise in realism—the production of indexical images—into an allegory of avisuality.[71]

The X-ray extended the boundaries of the visible world while revealing the limits of human perception.[72] Lippit identifies 1895 as a turning point in the history of vision, as the concurrent discoveries of the X-ray, psychoanalysis, and cinema "transformed the structure of visual perception, shifting the terms of vision from phenomenal to phantasmatic registers, from a perceived visuality to an imagined one. From visual to avisual."[73] While psychoanalysis concerned itself with the unconscious mind, photographs also shaped Sigmund Freud's thinking, and the photographic negative become an important metaphor to illustrate his theory of the unconscious.[74] The latent image imprinted on the photograph parallels the unconscious mind, where information lies below the day-to-day processing of conscious perception. Benjamin's notion of the "optical unconscious" uses psychoanalysis as an analogy to attune us to "the unknown, the unseen, and the uncontrolled" in "the making, circulation, and viewing of photographs."[75] He suggests that the photograph contains complex and unruly depths, well beyond what the photographer intended to portray.

When the bomb was dropped on Hiroshima, it was not immediately apparent to survivors that the bomb was atomic. The flash of light, which lasted one fifteen-millionth of a second, was believed by some survivors in Hiroshima to have been caused by a magnesium bomb, as magnesium emits a cloud of white smoke.[76] However, the detonation exposed a supply of X-ray film in a hospital, confirming to survivors shortly after the blast that the bombs were atomic.[77] The dispersed effects of radioactivity also surfaced not long after atomic testing in the United States. After the Trinity bombings, Dr. J. H. Webb, a scientist at Kodak, received complaints about spotting and fogging on film being used in Illinois.[78] Testing revealed cerium-14, a radioactive element not found in nature and produced only through nuclear fission. Kodak maintained control over the manufacturing process and materials used, so contamination was highly unusual. The exposure came from a radioactive contaminant in the strawboard material used to package the film in a mill in Vincennes, Indiana. Similar results were found in a mill in Tama, Iowa. Webb concluded that the contamination must have come from river water, indicating that the wind-borne radioactive fission products from the Trinity bombing were deposited into the watershed by precipitation. The bomb's effects were thus not contained to the blast site. Webb was aware of the potential political implications of the discovery and did not announce it until 1949. In both of these cases in Japan and in the US Midwest, photography—or, more specifically, photographic film—revealed human-induced radiation exposure that was officially unconfirmed. In addition to the role of photographic practices in identifying the initial facts of radioactivity present in naturally occurring materials, this later use of photosensitive materials to detect the consequences of the deliberate harnessing of radioactive properties was an uncanny echo of when prospectors in Canada's Northwest Territories rubbed photographic film on rock to locate pitchblende, as radioactivity would register a material trace.[79]

How does atomic light complicate notions of the photographic or the photographic surface? Here, it is helpful to return to Lippit's reflection that uranium "facilitated the realization of certain drives intrinsic to photography." Paul Virilio describes the bombs dropped on Hiroshima and Nagasaki as *"light-weapons,"* which inherited the experiments of Niépce and

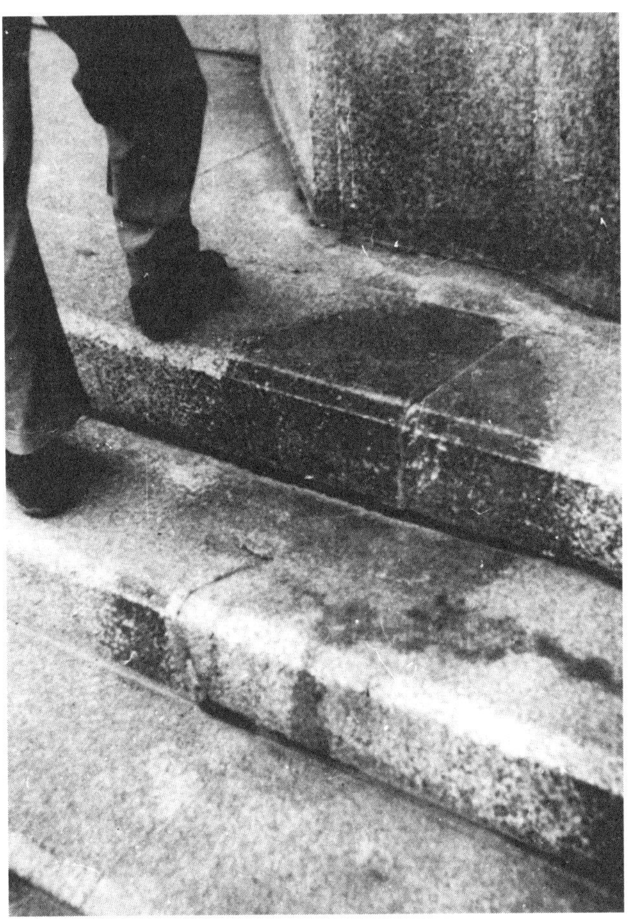

Daguerre, as well as the military searchlight.[80] This fusion of early photography with a militarized gaze pushes photography to the limit of one of its potential outcomes: violence, discipline, and surveillance. Indeed, atomic light formed a haunting perversion of the photographic process at Hiroshima and Nagasaki, as the heat from the blast burnt shadows of the objects in the path of the bomb. The evocatively titled *Human Shadow Etched in Stone* (1945) captures the shadow of someone waiting at the entrance of the Hiroshima branch of the Sumitomo Bank when the atomic bomb was dropped. The stone became a photographic plate

5.6 Yoshito Matsushige, *Human Shadow Etched in Stone*, Hiroshima, 1945.

through the sheer force of the concentrated form of "light" that hit them. The material traces of the bomb make literal the tensions between visibility and invisibility in the context of both immediate and slow violence.

Seeing a Future in Remedial Seeds

The light and heat of atomic bombs, in a sense, turned entire cities into photographic surfaces, fusing photography with spectacular, immediate violence. But what of photography's other potential roles in the context of atomic culture? I want to conclude this study of uranium, photography, and the visibility of slow violence by returning to Susanne Kriemann's *Pechblende* series. In the final exhibition of her series *Pechblende (canopy, canopy)* (2017), Kriemann uses heliogravure, a method of reproducing photographs with ink, to capture the plants false chamomile, wild carrot, and ox tongue in subtle, earthy shades of green and gray (plate 25). Kriemann harvested the weeds from Gessenweise and Kanigsberg in the former German Democratic Republic, where the mining company SDAG Wismut mined pitchblende. Unlike most gravure prints, however, the pigments that color this series include the uranium and other heavy metals that the plants depicted in the images absorbed from the soil, air, and water surrounding the mine.

These heliogravure prints follow an earlier set of photograms of the same subject. In the earlier series, the artist harvested false chamomile, wild carrot, and ox tongue, which she formed into loose arrangements and dried. The dried arrangements were placed on light-sensitive paper and exposed with a smartphone-camera flash. The original plants were displayed alongside the photograms, bringing together the physical object and its representation. The artist describes how the photograms entangle distinct temporal scales: the deep time of the uranium extracted from the earth, the hundreds of years it takes to reclaim polluted land, the seasonal life of a plant, and the flash of a smartphone in the darkroom. Beside the photograms, captions list the botanical information, the date, and the location of harvest. They also list the heavy metals and other elements found in the plants. The plants absorbed substantial traces of metals and minerals from the polluted soil: lanthanum, gadolinium, germanium, uranium, mercury, lead, nickel, zinc, aluminum, and copper. The captions are written in earthy-hued pigments produced from the pulp of the weeds. The pigments are, therefore, material evidence of the damage left behind by resource extraction. The pigments will eventually fade, making the text invisible to the human eye. They will, however, continue to emanate energy from the metals and minerals that the plants have absorbed, introducing a range of visual material that lies beyond human perception.[81]

Kriemann selected the three plants featured in the photo series because they are hyperaccumulator plants. Hyperaccumulators absorb, neutralize, extract, or store industrial contaminants in a process called phytoremediation. The artist Mel Chin initiated the first field study to use phytoremediation to restore polluted soil in 1991. Working in consultation with Dr. Rufus Chaney, a senior research scientist at the US Department of Agriculture and funded by an NEA grant submitted by the Citizens Environmental Coalition (CEC), Chin installed *Revival Field* in the Pig's Eye Landfill in Minnesota. Chin described the conceptual artwork as an "invisible aesthetic" where plants essentially "carved away" pollution so that "life could return to that soil," resulting in a "wonderful sculpture."[82] Chin's sculpture confirmed the scientific potential of this low-intervention, low-cost remediation method. Traditional industrial methods of remediation are faster, but they cause further disruption to land and ecosystems. *Revival Field* represents an example of how art can move beyond a symbolic exploration of environmental pollution toward work with science and nature to reclaim devastated territories. Phytoremediation is a synergistic process of detoxifying, absorbing, and transforming. At the same time, it challenges an instrumental relationship to land centered on its extractive objectification or subjectification of an abstract, all-encompassing nature. Phytoremediation as a process and metaphor imagines and enacts new ways of being in common. In doing so, it moves beyond the aesthetic into the possible.

Here, plants become "the archivists of contamination in fallout zones deserted by humans."[83] Archives are usually considered a uniquely human preoccupation, concerned with preserving history, text, and institutions. The potential archive formed by seeds orients to the changing seasons, undergoing a constant process of death and renewal. Changes in the geological layer form a different kind of archive: an ancient record of life, death, adaption, and—occasionally—cataclysmic destruction.

Phytoremediation combines these two distinct temporal registers: the deep time of metals found in the soil with the seasonal cycles of seeds and weeds. Art historian Jill Casid uses seeds as a metaphor for renewal in archives, writing that seeds have the potential to "materially rework how the matter of the archive is physically remembered, raking over and reseeding the ground of the past for the materialization of a different future."[84] Kriemann's weedy archives decontaminate and reseed. This is both a literal process of detoxification and a metaphorical means of addressing the past and recovering something hopeful or productive.

In contrast to the tendency to focus on the spectacular aspects of the atomic, Kriemann's archives center on the mine, the body of the worker, and the mining landscape. The ecological turn has directed attention to the natural world, but labor remains an essential area of analysis in environmental justice. Within Kriemann's archive, the geological, botanical, and human scales are woven together. *Pechblende (Volume)* and *Pechblende (canopy, canopy)* turn to plants grown in contaminated soil, but the archive that precedes them, *Pechblende (Chapter 1)*, centers on the miner (plate 26). In the installation, miners' tools are projected onto the gallery walls through a camera obscura. The spectrum of colors references the chromatic range of byproducts of pitchblende: bright green, fluorescent yellow, gray. The darkly lit exhibition space becomes an analogue for the darkroom and the photographer's labor.[85] The juxtaposition of the workers' tools and the space of the researcher point to tensions and interrelations between intellectual labor and physical labor. For example, Allan Sekula has observed that while Denis Diderot's *Encyclopédie* (1751–66) included intellectual and manual knowledge, the compendium reinforces a hierarchy of skills, for in Diderot's framing, intellectuals refine and improve the practices of artisanal workers.[86] The abstract equivalence proposed by Diderot between intellectual and artisanal labor hides the hierarchy that structures the perceived value of different kinds of labor. In linking the two sites, Kriemann suggests a complex equivalence. The mine, archive, and darkroom become analogous (if not identical), all involving immersion and extraction, forms of labor performed out of sight. While Sekula has argued that "archives are not like coal mines: meaning is not extracted from nature, but from culture," Kriemann proposes a less binary division of nature and culture, presenting an archive where the synthetic and natural, the human and extrahuman, the damaged and pristine, as well as intellectual, artisanal, and industrial labor, become impossible to separate.[87] The understanding of humans as a geological force means that the emphasis on the natural world does not displace human history;[88] rather, Kriemann identifies the body of the worker as a key site of interaction between human history and the natural world.

In *(Prologue)* and *(Chapter 1)*, the form of the archive is legible and familiar, linking images and text to produce meaning. In the final archive, however, a different orientation to the natural world introduces ways of knowing and healing that potentially exist outside of human knowledge structures. Kriemann orients our attention to the rhizomatic networks of root systems that elude human vision below the surface of the earth. While the subject matter documents the destruction wrought by radiation and its applications in military technology, an implicit hopefulness underlies the act of building an archive. Kriemann proposes that something productive can be recovered from the ruins of history. Climate change is often framed as what T. J. Demos calls an "emergency temporality but a forgotten history," which obscures the choices that have produced or exacerbated ecological breakdown.[89] These forgotten histories have wide-ranging implications and lessons. Archives are a key heuristic for Kriemann's intervention into legacies of radiation. For Jacques Derrida, the archive is "the very question of the future, the question of a response, of a promise and of a responsibility for tomorrow."[90] Kriemann develops a historical investigation of uranium extraction while using the archive to destabilize the inevitability of an extractive future. Kriemann's practice does reconnecting work: tracing the movement of uranium from the mine to its applications in cultural, medical, and military realms, revealing the complex networks connecting these histories. *Pechblende* invites the viewer to think critically about the material foundations of our world as they connect to environmental catastrophe, proposing potential new ways of seeing, relating, and organizing society.

At this point, I want to return to phytoremediation and the (slow) healing of a landscape damaged by mining. Phytoremediation proposes a way of understanding extractive landscapes that moves beyond damage narratives. Joseph Masco suggests that, following the Atomic Age, we cannot escape the reality of hybridization, a world of *"radioactive tumbleweeds, contaminated fruit flies, and toxic alligators."*[91] This perspective shifts the emphasis from environmentalism and conservation to questions of environmental justice. Environmentalism directs attention to the preservation of old-growth forests, rainforests, and oceans.[92] In this framing, other landscapes—such as a former uranium mining site—are dismissed as damaged and, therefore, beyond saving. As the anthropologist Anna Tsing argues in her study of former industrial logging forests in Oregon: "Industrial transformation turned out to be a bubble of promise followed by lost livelihoods and damaged landscapes. And yet: such documents are not enough. If we end the story with decay, we abandon all hope—or turn our attention to other sites of promise and ruin, promise and ruin."[93] To move toward energy transitions, sustainable futures, and other ways of valuing the natural world and each other, the stories we tell about our collective past and experiences of the present can't end in decay and ruin. In contrast to narratives of damage, phytoremediation is a reclamation process. As a means of reclaiming shattered landscapes, it operates through a markedly different notion of what is valued, to what end, and how to draw it out. On the surface, phytoremediation is not the most efficient means of extracting toxins. It demands close attention, low intervention, and time. It requires a certain letting go of human processes to trust natural ones. This method of remediation (or remedy) draws in notions of care.

Damaged landscapes contain "the small, partial, and wild stories of more-than-human attempts to stay alive. . . . The landscapes grown from such endings are our disaster as well as our weedy hope."[94] Weeds—unwanted, unvalued plants—are often hardy and durable, albeit less aesthetically uniform than the manicured lawns and showpiece flowers that line suburban streets. The monoculture of domestic lawns damages biodiversity, requires excessive water use, and serves as a perceptible source of carbon and nitrogen. In contrast, many species designated as weeds have been key to healing polluted landscapes through phytoremediation. Kriemann highlights the need to shift in what we value: there is an upturning of aesthetic hierarchies—just as we have to consider unwanted, "damaged" places as worth celebrating and saving. Using actual weeds as a "weedy hope," the history of the future proposed in *Pechblende* locates entanglements between the human and natural world. In doing so, it inherits difficult pasts while locating a yet-to-come in a shattered landscape, working toward resurgence and growth.[95] Kriemann's practice excavates the histories that got us to this moment and, critically, places an emphasis on resilience, healing, and connection. But the desire to construct a counter-archive reflects a belief that, in the face of the overwhelming scale and the perceived inevitability of climate change, possible futures can be rewritten and called forward from the ruins of failed histories. Kriemann works within a continuum that bears witness to histories of the past and their relevance for the future. In the process, such work calls a new future into being.

Kriemann's autoradiographs and lithographs bring into view forms of slow violence that are illegible to the human eye. Uranium's ability to extend human vision reveals the possibilities of material traces in moments where representation fails. Seeing is inadequate to trace and document slow violence, but these histories become legible through the physical imprint on the photographic plate. Uranium's applications in military technology enabled spectacular, world-altering violence accompanied by the slow poisoning of radioactivity. On the photographic plate, these histories are crystalized.

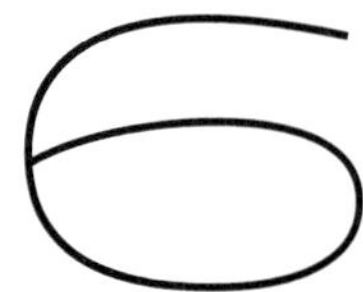

Rare Earth Elements
and De/Materialization

TEN THOUSAND DIGITAL IMAGES are scaled down in size, reduced to their base elements of chromatic value and density. The abstracted blocks of color and form transmute into a copy of Nicéphore Niépce's *View from the Window at Le Gras* (plate 27). Catalan conceptual artist Joan Fontcuberta's *Googlegram: Niépce* (2005) updates Niépce's heliograph for the digital age, replacing bitumen with bits, bytes, and pixels. The title shifts attention from the autogenic power of the sun (helio) to the invisible work of the algorithm (Google). A Google Image search for "photo" and "foto" gathered the images that the artist then entered into photomosaic software. The algorithm's selections become the base materials for the photograph. However, Google's archive is constantly in flux, undermining the idea of the archive as a site where knowledge is constructed and fixed.[1] Fontcuberta suggests the digital image is "an image without place and without origin:

deterritorialised, it has no place because it is everywhere."[2] Appropriately, then, the image itself is slippery. The pattern of the photomosaic image is only visible through distance, but a myopic view is required to see the granular thumbnails that make up the image. The excessive amount of data within the photograph results in informational abstraction, as meaning is lost while moving between scales. Here, the image stands in for a larger cultural shift: from the evidentiary value of the visual—seeing is believing—to the avisual truth of data. The fragmented, blurry image makes tangible how the digital world undermines photography's seemingly privileged evidentiary relationship to "the real." Fontcuberta's image directs attention to how the digital forms a new type of evidence primarily outside of the visible, challenging the authority of visual authenticity.[3] Nonetheless, through the re-creation of Niépce's heliograph, Fontcuberta traces a path from the early chemical experiments that defined photography to the pixelated present, questioning the relationship the digital image has to analog photography and the mines that made both possible. Following this prompt, this chapter turns to the digital to consider the extractive and visual image economies of the present.

Googlegram: Niépce is a digital c-print, or chromogenic print, a process that spans the analog and the digital. The image is made from a digital file rather than a negative, but light and silver halides remain key materials. The print is formed from three silver emulsion layers sensitized to the primary additive colors of light (red, blue, and green—typically colored by rare earth elements), which are added by lasers or LED lights. C-prints are a subtractive form of building color, where light-absorbing inks (cyan, magenta, yellow, and black) are layered over a white substrate. Finally, the silver is bleached in a chemical bath, resulting in a positive image in full color.[4] In the physical print, the interplay between silver halides, pigment, and light makes the photograph. But metals are, of course, also a critical part of the digital half of the process.

Both digital cameras and smartphones use minuscule amounts of many different metals and minerals, forming what critical theorist McKenzie Wark calls the "mineral sandwich in your pocket."[5] It is estimated that 84 percent of the stable elements on the periodic table are used in smartphones, perhaps *the* critical tool for many for both producing and accessing photographs today.[6] In the previous chapters, we used the qualities and social lives of one metal or fossil fuel as the guide for our investigation, though along the way, we were regularly reminded that all photographs depend on complex combinations of diverse materials. The digital image poses this complexity and analytic challenge anew, at a higher level.

Take for instance this cursory, nonexhaustive list of minerals used in even basic digital photography and image-viewing, as well as the places they are mined: rare earth elements add color and are used in lenses, magnets, speakers, and motors (China, United States, Australia); platinum group metals are found in optical fibers and liquid crystal display glass (South Africa, Russia, Canada); for circuitry and wiring, copper (Chile), silver (Mexico, Peru), and gold (China, Russia, United States, Canada, Burkina Faso, Australia); for rechargeable batteries, lithium (Australia, Argentina, Bolivia, and Chile), cobalt (Democratic Republic of Congo), and graphite (China, India); tellurium reduces corrosion (China, Canada, Japan); manganese is used in dry cell batteries and circuit boards (South Africa); silica sand, rare earth metals, and potassium harden glass screens (China, United States); mica forms electrical insulators (Namibia, India); bauxite provides LED backlighting and is used in casing (Australia, Guinea, and China); tin solders circuit boards (Burma, Peru, Indonesia, and China); indium is a component in transparent circuits in the display (Republic of Korea); arsenic is used in radio-frequency and power amplifiers (China, Chile, Russia, Mexico, and the Philippines); tantalite regulates voltage and audio quality (Brazil); and lightweight casing requires aluminum (Australia, China, and Africa) and petroleum-based plastics (Saudi Arabia, Russia, United States, Iran, Brazil, Venezuela, Canada, and China). The variety of materials used in smartphones make tracing the mining supply chains behind the photograph more complex by orders of magnitude. And yet, the need to develop a means to think through these violent interrelations is just as pressing as it is multifarious—as is the case in understanding any social phenomenon in a world riven by the racialized inequality and uneven development that capitalism requires and reproduces. As many people are at least vaguely aware, numerous minerals powering our smartphones are conflict minerals, including cobalt, coltan, and tin. If contemporary image-making, far from the dematerialization promised by New Economy rhetoric, is extremely resource-intensive, how can we think of the relation of new forms of image-making to extraction?

In light of these material realities, I do not position the analog/material *against* the virtual/demateral, as these boundaries are not easily demarcated and tend to crumble under scrutiny. Photography's reliance on mined minerals only heightens the ties between analog and digital image-making. In many ways, digital images are an extension of earlier discourses and problems in photography.[7] At the same time, the rise of the digital image in the twenty-first century marked a profound shift in the status of the image, both materially and conceptually. Acknowledging both continuity and rupture, this chapter explores the material costs of the digital world.

To consider how mining does and does not come into view in the twenty-first century, I turn to rare earth elements, minerals central to smartphone cameras—and most models for a green transition away from a high-carbon economy. I read the relative visual invisibility of rare earth elements against lithium, another technology metal that, unlike rare earths, is heavily photographed. From there, I follow metals through the data mining of Cloud infrastructure, which, as the artist Trevor Paglen shows, runs deep underwater in ocean cables. Then I turn to the eventual outcome of technologies as e-waste, with a focus on photographs by Pieter Hugo and Nyaba Leon Ouedraogo of Agbogbloshie, the e-waste dump in Accra, Ghana, where a secondary form of mining occurs as spent technologies that enable digital imagery are stripped down to their component parts. Finally, I consider the possibilities of ethical spectatorship in the image-saturated present. In emphasizing the materiality of the digital image, I push back against narratives that the digital is less material and, by extension, necessarily less environmentally damaging. The conceptual shift from a material object to dematerial representation, from the darkroom to Lightroom, from the archive to the Cloud, symbolically neutralizes the ongoing violence of extraction, hiding the environmental and human costs undergirding the digital photograph.

De/Materialization and Mining in the Rare Earth Age

Today, we often view photographs through screens. Transitory images on screens seem more ephemeral, somehow less material than, say, a silver gelatin print or a Polaroid. Popular conceptions describe such digital images as dematerial or immaterial—referring to their fleeting presence on the screen, of course, rather than the screen itself. In the parlance of economics, dematerialization names a reduction in the amount of materials used and the waste generated in production.[8] In essence, as firms dematerialize, less physical material is required to fuel economic growth. For example, digital music subscriptions have largely displaced the vinyl record or CD, marking a shift from the provision of physical products to digitally rendered services, where the "product" is access to the music, often via a subscription service. Less raw material is needed to make the individual unit—in this case, the experience of listening to a singular album. Dematerialization is also used more broadly to describe the increasing predominance of the digital world and the shift from physical objects to digital files on hard drives and screens. Rather than printing a physical photograph, for instance, an image taken with an Apple iPhone might be stored on the Cloud and posted to Instagram, where followers worldwide can immediately engage with the visual data. Companies like Apple, Google, and Instagram rely on a data-driven approach that bridges the object (the phone or camera) with services (Cloud storage and social media): both the physical stuff of the object and its networked existence depend on physical infrastructure. This, however, does not mean digital cameras and images necessarily have less of an impact materially. Largely, metals have replaced bulkier plastic components, allowing for sleek, streamlined hardware. Metals are resource-intensive to extract, so even trace amounts can have an outsized impact. Economists often qualify the term *dematerialization* as the reduction of materials for *one* unit (the commodity itself), which does not necessarily reduce the overall need for materials across the economic sector. In cultural conversations about production and consumption, this nuance is often missed. Phones are indeed smaller, but their material impact is not.

In visual culture, the naming of image software crystalizes the imaginary of immateriality. Images are stored on the Cloud, while Adobe's most popular image-editing program is Lightroom. Once again, our vision is directed to the atmosphere, not the mine. Lightroom proposes that photography's tactile, chemical histories—that relied on darkness—have been replaced by the ease and transparency of light. While moving from dark to light is associated with increasing visibility, photographic work classically happens in the dark. The darkroom is the most tangible site of human labor in the photographic process, and this workspace comes to be eclipsed in this rhetoric.[9] What are the implications of this reframing?

The digital image seems to represent the attainment of the desire to transcend materiality and labor, which, as we have seen, is woven throughout the history of photographic discourse. Media theorist Johanna Drucker argues that the perception of the digital as immaterial reflects a desire to separate information from matter, rooted in a longer intellectual tradition "that idealizes the immaterial, even placing it in a theological frame, above embodied knowledge." In seemingly transcending materiality, the digital is the culmination of one of photography's narratives. Despite this seductive framing, computational media is "overwhelmingly material" in both production and use, damaged nevertheless by time: aging materials, wear from poor storage, and imposed obsolescence blocking interoperability.[10] Narratives of digital immateriality hide the environmental and human costs of digital technologies and, by extension, the digital image.

When I first mapped out this book's structure, I planned to use rare earth elements (REE) as a guide for this chapter. Atomic numbers 21, 39, and 57–71, "rare earths" as a category describes seventeen chemically similar elements with magnetic and conductive properties. Rare earths are hard to classify—they are neither rare nor earths—and even metals such as uranium have been classified as rare earths at various points in time. Despite their slippery classification, rare earths are a suitable guide for this inquiry into the digital for two reasons. First, rare earth production is critical to green technology. The US Department of Energy calls rare earths "technology metals." As many of them have limited supply chains and are considered critical metals, REEs are prominent in geopolitical discourse. Commentators have called our current era the "Rare Earths Age," and geographer Julie Klinger argues that rare earths have been "vested with strategic significance."[11] Secondly, sixteen of the seventeen rare earth elements are used in smartphones, though the elements used in different models vary.[12] Qualities of rare earths enable faster, lighter, and more efficient technologies. They are the key to miniaturization, driving the widespread adaption of smaller, more portable smartphones and cameras. Critically, however, rare earths also play an essential role in mediating the image for both production and consumption. The display screen is perhaps the closest analogy to the physical print, the substrate of the photograph as an object, making rare earths a logical guide for our inquiry into the mining underlying the digital image. Lenses use lanthanum and yttrium while the vivid colors on screen displays are made from rare earths—europium and yttrium (red), cerium (blue), and terbium (green)—emitting light at wavelengths to which the human eye is sensitive, forming the chromatic building blocks of the image. The demand for rare earths as colorants is significant: 8,750 tons of rare earth phosphors and pigments were used in 2012, which is remarkable given that rare earths are found and used almost exclusively in trace amounts.[13]

While discussions of rare earths as critical metals have become common in public discourse, rare earth mining itself is rarely photographed.[14] As a result, the labor and results of such strategically important mining rarely enter into the realm of visibility. What accounts for rare earths' relative out-of-sightness? To answer this question, it is helpful to turn to China, which has dominated rare earths production for decades. In 2010, China accounted for 97.7 percent of global rare earths production, mainly from the Bayan Obo rare earths mine, just north of Baotou, Mongolia, the rare earths capital of the world.[15] Rare earths are extremely uncommon in pure form and often found only in trace amounts in iron, phosphate, or copper-gold deposits or bonded to toxic materials like arsenic, uranium, thorium, and fluoride. Separating rare earths is technically challenging while socially and environmentally responsible extraction is especially costly. The elements are seldom found in sufficient quantities to make mining them profitable: to use corporate jargon, there is very little "economic viability" in rare earth mining. Rare earths are not mined in "remote" places simply because that is where the scarce metal happens to be found but rather because undertaking such mining profitably at scale requires the deliberate construction of sacrifice zones, given the toxicity of extraction and processing according to current methods.[16] Exposure to rare earths can cause organ damage and corrode skin while promethium, gadolinium, terbium, thulium, and holmium can cause radiation poisoning. Separating rare earth ores requires several dozen chemical processes using acids and toxic chemicals that are highly carcinogenic. As Klinger summarizes, one metric ton of rare earths results in "approximately one tonne of radioactive wastewater; seventy-five cubic meters of acid wastewater; 9,600 to 12,000 cubic meters of waste gas containing hydrofluoric acid, sulfur dioxide, and sulfuric acid; and approximately 8.5 kilograms of fluorine."[17] In Baotou, many of these byproducts are dumped in a tailings pond that holds two hundred million metric tons of radioactive slurry.

Echoing a trajectory played out in mining regions around the world, the exploitation of minerals has transformed life in the area. In 1950, before rare earth mining started in earnest, the city had a population of 97,000. Today, the population is more than 2.5 million, rapid growth resembling the transformations of Potosí. Disease rates are very high due to the toxicity of rare earth extraction. The standard ratio of people diagnosed with cancer nationally is two in one thousand. Rates in Baotou are one in seven.[18] Eventually, the Chinese government moved to limit production due to the high environmental costs, creating what markets experienced as the rare earths crisis of 2010–2014.[19] In response, the United States, Japan, and the European Union filed a World Trade Organization (WTO) lawsuit against China in 2012 for lowering export quotas of a critical mineral. The WTO ruled against China in 2014, but in the meantime, trade restrictions and rising costs resulted in rare earth mines opening up in other countries. By 2021, China's percentage of production had dropped to 60 percent. The environmental impact of REE remains a fraught discussion, however. At the time of writing, the United States has one rare earths mine, the Mountain Pass Mine in California, which reopened in 2018. However, the minerals are still shipped to China for the particularly toxic separation process.

Like other sacrifice zones, Baotou is not a site that often enters into the realm of visibility. Foreign visitors are prohibited near the Bayan Obo mines, which limits the production of images for international consumption.[20] As I researched this chapter, I found some aerial images from the Environmental Justice Atlas and a few photos taken by Reuters or other news outlets. But there were not enough images to draw any kind of conclusion about the visual culture of rare earths except that these mines stay largely out of sight. By way of contrast, I briefly turn to lithium, another metal crucial to recent technological developments, which is much more prominent in visual culture.

The aesthetic beauty found in lithium mining has perhaps allowed it to dominate the visual culture of mining in the twenty-first century. Consider an aerial-view photograph of the SQM lithium mines in the Salt Flats of Chile's Atacama Desert by photographer Edward Burtynsky (plate 28). The Andes rise in the background, framing the scene. The staging of the image is striking, presenting the mine as neat, contained, and colorful. The bright colors of lithium brine ponds at various stages of evaporation resemble a paint box set in the desert landscape. The cheerful colors, from yellow-green to cerulean, are a byproduct of brine mining, the primary method of extracting lithium. In Atacama, lithium is found in brine reservoirs under dried lake beds, and water is pumped deep underground to push the lithium to the surface. Atacama is the driest place on Earth, making it well suited for the slow evaporation required for lithium extraction. A stew of manganese, potassium, borax, and lithium salts is dissolved in water, which is then evaporated, filtered, and moved into different pools. In some ponds, chemicals like hydrochloric acid have been added to isolate certain products. After twelve to eighteen months, the mixture has been sufficiently filtered to enable lithium carbonate extraction.

Lithium is a crucial material in rechargeable batteries and is central to most plans for a green energy transition that moves beyond fossil fuels, a fact that has created a contemporary "gold rush for lithium."[21] Water mining is incredibly resource-intensive, especially in desert regions where water is scarce. In Salar de Atacama, mining consumes 65 percent of the region's water, as one ton of lithium requires approximately five hundred thousand gallons of water. Guillermo Gonzalez, an expert in lithium batteries at the University of Chile, summarizes, "Like any mining process, it is invasive, it scars the landscape, it destroys the water table and it pollutes the earth and the local wells. This isn't a green solution—it's not a solution at all."[22] Lithium mining produces toxic byproducts, including polluted water, tailings ponds, and slag heaps of toxic waste. But despite the environmental consequences, lithium mining *looks* environmentally friendly. The green and blue ponds signal "nature"; they are colors symbolic of a healthy earth. Visually, lithium mining appears to promise a new era of extraction: aboveground, contained, and ecofriendly. The healthy glow of a postcarbon world.

Seemingly without irony, Burtynsky writes "We will need it all!" alongside the photograph in an article on his website, titled "A Good Anthropocene," highlighting progress in addressing climate issues.[23] In the article, Burtynsky lays out his intentions to raise consciousness about environmental issues but not vilify or blame. The limitations to this framework become apparent as Burtynsky simplistically celebrates lithium as a sustainable solution to fossil fuels. After outlining that *all* of Chile's lithium will be needed to fuel the green energy transition, Burtynsky goes on to say: "Closer to home here in Canada, we have Big Lonely Doug on Vancouver Island. The second-largest Douglas fir tree in the country and approximately 1,000 years old, BLD stands alone in a clear-cut at 66 metres tall and 12 metres in circumference. BLD is a stark and powerful symbol of the need to protect our old-growth forests."[24] The peculiar unevenness of Burtynsky's vision for development situates one ecosystem as marked for protection while another is justified as a necessary sacrifice zone. A neocolonial grammar orders this vision of protecting trees in Canada—a nation-state home to many mining firms ravaging Latin American landscapes—while intensifying apparently justified extractivism in Chile. While the shift from catastrophism is a welcome acknowledgment that there is still time to avert some of the more damaging aspects of climate change, here we see the troubling tendency to ignore questions of inequality, power, and historical causation, a tendency that inflects much of the contemporary discourse around the Anthropocene. This approach addresses the symptoms and perhaps proximate triggers of climate crisis rather than probing and restructuring its fundamental social causes.

Burtynsky's evocation of a "Good Anthropocene" connects to a larger movement that, in its most extreme manifestations, argues climate breakdown is, in fact, a kind of economic opportunity. Emblematic of this framing is the Good Anthropocene articulated in "An Ecomodernist Manifesto" (2015).[25] Collaboratively written by eighteen leading scientists, the manifesto challenges the environmentalist emphasis on limits to growth, proposing that efficiency and technological solutions can solve climate change. Spearheaded by the Breakthrough Institute—an environmental research center situated in Silicon Valley, California—and its cofounders Ted Nordhaus and Michael Shellenberger, the manifesto's pragmatic defense of economic growth and technological intensification has reached a broad audience. The manifesto summarizes, "Knowledge and technology, applied with wisdom, might allow for a good, or even great, Anthropocene." Critics have described the institute as the "leading big money, anti-green, pro-nuclear" think tank; however, the idea that human ingenuity and technological innovation alone can fix climate breakdown is seductive.[26] Not mentioned in the manifesto are questions of environmental justice, capitalist imperatives, spatial and generational equity, racism's global organizing role, or neocolonialism. Instead, the Breakthrough Institute's mission is to find technological solutions—and economic profits—to pressing ecological issues. T. J. Demos argues the institute reflects the influence of Big Tech in green economics, which is shaping a "neoliberal Anthropocene" haunted by the "threat of white supremacist tendencies and colonial, extractive futurism."[27] The narrative of a Good Anthropocene suggests that we can simply move beyond the legacies of colonialism and extractive capitalism that produced our current moment of crisis with a few technocratic tweaks, bypassing any need for structural change to systems of production, reproduction, and distribution. However, technological fixes without structural changes are insufficient to address the environmental challenges caused by extraction, particularly given the imperatives to continual cycles of growth, crisis, and sacrifice built into the capitalist mode of production.

The smartphone is a tangible example of an object that has gotten smaller, lighter, and faster at the same time as it has grown ubiquitous. The increasing use of technology metals such as rare earths and lithium has indeed reduced the physical stuff needed to make a phone in terms of mass. However, we use smaller and more efficient phones much more, both individually and across the population, which stimulates rising demand for these particular raw materials. This counterintuitive dynamic of scaled-up consumption following dematerialization has been identified before. In 1865, the economist William Stanley Jevons, concerned with resource depletion, published *The Coal Question*, in which he announced a strange finding: higher efficiency in energy production *increased* the amount of coal consumed. Jevons demonstrated that when coal was used more efficiently, it lowered prices, stimulating increased use, further depleting supplies. The Jevons paradox poses a significant challenge to arguments that increased efficiency in production will eventually reduce the impact of human activity on the natural world. As environmental scientist Vaclav Smil shows in his comprehensive study of dematerialization, the Jevons paradox remains true in our present moment, as the combination of cheaper energy and less expensive raw materials has led to higher consumption. As Smil writes, "Less has thus been an enabling agent of more."[28] Indeed, while significant technological innovations in recent decades have increased energy efficiency, demand for electricity and fossil fuels has expanded. This is evidenced by the fact that extractive activity is escalating the world over despite our increasingly networked, seemingly immaterial world. Rising consumption and the reality that technology is increasingly metals-intensive (particularly but not exclusively in terms of REEs) drive extraction. While Burtynsky's photograph implies a certain green-extractive futurity, lithium photographs stubbornly show some of the materiality that underpins the digital world. We should not be lulled by the fairy tale that, under capitalist pressures to reduce cost and expand production to maximize profitability, we will accidentally innovate ourselves out of our destructive consumption of the earth's minerals, any more than workers should believe that a labor-saving innovation at work will grant them more leisure time.

The Cloud

Once extracted from the earth, minerals and metals are transformed into the infrastructure that connects the networked present. The data mining of Silicon Valley is inextricably linked to the physical mining of raw natural resources. In the Cloud, we find that the visual lies at the center of our use of the products of the mine: according to the Shift Project, a French think tank advocating for a postcarbon economy, 80 percent of all data is images or video.[29] Still and moving images flow through network cables and mobile network antennae to remote software housed in data centers worldwide. An increasing quantity of these images are ultimately stored in Cloud infrastructure. A data storage system off of individual consumers' devices, the Cloud enables remote access to data, a key node in the networked present, as data is liberated from the device's hardware. The Cloud promises immateriality: the ephemerality of atmosphere. This is heightened by the aesthetics of the Cloud—what media studies scholar John Durham Peters describes as the "fluffy, benign cumulus clouds" that signify online storage.[30]

To consider the materiality of the Cloud, allow us to trace our path once again from the surface to the underground to the atmosphere—though this particular Cloud moves its data primarily through undersea cables. How does the digital photograph look from the ocean floor? Here, we can think of Trevor Paglen's photographs, which show underwater internet cables in an attempt to make a material backbone of our networked world visible. In a painterly c-print photograph, *NSA-Tapped Undersea Cables, North Pacific Ocean* (2016), cables run along the ocean floor, immersed in deep but hazy shades of blue (plate 29). The photograph directs our attention to the reality that there are over three hundred active fiber-optic megalines spanning over half a million miles and connecting every continent besides Antarctica.[31] Just as many highways follow old rail routes, fiber-optic cables follow paths built for telegraph and telephone lines. As Paglen tells the viewer, this physical infrastructure can also be easily monitored: many of the cables meet in one place, which makes them easier to tap—in this case, by the US National Security Agency. According to Paglen, these "deep-sea geographies" challenge the conventional emphasis on horizontal geographies, orienting attention to "the vertical dimensions of human world-making."[32] It seems counterintuitive that the data stored on the Cloud moves under the ocean rather than through the atmosphere, pinged off satellites. However, over 99 percent of the world's data travels through fiber-optic cables that snake through the sea, using light to encode information since it can move data faster and cheaper than satellites can.[33] Undersea cables conflict with cultural narratives of communications and media. As media studies scholar Nicole Starosielski states, the communications infrastructure we use is often wireless, shifting "our attention above rather than below," seemingly confirming longstanding imaginaries of communications technologies as transcending material limits.[34]

Once the technological device—end console and camera in one—is purchased, the demand for the resources it spurs continues. As more services have shifted to online storage, the energy and infrastructural demands for these services have skyrocketed. Calculations of the carbon footprint of the Cloud vary considerably, but it is estimated that digital technologies emit more greenhouse gases than aviation. The endless circulation of images consumes significant amounts of electricity and, typically, produces CO_2 emissions.[35] Take one striking statistic: in 2018, Luis Fonsi's music video "Despacito" hit five billion views on YouTube, using as much energy as forty thousand homes in a year in the United States.[36] The data required to stream "Despacito" moved from data centers to end users through fiber-optic cables. Data centers are the United States' most significant and fastest-growing consumers of electricity and water needed to cool the heat-generating servers.[37] Despite the outsized attention given to the environmental impact of Bitcoin mining, even more prosaic server farms have intensive needs for materials and energy. A midsize data center uses between 80 million and 130 million gallons of water annually, equivalent to the water used in three hospitals in a year.[38] The infrastructure of the Cloud continues to expand, and data centers continue to grow in both size and number, introducing new environmental pressures at every step. Despite this, technology companies deploy the rhetoric of green energy, hiding the reliance of even their immaterial services on resource exploitation. Following the withdrawal of the United States from the Paris Climate Accord, Facebook CEO Mark Zuckerberg stated, "Withdrawing from the Paris climate agreement is bad for the environment, bad for the economy. . . . We've committed that every new data center we build will be powered by 100 percent renewable energy. Stopping climate change is something we can only do as a global community, and we have to act together before it's too late."[39] Google posted on its website: "We're a data-driven company. The science of climate change tells us that building a carbon-free electrical grid is an urgent global priority."[40] Big Tech companies' postcarbon messaging downplays their businesses' consequential material impacts, as their rhetoric of technological innovation obscures these earthly realities.[41]

Narratives of the Cloud draw us back to our earlier discussions around the politics of atmosphere. In chapter 3, I outlined how silver-based photographic processes were materially impacted by air pollution, as different elements mined from the earth for human use encountered each other on the surface of the physical image. Most photographs produced today are digitally produced, which raises a different set of questions about how the photograph interacts with particulate matter in the atmosphere and in our changing climate. During the 2020 wildfires in California and Oregon, people reported that smartphone cameras changed the orange sky to gray. The *Atlantic* ran an article titled "Your Phone Wasn't Built for the Apocalypse," reflecting how it was "as if the smartphones that captured the pictures were engaged in a conspiracy to silence this latest cataclysm."[42] In analog photography, the film's chemical color tone determines the photograph's range. In digital photography, camera sensors are color-blind, registering brightness alone and using a process called white balance to build color. This process relies on calibrating the visual field against a reference point, using white as a marker to determine what objects in the photograph should look like. Smartphones do not capture color directly; it's always constructed. Pointed at the hazy sky, the camera removed orange, as the white-balance process identified such a vast orange field in context as unrealistic. The surreal orange sky caused by wildfire smoke over Santa Cruz, California, exceeds the representational possibilities imagined by the camera engineers. This does not suggest that analog photography is a privileged medium to document moments like this, as film too has limited tonal ranges.[43] Instead, it reminds us that no photographic process is a neutral window onto the world but is always somehow mediated and limited in what it can reveal.

The uncanny orange skies and haunting reddish shadows that follow wildfires highlight how pollution and disaster impact how we see. In the face of ecological devastation, we are met with what Demos has called "the insufficiency of the image."[44] In his analysis of photographs of catastrophic wildfires that circulate in the news and on social media, Demos highlights how images of disaster, many of which can be broadly classified as sublime, circulate to sensationalize crisis.[45] The repetition of these visual forms can have the effect of normalizing—or aestheticizing—catastrophe. The discourse around climate change too often fails to address or name the causes of environmental catastrophe and attritional violence, which consigns climate change to the abstraction of "disaster," a category with limited political potential. For example, less than 4 percent of mainstream news coverage of the deadly California wildfires in 2018 mentioned climate change.[46] By the time of writing in 2021, it seems like this discourse has decisively shifted, with the development of more widespread recognition across mainstream media reports of the role of climate change in exacerbating natural disasters. Still, the emphasis remains on the spectacular nature of catastrophe rather than moving the political conversation toward mitigation and adaptation.

As we have seen, photography is limited in its ability to reveal the structural causes that have contributed to the "catastrophic convergences" of our present.[47] It is, of course, not alone as a medium in its inability to reveal the full scope of the problem. Linking environmental issues to their proximate and fundamental causes, and outlining the cumulative nature of their effects, is a difficult task that poses challenges for environmental organizing.[48] There are significant limitations of photographic representation in visualizing experiences and analyses of climate change, let alone what can be done to stop it. Yet, as Demos states, "Critically reading these images does some work to restore hopefulness—provided by research, interpretation, writing, teaching, learning, building community. It grants new life, against all odds, even if against optimism and its cruelties, perhaps resulting in something like undefeated despair."[49] This is the spirit with which I turn our attention to the material consequences of our apparently dematerializing image culture—not to suggest that there is no way out but rather to highlight how many places we can possibly intervene.[50]

E-Waste Mining and the Digital Image

A discarded keyboard, missing keys, and other detritus are partially covered in soil. Loose wires scatter nearby. The image is tightly framed, revealing very little about the surrounding scene. It resembles a still from a postapocalyptic film, an elegy for the remnants of a past society. The enigmatic image opens the South African photographer Pieter Hugo's photo book *Permanent Error* (2011), which centers on Agbogbloshie, an e-waste dump in Accra, Ghana. In the second image, a tangled web of wires, circuit boards, and metal are in flames. Dark smoke billows out of the bright, hot fire. The wider frame comes into view in the third image in the sequence. A crowd of people is framed by thick smoke pouring out of many fires, merging into the flat gray sky. Hugo situates the viewer in Agbogbloshie, where computers, smartphones, circuits, transistors, capacitors, and semiconductors are stripped down to base metals. Young people working without masks, gloves, or any safety materials break down old technologies at the dump, which spans ten kilometers. The work is done by hand or with makeshift tools found among the detritus. Plastic or rubber is melted down to isolate the valuable metals. Hugo describes the waste recovery as darkly but "beautifully alchemical," relying on fire for material transformation. Due to the omnipresence of fire and toxic smoke, Agbogbloshie is nicknamed "Sodom and Gomorrah," referencing two legendary biblical cities destroyed by God for their wickedness.

Electronics are the fastest growing portion of global garbage production, totaling an estimated fifty million tons of electronic waste each year, the discarded products of innumerable forms of extraction.[51] In e-waste dumps, at the other end of the technology life cycle, a different kind of mining occurs. This secondary stage of extraction, in which scrap or excess metals are recovered, has a long history. Eastman Kodak's silver recovery plant, stabilizing their silver supply, is of course one forebear, as is the mining for uranium and later radium of waste piles left behind after silver extraction in Joachimsthal in Germany. Many of the digital age's most critical metals and minerals, like rare earth elements and lithium, cannot be separated using rudimentary refining methods like burning. The existing methods for recovering these valuable elements are expensive and complex. While many countries have regulations on electronics recycling, most electronics are discarded improperly. For instance, in 2012, $206 billion was spent on consumer electronics in the United States, but only 29 percent of e-waste was recycled.[52] Most spent electronics are sent to Africa, China, India, the Philippines, and Vietnam in shipping containers, typically labeled as second-hand goods rather than waste. However, in his study of e-waste in Lagos, Nigeria, Jack Sullivan shows that half of the materials cannot be resold.[53] E-waste destined for Agbogbloshie arrives in shipping containers through the Port of Tema, where wholesalers purchase the containers and send the materials to Accra, where they are broken down, stripped, and separated. Copper and other valuable materials are then resold. The trace amounts of rare earth elements and lithium, among other materials, become waste. In the decade since Hugo took these photographs, electronics production has rapidly increased: electronic waste increased 20 percent in volume between 2015 and 2017 alone.[54] Planned obsolescence, even of relatively dematerialized goods, likewise ramps up the flow of waste, posing significant problems of environmental justice.

Hugo's stark photos attempt to make toxicity visible by emphasizing the burning waste and acrid smoke over a flat gray sky. The square photographs, muted color range, and visual repetition build a narrative arc throughout the series. People typically look back at the viewer, aware of the encounter with the camera. In *Al Hasan, Agbogbloshie Market, Accra, Ghana* (2009), a worker stands behind a wheelbarrow on the ground, full of electronic wires (plate 30). Al Hasan stares at the camera, framed by smoke. Hugo writes:

> Notions of time and progress are collapsed in these photographs. There are elements in the images that fast-forward us to an apocalyptic end of the world as we know it, yet the alchemy on this site and the strolling cows recall a pastoral existence that rewinds our

minds to a medieval setting. The cycles of history and the lifespan of our technology are both clearly apparent in this cemetery of artifacts from the industrialised world.[55]

Through Hugo's lens, Agbogbloshie slides out of time, as the formal beauty of the subjects in the frame contrast with the hellish environment of smoke and fire, which curator Karen E. Milbourne describes as "a setting imagined by Goya."[56] The wheelbarrow, a tool for building and planting, is repurposed for a toxic form of harvesting. Standing amid the wreckage of the networked world, Al Hasan is situated in a lost Eden. The scene seems to draw from the extended, even imaginary temporalities of painting rather than the singular moment of the photograph.

Hugo's unsettling portraits are controversial and have been criticized for their aestheticization of poverty, compounded by the power dynamics of a white photographer documenting the abject working conditions of Black Africans for an art-world audience. However, literary scholar Cajetan Iheka has argued that despite the extractive dynamics that inform the creation of the images, Hugo's images bring scenes such as this into a wider public consciousness, allowing for a critique of consumption, waste, and the racially unequal toxic burdens that structure the global flow of materials.[57] The images ask the viewer to consider racial capitalism's environmental sacrifice zones, encompassing extractive frontiers and sites to dispose of toxic waste.[58]

We might set Hugo's photographs alongside another group of images of Agbogbloshie, taken by the Burkina Faso–born photographer Nyaba Leon Ouedraogo in 2008. Ouedraogo took some straight portraits, but we nevertheless often see people at work. In one, a male figure leans forward, his foot close to the camera lens, surrounded by singed wires and a tire. The physicality of the backbreaking labor stands out as he bends to pick up scraps, framed by billowing black clouds in the background and the charred refuse in the foreground. Shot close to the ground, the photographer situates the viewer near the work of gleaning scraps for something valuable. A leg spans the right side of the frame, adding to the impression of photojournalistic realism. Ouedraogo's dramatic framing and high contrast are striking, but Ouedraogo described his intention as less formal and more political: "Not showing my images for what they depict, but for what they transmit"—namely, the human and environmental costs of waste.[59] Ouedraogo's invocation of the concept of "transmission" conjures the networked world of technology but also summons the specter of disease, of transmissible viruses and carcinogens.

What is transmitted at Agbogbloshie? The site has dangerous levels of carbon monoxide, mercury, dioxin, cadmium, antimony, lead, mercury, thallium, hydrogen cyanide, polychlorinated biphenyls (PCBs), chlorobenzenes, polybrominated diphenyl ethers (PBDEs), triphenyl phosphate (TPP), and brominated flame retardants (BFRs). Some of these materials migrate to nearby regions in the particulate matter of smoke. Others enter the soil, which drifts into toxic dust. Pollutants seep into waterways as the Odaw River flows to the Korle Lagoon, which feeds into the Gulf of Guinea. The poisonous materials pose heightened risks to people living in and around Agbogbloshie, but as toxins enter the air, soil, and water, toxicity becomes hard to contain. The toxins enter the food chain and become toxicants, increasing in concentration.[60] Burns, headaches, respiratory problems, and nausea are immediate effects of exposure to toxic waste, but the slow carcinogenic mutations give way to cancer and other industrial diseases. In addition to its material realities, toxicity is also a problem of representation, for it is largely invisible, as we have seen. In both projects, the photographs push the conventions of portraiture as the image of a singular individual, staging an ecological encounter with an entire scene immersed in a toxic haze. While the smoke suggests the permeability of landscapes and bodies, in the photographs, we cannot see how these materials enter the lungs of the workers, who face serious health risks.

While the images struggle to show us toxicity, they make visible the human and ecological costs of the value extracted from the global majority, on both ends of commodity chains, to fuel the insatiable demand for technology. It shows how the extraction of resources, even through such artisanal materials recycling, continues to be driven by the extraction of energy from workers. Or, in Ouedraogo's words, "The garbage of the rich poisons the children of the poor."[61] In another 2008 photograph by Ouedraogo, three of these children are shown from behind, sitting on old computers. Beside them is a worn-out soccer net. The sky is blue; fire and smoke are not present in this scene. The more quotidian framing shifts from Hugo's toxic sublime to a more contemplative but equally devastating indictment. Ouedraogo brings into view the fuller lives that the global waste trade has damaged. The mining of metals at Agbogbloshie is part and parcel of the digital world, a critical aspect of the labor that allows the digital to happen.[62]

By photographing Agbogbloshie, Hugo and Ouedraogo demand we look at the toxic jobs that underpin our networked world. The fact of massive concentrations of e-waste among poor communities in the Global South is a major problem of environmental justice and undermines the marketing of new technologies as sleek, efficient, and often greener. Consider, for instance, a video advertisement released in 2022 for Apple's iPhone 13, titled "Now in Green." Abstracted organic forms like plants and the crystalline form of rocks fold into the sleek form of the green iPhone. The ad breaks from the clinical white minimalism of many Apple ads by proposing a merging of the organic and the technological; the phone both emerges from and then swallows pixelated and distorted images of the natural world. "Now in Green" describes the color of the phone but also gestures to "green" culture more broadly,[63] a bitterly ironic claim given the massive waste produced by the planned obsolescence built into Apple's business model.[64] As a strategy to increase sales, planned obsolescence relies on three methods: hardware with component parts that are unreliable or hard to repair, software upgrades that are incompatible with older hardware or that are programmed to stop working after a certain number of actions, and marketing to push newer models even if older devices still work. Apple deploys all of these methods. One of the primary drivers of new smartphone sales is increasing camera quality, even though the camera is typically the most expensive single function in a smartphone, as external suppliers make image sensors and lenses.[65] Here, we see the significance of images of places like Agbogbloshie: they show what underpins the glossy, green marketing of Big Tech. Merely recycling correctly cannot reduce waste on the scale required, let alone without exposing entire communities to toxins in the process. Instead, the political-economic emphasis must be on preventing overproduction and planned obsolescence, producing to meet needs rather than to profit through the production of excessive trash.[66]

It's worth emphasizing that planned obsolescence of this kind is not simply a matter of consumers wastefully following fashions. Consumers are given limited options to viably keep older electronics functioning, as software updates make it difficult to keep using an older phone while newer models are harder to repair. To participate in the increasingly connected 24/7 world of global capitalism, for both work and for social life, one must keep upgrading. The constant upgrading also impacts our ability to opt out, for as art historian Jonathan Crary observes, "catching up" becomes impossible when individuals become "estranged and disempowered because of the velocity at which new products emerge and at which arbitrary reconfigurations of entire systems take place."[67] We have been considering the material underpinnings of the so-called dematerial image—from the necessarily remote rare earths mining zone, to the fiber-optic cable on the ocean floor and its resource-intensive server farms, to the e-waste dump animated by people trying to squeeze a living out of salvaging materials from their premature destruction. However, as Crary shows, these physical infrastructures have decisive cultural lives—the immaterial reacts back on the material.

The Work of Art in the Age of Dematerialization

Throughout this book, I have argued that photographs are material objects, and their materiality is essential to understanding their extended social life. However, photographs also participate in larger image economies. Since its inception as a medium, there has been a tension in photography between the image-object and the photograph-as-representation, most famously theorized by Walter Benjamin in "The Work of Art in the Age of Mechanical Reproduction," which described the endless circulation of images disconnected from their source, a phenomenon that characterized the image-worlds of the early twentieth century. Benjamin's emphasis on circulation has an even more intense resonance today. We live in a hypervisual world. It is estimated that, in 2017, 1.2 trillion photos were taken,[68] and a conservative estimate places image uploads in 2018 at thirty million images posted *daily* to Twitter, fifty-two million to Instagram, and 350 million to Facebook.[69] This volume of images far exceeds our collective capacity to engage with them.

If the photograph tells us something—however incomplete—by freezing time, the circulation of contextless and unorganized images interrupts our ability to engage with the visual. Algorithms now process and organize photos for our consumption, weeding through a potentially altogether random stream, and the image is reduced to content. Programs like TikTok and Instagram run off algorithms that reward regularity, standardization, and quantity. Further, the algorithms and artificial intelligence that sort and catalog images often come to reflect political and racial biases present in the societies and firms that produce both their raw material and their strategic tendencies.[70] At the same time, the overwhelming flow of images results for many in a loss of meaning, as the singular image dissolves into a flow of disjointed information.[71] Media studies scholar Sean Cubitt draws attention to the reality that the amount of data produced now exceeds storage capacity, requiring compression of files and partial data loss. "The past is compressed, while the power to extrapolate future states from recent data expands, to the extent that, from the standpoint of capital's eternal present, the onward growth of capital and the collapse of the planet appear preordained. The mass image does not simply accelerate: it generates a new kind of time."[72] The "eternal present" of the digital image alters the photograph's relationship to time. Only the algorithm can sort the mass image, making the "when" of the mass of digital photos relatively irrelevant. They all exist in a potentially perpetual present, but a present where the now is "always already over."[73] Or, as Geoffrey Batchen describes, the digital image is "in time, not *of* time."[74] The singular moment marked in a photograph becomes one moment in a larger data aggregation that overwhelms the isolated moment.

As Benjamin noted a century ago, the endless flow of decontextualized images impacts our politics, ethics, and sense of time. Smartphones have escalated the acceleration Benjamin observed, the bleeding of work into daily life, and the rise of the attention economy. Crary describes the "immense incapacitation of visual experience" that comes with capitalism's nonstop work cycle, which "disables vision through a process of homogenization, redundancy, and acceleration."[75] Ethical spectatorship is incredibly challenging within the conditions of the attention economy. Further, data mining underpins the anesthetizing flood of images. If the twentieth century was marked by what Guy Debord called "spectacle," the shift to platform capitalism or computational capitalism has changed the political economy of image-making. As Trevor Paglen and Kate Crawford argue, "Images are no longer spectacle but they are in fact looking back at us, being actors in a process of massive value extraction."[76] Just as no one can claim to occupy an ethical standpoint wholly outside of complicity with extraction, involvement in this economy of visual representation is somewhat inevitable to participate in social life. Accepting this state of compromised embeddedness is the starting point for a more complex conversation about what images can and cannot do to shape ethical engagement, enabling the possibility for less extractive forms of collective life.

While the rise of the digital world wasn't the death of photography, it seems that, in some senses, photography has failed.[77] In this failure, however, there is possibility. Now, bear with me as I attempt to square a circle. Criticisms of photography's political potential have, for over a century, centered on the way the medium removes a scene from its social and political context. I have argued that images are nevertheless always rooted in the conditions of their production and thus proposed that asking a different set of questions allows us to situate photographic interventions within complex chains of production and consumption, allowing political histories of the social life of photography to return to the center of the frame. Understanding these histories is significant because the visual has long been an important site for constructing meaning and making sense of the world, which are fundamental steps toward changing it. In particular, photography is a privileged tool for contemplation. In the ability to isolate a single moment for reflection, there might be a lesson we need urgently for our present. In a sense, photographs resist acceleration by freezing time, by slowing it down. They can, potentially, give us pause.

In photographs, there is an opportunity for a kind of slow looking that can initiate ethical spectatorship. Through the complex relational webs enabling the photograph, they also invite us to trace back the structural causes that make the image possible and, in many contexts, have also contributed to our current moment of crisis. Standard critical narratives of climate change provoke panic, which can be a recipe for inaction. Photographic contemplation, however, can provide a productive invitation to think slowly, from a nonpurist place of implication, recognizing that we need to stop, think, and make decisions based on a longer, more relational view to combat either slow or spectacular violence. To undertake that transformative struggle, society must fundamentally shift what it values.

Photography, of course, cannot enact this shift alone. More often than not, photography abstracts and anesthetizes. An assumption underlying this book is that images often teach us to see nature and people extractively. Extraction is, in part, a problem of vision. But as Joanna Zylinska argues, photography "slows down time and can teach us humans to look at ourselves and our environment differently."[78] Seeing differently is urgently needed, particularly if we are going to act collectively, differently. Is there a politics that emerges from looking at photographs of extraction? Not necessarily, but perhaps. The photograph can shift attention from the horizon of the future to the here and now. And, as climate crisis has shown, we cannot think about the future without directly intervening in the present. To tell us something about how we got here and where we need to go, the image needs to be resituated within the complex networks of human and nonhuman labor that bring it into being. The archive of images that composes this book reckon with history, reassert the presentness of the present, and, in doing so, reimagine the future. This, I argue, should be the task of photography in the Anthropocene.

Conclusion

Discussions around climate crisis rarely address the reality that without reducing extraction in all of its scope, we cannot meet the ecological challenges of our present. Instead, the search goes on for ever-more-distant sites of extraction. For example, in 2019, NASA announced a planned probe mission to 16 Psyche, a metal asteroid formed of gold, iron, and platinum, located between Mars and Jupiter. The relative scarcity of platinum, and its centrality to both emissions reduction and health care, is one of the justifications for exploring asteroid mining. Led by Arizona State University and using SpaceX's launch services, the NASA probe was quickly picked up in the press. Fox News proclaimed, "NASA Headed towards Giant Golden Asteroid That Could Make Everyone on Earth a Billionaire." Newspapers echoed this gold-rush language: "NASA Chasing GOLDEN Asteroid That Could Make Everyone on Earth a Billionaire," "Hubble Examines 16 Psyche, the Asteroid Worth $10,000 Quadrillion," and "Hubble Examines Massive Metal Asteroid Called 'Psyche' That's Worth Way More Than Our Global Economy," to name a few. Leaving aside the absurd premise that the wealth would be distributed equally—something unheard of in the history of capitalist property relations—the ongoing fascination with new commodity frontiers is clear, as is the unwillingness to recognize that we need to extract, and therefore consume, less.

The extractivist futurity prefigured in mining asteroids poses a seductive answer to a problem that has become pressing: the limits to growth. According to Franco "Bifo" Berardi: "Modern capitalism is based on an idolatry of energy. It is based on an obsession with growth, expansion, productivity, acceleration."[79] With the combustion of coal into energy, society seemed to move beyond the tangible limitations to growth imposed by the productivity of land. Describing the changes enacted by coal, Marx made an analogy between land and bodies, describing how capitalism "simultaneously undermin[es] the original source of all wealth: the soil and the worker" through a prioritization of short-term growth that robs the soil of its fertility while "laying waste and debilitating labor-power."[80] In this, Marx reminds us that rapid growth is based on the extraction of energy, which comes with costs: exhaustion. As the Anthropocene has revealed, there are not only limits to growth but consequences. Despite this, extraction continues unabated. Asteroid mining and space exploration seem to offer a way off "spaceship Earth," as global temperatures tick up entire degrees, causing rising sea levels, wildfires, failing power grids, and extreme weather—a narrative of escape that has purchase on both the Left and the Right.[81]

In this book, I have charted the extraction of several different metals and minerals and how they were and are used in photography. While the photographic processes have changed, all of these materials are still extracted. The role of images in mass culture has shifted with the advent of the digital image; its roots, however, remain fundamentally material. This is not how we tend to imagine our networked present. In Sebastião Salgado's *Workers: An Archaeology of the Industrial Age* (1993), the photographer describes how his work captures an era "when men and women at work with their hands provided the central axis of the world."[82] The use of the past tense "provided" signals that for Salgado, this world driven by human labor is coming to an end, a common narrative in contemporary discourse about economic change.[83] The perceived dematerialization of labor is directly tied to the perceived dematerialization of infrastructure— and just as erroneous. Smartphones and cameras are one area where relative dematerialization demonstrably occurred while the mass of the individual objects has decreased. However, the overall reduction of mass relies on more metals, which have a greater impact than the masses of plastic that formed earlier, bulkier models.[84] Further, as smartphones have gotten smaller, there has been skyrocketing demand. This would suggest that, to address the challenges of our present, technological innovation without structural change is insufficient. Without shifting larger patterns of consumption and ways of relating to the natural world, technological solutions will only finesse our march toward collective climate disaster.

As the demands for technologies grow—the volume of data stored globally is expected to grow sixfold from 2018 to 2025—addressing these environmental and political problems is urgent.[85] Here, the reorientations of vision that the photographs and analysis in this book enact become significant. We need to learn how to see and value differently. Climate change is a crisis that draws our material relations with each other and the world into the center of choices we must make collectively and deliberately, independent of capitalism's chaotic and implacable demand for profitable growth. The decision facing us in the present is whether we want that relationship to be rooted in extraction—with its attendant consequences of environmental damage and inequality—or in reconfigured engagements with the natural world and each other based in reciprocity.

Conclusion

ALL THAT IS SOLID MELTS INTO AIR

> There is a crack in everything, that's how the light gets in.
> Leonard Cohen, "Anthem" (1992)

FIVE LARGE BLOCKS OF ICE MELT slowly in the Texas sun, causing water to pool on the hot pavement. Spiderweb cracks fracture through the ice blocks, which eventually shatter and collapse. Encased within the ice were portraits taken in the high Arctic by Canadian photographer Louie Palu (plate 31). The ice blocks formed a sculptural assemblage called *Arctic Passage*, installed outside the Harry Ransom Center at the University of Texas at Austin during the SXSW festival in 2019. The ice casts a cool glaze over the distinctive tonal range of the portraits, which capture the crisp and saturated shades of the Arctic landscape. The portraits document Inuit people, indigenous to northern Canada and parts of Greenland and Alaska, whose traditional way of life is acutely threatened by climate breakdown. The other portraits capture Canadian Rangers and US Marines who are deployed in the Arctic, as well as members of the US, UK, and French militaries conducting Arctic training exercises. The portraits are straight frontal shots: the subjects stare directly back at the viewers of the installation. However, the ice forms a physical barrier, undermining the promise of an unmediated encounter with the subjects in the frame. As exposure to the elements melt the ice, the portraits become harder to see. The melting ice evokes the thawing Arctic sea ice, a major link in the chain of climatic changes leading to extreme weather patterns. The water breaks down the image, but the water also flows off. As the water disappears, it destroys the coherence of the work. The dissolution of the work becomes the agent of meaning as the melting ice materializes the vulnerability of both landscapes and bodies to anthropogenic climate change.

Palu's photographs have a surreal quality, rooted in the striking visual contrast between the arctic landscape and the intrusion of militarized force. The focus on the military presence in the North American Arctic reveals the links between climate change, geopolitical power, and militarized zones of extraction. The political and strategic dimensions of climate crisis become impossible to ignore, as the work draws sharp attention to the military response and contributions to the impacts of climate change on the Arctic, which is warming faster than any other site on the planet. Geopolitical strategists for a variety of states have argued that the ravages of climate change have increased the risk of conflicts resulting from resource shortages, access to water, and natural disasters. In a cruel twist, glaciers melting from the ecosystemic change fueled by our hyperextractive economy have revealed mineral deposits previously inaccessible for extraction, thus opening up the possibilities for drilling for oil and natural gas and potentially stimulating further extraction. Palu's body of work documents the intense and growing interest in the Arctic region and the ways in which states are using military force to secure resource-rich regions and strategic new shipping routes. It is crucial to recall that the military itself has a massive ecological footprint: the US Department of Defense has a higher annual carbon footprint than most countries on Earth, and the Pentagon is the largest user of petroleum outside of entire national economies.[1]

While we commonly encounter fine art photography behind glass and in climate-controlled environments, these portraits are intended to materially change through exposure to the elements. Glass and ice have a visual resonance, and glass and ice can both protect in the right environments, but here under the Texas sun, ice is an agent of disintegration. The emphasis on water also draws a connection to photographic processes. The photographer Jeff Wall has famously described how the liquid used in the darkroom disrupts the predictable mechanical functioning of the camera. As the photographer develops the image, it is soaked in fluids, an "immersion in the incalculable." The use of liquid introduces the possibility of something unexpected surfacing within the developed photograph: this immersion in the incalculable "appears with a vengeance in remote consequences of even the most controlled releases of energy."[2] Wall parallels the unpredictability of the film developing process with environmental change, reflecting that "the ecological crisis is the form in which these remote consequences appear to us most strikingly today."[3] As climate change has demonstrated, the belief that parts of our environment can be contained, isolated, separated, and individually controlled ignores the indissoluble connections that link ecosystems.

Watching Palu's photographs transform and collapse through their material encounter with the elements reminds us that the climate transformations we are currently experiencing can likewise not be isolated or contained. The intentional impermanence of Palu's photographs updates photography's relationship to time in a period of climate crisis. The image-object is intended to be transitory, unfixing the index through exposure to sunlight—the very element that brought photography into being. The power of sunlight both makes and unmakes the photograph, which is analogous to the productive and destructive power of solar energy and carbon in the era of the greenhouse effect. The sun sustains life based on carbon, but the two increasingly come together in sufficient concentrations to threaten it. In Palu's installation, photography leaves the protected confines of the gallery and museum to face the realities of the present. The location adds an evocative dimension, for Niépce's original heliograph is on display at the Harry Ransom Center. The home of the first successful fixing of the image becomes the context where the image is undone. Palu's ice block photographs acknowledge their own vulnerability in an era of melting ice caps and mass extinction.

Palu's ephemeral images consider the fragile precarity of both image-making and of all life in the context of climate crisis. They provide a helpful pivot to consider the consequences of extraction and the stakes of thinking environmentally about photography. While this book has endeavored to enable our focus on the multifarious ways that the extraction that enables photography has contributed to our current reality of climate breakdown, we have dwelt significantly less on where this leaves us. By way of conclusion, I briefly consider two different approaches to the delicate vulnerability of life, ecosystems, and the image that Palu brings into focus: corporate attempts to preserve and own images, on the one hand, and a web of interconnection that reminds us of our embeddedness within both extractive capitalism—and a broader web of life beyond it—on the other.

Returns, Re-visions

By redirecting attention to the importance of mining in making images, I have argued that photographs materially *and* metaphorically emerge from the underground. What might it mean for photography to return to the underground? Alfredo Jaar's *Lament of the Images* (2002) focuses on seventeen million images that were buried 220 feet below the surface in a former limestone-mine-turned-corporate-bomb-shelter at Iron Mountain in western Pennsylvania. The images were buried by Corbis, a stock photography company owned by tech mogul Bill Gates. In 1995, Corbis purchased the Bettmann Archive, a significant photography collection built by the German curator Otto Bettmann, who began acquiring images in 1936. Over time, the images began to chemically deteriorate, and Corbis moved the images into cold storage deep underground while likewise beginning to digitize the images. Repurposed as a corporate safe haven, the mine becomes an escape hatch from the consequences of extractive capitalism and militarism. The images return underground.

Corbis celebrated the archive as "captur[ing] the entire human experience throughout history," making Gates, as Allan Sekula notes, "the new paradigm of the global archivist."[4] In 2016, Corbis rebranded as Branded Entertainment Network (BEN), selling their image-licensing wing to Visual China Group, where they were licensed to Getty Images. The Bettmann Archive remains stored in the Iron Mountain facility, and

Pennsylvania, U.S.A., April 15, 2001.

It is reported that one of the largest collections of historical photographs in the world is about to be buried in an old limestone mine forever. The mine, located in a remote area of western Pennsylvania, was turned into a corporate bomb shelter in the 1950s and is now known as the Iron Mountain National Underground Storage site.

The Bettmann and United Press International archive, comprising an estimated 17 million images, was purchased in 1995 by Microsoft chairman Bill Gates. Now Gates' private company Corbis will move the images from New York City to the mine and bury them 220 feet below the surface in a subzero, low-humidity storage vault.

It is thought that the move will preserve the images, but also make them totally inaccessible. In their place, Gates plans to sell digital scans of the images. In the past six years, 225,000 images, or less than 2 percent of them, have been scanned. At that rate, it would take 453 years to digitize the entire archive.

The collection includes images of the Wright Brothers in flight, JFK Jr. saluting his father's coffin, important images from the Vietnam War, and Nelson Mandela in prison.

Gates also owns two other photo agencies and has secured the digital reproduction rights to works in many of the world's art museums. At present, Gates owns the rights to show (or bury) an estimated 65 million images.

it now has over one hundred million images, requiring immense physical infrastructure for file storage, transmission, and metadata, though many of the archival images are still not available digitally.

Jaar's *Lament of the Images* reflects on the implications of the archive being bought by Bill Gates, who "owns the rights to show (or bury) an estimated 65 million images." Jaar's oeuvre is marked by an interrogation of the ethics of images, suggesting that visibility does not necessarily result in action, nor even ethical spectatorship, on the part of viewers. However, Jaar

C.1   Alfredo Jaar, *Lament of the Images, Pennsylvania, USA, April 15, 2001*, 2002. Illuminated texts mounted on Plexiglas, light screen. Courtesy of the Galerie Lelong & Co., New York; the Louisiana Museum of Modern Art, Humlebæk; the Museum of Modern Art, New York; and the artist, New York.

highlights that the corporate ownership of images as property, their enclosure, limits the stories that can be told with images. Gates purchased a degree of control over images that could potentially shape our understanding of the world. At the same time, the idea of what it means to "own" an image has proven unsteady as the slippage between the image-object and its representation is crystalized here, caught "between singularity, rarity, authenticity and multiplicity, ubiquity, equivalence."[5] Significantly, the circulation that defines images is curtailed by this kind of regime of property rights; they are metaphorically put on ice. As Sekula reflects, "These pictures wait like slabs of dried cod for the revivifying water of the gaze, for the laser beam of the scanner. Their rediscovery is unlikely. Researchers are forbidden to enter. Specialists in conservation applaud the care and thoroughness of the operation."[6] The rhetoric of digital images—and their economic potential, when monetizable—draws on immateriality and mobility, linking Corbis's rhetoric to Holmes's celebration of the possibility of divorcing form from matter.[7]

If Palu's ice block images enter into the world to decay, storing these images underground signals an aspiration to eternity. Corbis suggests that the images, stored in subzero temperatures, could be preserved for up to five thousand years. If the originals are vulnerable, the digital promises (but of course cannot actually deliver) immortality. The impulse to preserve is significant and in many ways built into the institutional structures of art history and art museums. The desire to protect an important archive in some ways parallels, from one angle, the desire to preserve ecosystems. Environmental discourse, particularly in its mainstream form, has often focused on preservation and conservation, rooted in a desire to protect and maintain ecosystems. Our attempts to preserve, it might seem, often end up harming that which we seek to protect. For instance, our attempts to protect nature have created unnatural ecosystems like national parks, which have exclusionary, racist histories and often make the larger ecosystems more vulnerable.[8] Fire-suppression policies create the conditions for more extreme wildfires.[9] Likewise, the targeted killings of species to protect other species, either by culling or through poisons, often have unexpected cascading impacts. In the Corbis example, protecting the images removes them from general circulation and, by extension, limits their applications within public historical memory. The fixation on protecting certain things comes at the expense of other entities. As geographer

Jason W. Moore suggests, we need to shift from protecting ecosystems to valuing them.[10] Without asking more complicated questions about structural causes and being in relation, preservation might do more harm than good. If photographs are going to play any role in helping us understand the choices that led to present crises or in seeding a different future, they need to be in the world.

I want to return one final time to Cariou's practice, to a petrograph that shows an intricate network of the spider's web, gleaming in the golden hues of bitumen (plate 32). The slow fixing of bitumen by sunlight draws to mind the slow accrual of geology and the slow and patient building of the spider's web. The filamentary threads of the web also evoke the slow and patient coalition-building of environmental justice organizing—the resilience required to build sustainable and meaningful change. The tensile strength of the spider's silk is remarkable: a golden-orb spider weaves threads that are stronger than steel yet fifty times as light. As a metaphor, the web is ambiguous: it stands in for cunning traps and secrecy as well as creativity and interconnectedness.[11] The trap set by the spider is not invisible, but it is hard to see; it resists coming into view, like the image of extraction. Webs abound in ecological thought as they propose interconnection—hence the utopian framing of the "world wide web," which gave way in actual fact to simultaneous fragmentation and centralized control by tech monopolies. The slow, patient building of the spider's web challenges a static, stratified history by pointing out this interconnectedness. We see the impossibility of separating relations: we are bound up with bitumen and the structural forces that weaponize it. Alberto Corsín Jiménez proposes the spiderweb as an apt metaphor for a precarious world that "tenses itself with violence and catastrophe but also grace and beauty."[12] Reflecting on the use of bitumen in his photographic practice, Warren Cariou reflects, "Often when pondering bitumen, I return to a kind of fluctuational thinking: between creation and destruction, medicine and toxin, pleasure and repugnance, euphoria and grief. My mind can't settle. I've come to believe that's as it should be."[13] A similar tension animates *Camera Geologica*. Mined materials—bitumen, silver, platinum, iron, uranium, rare earth elements, and myriad other metals—are in a sense magical: they build cities, fertilize crops, provide the energy that lights up homes, and, in artistic practice, allow us to document our experiences and express our hopes, dreams, and fears. They enable the highest forms of human expression. The extraction of these metals, however, causes unimaginable harm. The consequences and power of extraction cannot be ignored.

Thinking seriously about extraction and the myriad ways it is woven through our existence highlights the compromised ethics of our present. Extraction necessitates recognizing that there is no straightforward answer or simple solution to the problems we face: there is no quick fix. Rare earth elements, lithium, cobalt, and the other technology metals will only be a temporary fix to the problems caused by burning fossil fuels, bringing in their stead new challenges. Reckoning with extraction demands an inheritance of hard histories, of loss, and of our uneven implication in these histories. From there, we can, as Donna Haraway proposes, "stay with the trouble." We cannot be seduced by the magical thinking of a quick fix to problems that have been centuries in the making, but we also cannot succumb to climate grief and inaction: too many debts are owed to the present and the future to do so. Instead, it requires us to be more conscious about what we do and how we do it, inviting us to think a bit more reflectively about how we move, collectively, through the world. However, perhaps living less extractively is not a question of negation or lack, of what we have to give up, but, rather, of what we can build. What relationships can we foster, what engagements of reciprocity can we enter into? How can we redistribute resources to social relations that actively build better worlds? Shifting from extractive- and accumulation-based practices invites relationship-building, reciprocity, the rematriation of Indigenous lands and waters, and centering responsibility and solidarities to each other. It means developing economic systems of interacting with each other and the land that both sustain life *and* allow us to make conscious decisions about the unevenly distributed costs of doing so.

Understanding the ecology of photography and its possibilities in the context of climate breakdown requires holding the complex material realities that underpin the image in tension with the ways of seeing it makes possible. If we want to address the challenges of the present, the past holds so many potential answers for us. We can learn resilience and resistance to colonial-capitalist modes of hyperextraction by entering into better relations with our own histories and our own responsibilities in the present. Recognizing the patterns and values that underlie our relationship with the natural world is the first step to transforming our relations.

NOTES

Introduction

Parts of the introduction have been revised from "Mining the History of Photography," in *Capitalism and the Camera*, ed. Kevin Coleman and Daniel James (New York: Verso, 2021), 55–73.

1   Hoskins, "People Like Us," 17.

2   Marien, "Imaging the Corporate Sublime," 1.

3   Sawyer, "Woodruff v. North Bloomfield Gravel Mining Co."

4   Hult-Lewis, "The Mining Photographs of Carleton Watkins," 172–74.

5   Hoskins, "People Like Us," 14.

6   Scott, *Photography and Environmental Activism*, 40.

7   Marx, *Economic and Philosophic Manuscripts of 1844*, 37.

8   Azoulay, *The Civil Contract of Photography*, 129; Benjamin, "A Little History of Photography."

9   Holmes, "Doings of the Sunbeam."

10 Holmes, "Doings of the Sunbeam."

11 Gordon, *Book of Film Care*.

12 Dootson, *The Rainbow's Gravity*, 128.

13 Eastman Kodak Company, *The Home of Kodak*, 23. Writing on the importance of photographic paper includes Mintie, "Material Matters"; and Messier, "Image Isn't Everything."

14 Sheppard, *Gelatin in Photography*, 25.

15 Rogers, *Chemistry of Photography*, 17.

16 Mees, "The History of Sensitisers," 104.

17 Shukin, *Animal Capital*, 104.

18 Lovejoy, "Celluloid Geopolitics."

19 Eastman Kodak Company, *The Kodak Park Works*.

20 "Gun Cotton and Collodion," *Scientific American*, 109.

21 Holmes, "The Stereoscope and the Stereograph."

22 Schaefer, "Photographic Ecologies," 45.

23 Belting, *The Invisible Masterpiece*.

24 Holmes, "The Stereoscope and the Stereograph."

25 Mumford, *Technics and Civilization*, 25.

26 Holmes, "The Stereoscope and the Stereograph."

27 International Resource Panel, *Global Resources Outlook 2019*, 7

28 Hellweg is the coauthor of a comprehensive study produced by the United Nations on climate change. Watts, "Resource Extraction Responsible for Half World's Carbon Emissions."

29 Gómez-Barris, *The Extractive Zone*.

30 As curator Royce K. Young Wolf (Hiraacá, Nu'eta, and Sosore) explains, materiality and representation disconnected from land are incomplete histories. Young Wolf's careful reconnecting of objects in museum collections with the material, embodied histories of land is an important addition to art historical methodologies, and conversations with Young Wolf have influenced my thinking on photographs as emplaced objects. Throughout, I attempt to resituate objects within the landscapes of their making, though some attempts are more successful than others.

31 Coleman and James, *Capitalism and the Camera*, 11.

32 Malm, *Fossil Capital*.

33 Wrigley, "The Supply of Raw Materials in the Industrial Revolution"; Mumford, *Technics and Civilization*, 156–57.

34 Ziolkowski, *German Romanticism and Its Institutions*, 23.

35 Marx and Engels, *The Communist Manifesto*, 225.

36 Taylor, "Auras and Ice Cores," 77.

37 Ware, *Mechanisms of Image Deterioration in Early Photographs*.

38 "Fixing and Washing Silver Prints," *Photographic News*, 150.

39 For an analysis of extraction, ways of seeing, and world-making, see Gómez-Barris, *The Extractive Zone*; de la Cadena, "Uncommoning Nature"; Goffe, "Human Resources."

40 Rudwick, "The Emergence of a Visual Language for Geological Science"; Sekula, "Photography between Labor and Capital," 203.

41 Agricola, *De Re Metallica*, 7.

42 Merchant, *The Death of Nature*, 29.

43 Agricola, *De Re Metallica*, 19.

44 Agricola, *De Re Metallica*, 19–20. Agricola's treatise influenced the experiments of Georg Fabricius, who first identified silver chloride, though Fabricius did not observe the impact of light on silver chloride. Eder, *History of Photography*, 25.

45 Mumford, *Technics and Civilization*, 74–77.

46 Lorke, *Liminal Semiotics*.

47 For instance, Johann Wolfgang von Goethe's experience in silver and copper mines inspired the poem *Ilmenau*. Ziolkowski, *German Romanticism and Its Institutions*, 33.

48 E. Miller, *Extraction Ecologies*, 4.

49 Blake, *Milton*.

50 Gómez-Barris, *The Extractive Zone*, xvi.

51 Coulthard, *Red Skin, White Masks*, 14.

52 Lisa Lowe traces the connections between European liberalism, settler colonialism in the Americas, the transatlantic African slave trade, and the imperial trade with the East Indies and China in the late eighteenth and early nineteenth centuries, showing how "differentially situated histories of indigeneity, slavery, industry, trade, and immigration give rise to linked, but not identical, genealogies of liberalism." Lowe directs attention to "relation across differences" and "the convergence of asymmetries." Following Lowe, I want to think about these histories as entangled but not equivalent. Lowe, *The Intimacy of Four Continents*, 11.

53 Some might conceptualize this using Heidegger's ideas around "the conquest of the world as picture" as well as the understanding of nature as a standing reserve of resources. See Heidegger, "The Age of the World Picture," 67.

54 Mirzoeff, "Visualizing the Anthropocene," 217.

55 W. Mitchell, *The Last Dinosaur Book*.

56 Sontag, *On Photography*, 14; Solomon-Godeau, "Who Is Speaking Thus?," 181. Scholars have extended this analysis to consider the links between

photography and hunting. See, for instance, Braddock, "Poaching Pictures"; Brower, "Trophy Shots"; Ronan, "Capturing Cruelty."

57 Szeman and Wenzel, "What Do We Talk about When We Talk about Extractivism?" See also Raab, "Landscape and the Risk of Metaphor."

58 Gómez-Barris, *The Extractive Zone*, 1; Riofrancos, *Resource Radicals*.

59 Azoulay, *The Civil Contract of Photography*, 129; Benjamin, "A Little History of Photography," 512.

60 Gilmore, *Abolition Geography*, 79.

61 The often-contradictory aims of national parks and their shifting relationships to the natural world have been explored by Berger, "Overexposed"; M. Barringer, *Selling Yellowstone*; Campbell, *A Century of Parks Canada*; Cronin, *Manufacturing National Park Nature*; McLaren, *Culturing Wilderness in Jasper National Park*; Gaudet, "Patterns of Possession."

62 Kelsey, "Sierra Club Photography and the Exclusive Property of Vision," 12.

63 Hodgins and Thompson, "Taking the Romance out of Extraction," 395.

64 Demos, "The Agency of Fire."

65 Tiffany Lethabo King uses the concept of ecotone to think through shoals, formations that are "liminal, indeterminate, and hard to map," locating potential in the friction produced by these indeterminate spaces and the new topographies they shape. King, *The Black Shoals*, 3–4.

66 Marx, *The German Ideology*, 36.

67 Marx, *The German Ideology*, 26.

68 Benjamin, "Theses on the Philosophy of History," 254.

69 Ray 2015, "Interview with Simon Starling."

70 Benjamin, "Theses on the Philosophy of History."

71 Benjamin, "A Little History of Photography," 511.

72 Berger, *About Looking*, 40.

73 Brecht, "The Threepenny Lawsuit," 469.

74 Brecht, "A Short Organum for the Theatre," 192.

75 Fisher, *Capitalist Realism*, 2.

76 Brecht, "A Short Organum for the Theatre," 194.

77 Gilmore, *Abolition Geography*, 475.

78 Nixon, *Slow Violence*.

79 Lee, *Forgetting the Art World*, 19, 25.

80 Sekula, "Between the Net and the Deep Blue Sea," 33.

81 Stephanie LeMenager shows how the cheap energy facilitated by oil made the expansion of the US and Canadian middle classes possible, and many

things considered central to liberal values like public education and activist movements like feminism, antiwar activism, and environmentalism are underpinned by access to cheap energy. LeMenager, *Living Oil*, 3.

82   I use *responsibility* in the sense of what Donna Haraway calls being "response-able": having "the capacity to respond." Haraway, *Staying with the Trouble*, 78.

Chapter 1. Bitumen and a Reorientation of Vision

1   Treaty 8 (between the Crown and First Nations of the Lesser Slave Lake area) was initiated following the discovery of gold in the nearby Yukon and the subsequent Klondike Gold Rush, and the boundaries of the treaty reflect mining regions. It has a different history than the other numbered treaties in the Canadian West, which emerged following the collapse of the bison population as Indigenous communities in the West faced starvation, compounded by successive waves of disease. Beginning in 1878, the dominion under Prime Minister John A. Macdonald pursued a policy of state-supported starvation to force Indigenous communities to sign what came to be known as the numbered treaties. On what eventually became Treaty 8 lands, Indigenous communities in the region requested a treaty but the region was considered too northern to be suitable for settlement or development (see Daschuk, *Clearing the Plains*). After the discovery of oil in 1891, the Crown began to prepare a treaty, but it was not implemented. It was with the gold rush and the influx of white prospectors into the region that the government initiated the treaty process. As a retired Indian agent explained to Clifford Sifton, superintendent general of Indian Affairs, "They will be more easily dealt with now than they would be when their country is overrun with prospectors and valuable mines be discovered. They would then place a higher value on their rights" (Hall, *Clifford Sifton*, 272).

2   Leahy, "This Is the World's Most Destructive Oil Operation—and It's Growing."

3   Sierra Club Canada Foundation, "Tar Sands."

4   Niépce, "Heliography."

5   Cariou, "Petrography."

6   This photograph is the oldest camera-made photograph, but a photomechanical reproduction of a seventeenth-century Flemish print was made in 1825.

7   Benjamin, "The Work of Art in the Age of Mechanical Reproduction."

8   Mathiot, "The Capability of Photography," 46.

9   Barrett and Worden, "Introduction," xix.

10   See, for instance, Ghosh, "Petrofiction"; Wilson, Carlson, and Szeman, *Petrocultures*; and LeMenager, *Living Oil*.

11   The use of bitumen in mummification processes likely reflected changing religious beliefs and funerary practices as bitumen was not used consistently in embalming but it was associated with sacredness and divinity. See Clark, Ikram, and Evershed, "The Significance of Petroleum Bitumen in Ancient Egyptian Mummies," 2.

12   Temple, *The Genius of China*, 54.

13   Bitumen was discovered in the Athabasca region in 1848, but production didn't begin until 1967 with the founding of the Great Canadian Oil Sands Company. The Canadian state funded decades of research to find commercial uses for bitumen, including developing asphalt, roof tiling products, and, eventually, refining bitumen into a shippable form of crude. The decades-long investment in making the extraction of this oil viable suggests that there is nothing inevitable about the large-scale extraction of bitumen as an energy source. Spady and Angus, "Histories of the Present."

14   Barnes, *Keeping an Eye Open*, 39.

15   Finlay, *Color*, 104–6.

16   Waterhouse, "Nicéphore Niépce's Early Photographic Work with Bitumen," 145.

17   Waterhouse, "Nicéphore Niépce's Early Photographic Work with Bitumen," 145. Waterhouse writes that Niépce may have tried some local bituminous products before turning to bitumen of Judea.

18   Waterhouse, "Photo-Lithography and Photo-Zincography," 587.

19   Waterhouse, "Photo-Lithography and Photo-Zincography," 725.

20   Waterhouse, "Photo-Lithography and Photo-Zincography," 529.

21   Waterhouse, "Photo-Lithography and Photo-Zincography," 588.

22   Waterhouse, "Photo-Lithography and Photo-Zincography," 479.

23   Falconer, *The Waterhouse Albums*.

24   Waterhouse, "Photo-Lithography and Photo-Zincography," 479.

25   Ruffles, "Waterhouse, James," 1475.

26   Todhunter, *William Whewell*, 125.

27   Yusoff, *A Billion Black Anthropocenes or None*.

28   Wilson, Carlson, and Szeman, *Petrocultures*, 6.

29   Cirlot, *A Dictionary of Symbols*, 57.

30   Huber, *Lifeblood*, 70.

31 H. Davis, *Plastic Matter*, 74.

32 McLaurin, *Sketches in Crude Oil*.

33 T. Mitchell, *Carbon Democracy*.

34 LeMenager, *Living Oil*, 4–6.

35 McLaurin, *Sketches in Crude Oil*, 372.

36 Chakrabarty, "The Climate of History," 208.

37 LeMenager, *Living Oil*, 3.

38 Wilson, Carlson, and Szeman, *Petrocultures*, 6.

39 Smil, *How the World Really Works*, 42.

40 Smil, *How the World Really Works*, 42–43.

41 Rich, "Everything You Thought You Knew, and Why You're Wrong."

42 Tsēmā, "Artist Statement."

43 Tsēmā, "Black Gold."

44 LeMenager, *Living Oil*, 7.

45 For an analysis of oil and life, see LeMenager, *Living Oil*; Huber, *Lifeblood*.

46 Nickel, "Talbot's Natural Magic"; Cirlot, *A Dictionary of Symbols*, 317–19.

47 Batchen, "The Naming of Photography."

48 Holmes, "The Doings of the Sunbeam," 13.

49 Miles, "The Burning Mirror."

50 Batchen, "The Naming of Photography," 24.

51 Batchen, "The Naming of Photography," 28.

52 Recent studies have nuanced the emphasis on solar light. See, for instance, Flint, *Flash!*; and Dinkar, "'Our Best Machines Are Made of Sunlight."

53 Talbot, *The Pencil of Nature*.

54 S. Edwards, *The Making of English Photography*, 36.

55 Hutton, *Theory of the Earth*, 200.

56 J. Peters, "Space, Time, and Communication Theory," 402.

57 Peirce, *The Collected Papers of Charles Sanders Peirce*, 391.

58 Batchen, "Ectoplasm," 18.

59 Barthes, *Camera Lucida*, 87.

60 Barthes, "The Rhetoric of the Image," 44. As Edwards and Hart observe, Barthes's famous description of the Winter Garden photography engages with the image as a material object, "a photograph that carries on it the marks of its own history, of its chemical deterioration" ("Photographs as Objects," 1).

61 S. Edwards, *Photography*, 119.

62   Michaels, "Photographs and Fossils," 431; Parikka, *A Geology of Media*.

63   Mirzoeff, *How to See the World*, 22.

64   Sugimoto, "Pre-Photography Time-Recording Device."

65   Sugimoto, "Pre-Photography Time-Recording Device."

66   Batchen, *Burning with Desire*, 183.

67   Cadava, *Words of Light*, 5.

68   Wilder and Kemp, "Proof Positive," 363.

69   Zylinska, *Nonhuman Photography*, 104, 111.

70   Langley, "The Sun's Energy," 28.

71   Michael Robinson suggests that camera-generated heliographs could require exposure times of several days. Robinson, "The Techniques and Material Aesthetics of the Daguerreotype," 76.

72   Gaudreault, "One and Many," 32.

73   Silverman, *The Miracle of Analogy*, 59, 60.

74   E. Edwards, "Photographs as Objects of Memory," 230.

75   The haptic encounter with the object has been a subject of much research in photography studies; for instance, Campt, *Image Matters*; Olin, *Touching Photographs*; Edwards and Hart, *Photographs Objects Histories*.

76   Albers, *The Night Albums*, 109.

77   Silverman, *The Miracle of Analogy*, 104.

78   Both Gernsheim and the Getty technicians suggest that the underexposure of the original plate is the cause of the image's irreproducibility.

79   Pinson, *Speculating Daguerre*, 12.

80   Here, we can think of Barthes's discussion of photography's double temporality as well as Esther Gabara, who argues that double exposures challenge the viewer through the merging of multiple temporal moments, reorienting attention to multiplicity. See Barthes, *Camera Lucida*; Gabara, *Errant Modernism*, 114–15. I use simultaneity more broadly to acknowledge the coexistence and entanglement of multiscalar levels of experience.

81   Brecht, "On Form and Subject-Matter," 30.

82   Hunter, *Painting with Fire*, 41, 174–75.

83   Marx, *Capital*, 163.

84   Heise, *Sense of Place and Sense of Planet*.

85   Roberts, *Scale*, 12.

86   Farrier, "How the Concept of Deep Time Is Changing."

87   Burtynsky, "Life in the Anthropocene," 190.

88 Burtynsky, "Oil: Artist's Statement."

89 AWG research categories are Anthroturbation (i.e., disturbance of the crust of the earth, as in mining), Species Extinction, Technofossils, Boundary Limits, and Terraforming.

90 Nixon, *Slow Violence*, 14.

91 Zuromskis, "Petroaesthetics and Landscape Photography," 292.

92 The Anthropocene has entered into somewhat common usage in the sciences and humanities, but its implications and theoretical commitments have been criticized by many scholars. In ascribing responsibility to humans as a universal category, the systems of power that produced climate crisis and the unequal burdens communities have faced are erased. Others have criticized the hubris of naming a geologic era after humans, as being a continuation of the anthropocentrism at the heart of climate crisis. Several alternatives to the Anthropocene have been proposed, the most prominent of which is the Capitalocene, which was first used by Andreas Malm and popularized by Jason W. Moore. Other terms include the Plantationocene, the White Supremacy Scene, and the Misanthropocene. Donna Haraway has helpfully suggested we consider the Anthropocene a boundary, not an epoch. For an analysis of how capitalism has fueled climate change, see Demos, *Against the Anthropocene*; Haraway, "Anthropocene, Capitalocene, Plantationocene, Chthulucene"; Klein, *This Changes Everything*; Malm, *Fossil Capital*; Moore, *Capitalism in the Web of Life*. Other scholars have documented the disproportionate burdens faced by low-income, racialized, and Indigenous communities in the face of climate change. See Davis and Todd, "On the Importance of a Date"; Mirzoeff, "It's Not the Anthropocene"; Yusoff, *A Billion Black Anthropocenes or None*.

93 Burtynsky's work is often used to critique extraction; for instance, it is on the cover of Coleman and James, *Capitalism and the Camera*. Burtynsky is also well represented, however, in corporate collections: guided tours of the Bank of Montreal Corporate Art Collection in Toronto begin with *Silver Lake Operations #1 and 3, Lake Lefroy*, 2007. The tour guide of the BMO collection described why they began the tour of the corporate collection with this image: "We like to remind ourselves of where all of this comes from." While this observation is literally true—mining provides the raw materials for banking—in this context, Burtynsky's indictment of industrial development becomes abstracted into formal beauty and distanced contemplation. For an analysis of Burtynsky's work, see, for example, Emery, "The Mirror and the Mine"; Hackett, Kunard and Stahel, *Anthropocene*; Schuster, "Between Manufacturing and Landscapes."

94 Sekula, "Reading an Archive," 448.

95   Ghosh, "Petrofiction," 30–31.

96   Sekula, *Fish Story*, 12.

97   Szeman and Wenzel, "What Do We Talk about When We Talk about Extractivism?," 7.

98   Nixon, *Slow Violence*, 41.

99   Coleman and James, *Capitalism and the Camera*, 10.

100   Cowen, *The Deadly Life of Logistics*, 8.

101   Galison and Jones, "Representations of Oil Spills."

102   Thank you to Patrick DeDauw for this formulation.

103   Haraway, *Staying with the Trouble*, 1.

104   Mackert, "Sisyphus's Prestige," 153.

105   Energy Mix, "Canada Boosts Fossil Subsidies to $14.3B Per Year."

106   LeMenager, "Petro-Melancholia," 31.

107   Ahooja, "Alberta."

108   Barthes, *The Neutral*, 51.

109   Todd, "Fish, Kin and Hope," 104.

110   The term "fossil fuel" was first used in 1759, just before the invention of "economy" and "photography." Coleman and James, *Capitalism and the Camera*, 2.

111   Liboiron, *Pollution Is Colonialism*, 7.

112   See, for example, Azoulay, *Civil Contract of Photography*; Campt, *Image Matters*; Knight, "A Relational Ecology of Photographic Practices"; and D. Palmer, *Photography and Collaboration*.

113   Cariou and Gordon, "Petrography, the Tar Sands Paradise, and the Medium of Modernity," 13.

114   Le Guin, "The Carrier Bag Theory of Fiction."

115   Kimmerer, *Braiding Sweetgrass*, 26.

116   Kimmerer, *Braiding Sweetgrass*, 26.

117   Cariou, "Portfolio," 254.

118   Cariou, "Portfolio," 253.

Chapter 2. Silver and Scale

1   The US Geological Survey places the number at 28 percent in 1999, the year that the demand for silver peaked in photographic industries, while D. S. Cameron suggests that photographic industries account for roughly 60 percent of all silver produced, though this number seems unusually

high. See Butterman and Hilliard, "Mineral Commodity Profiles"; and Cameron, "Recycling."

2   Eastman Kodak Company, "Silver," 17.

3   Eastman Kodak Company, "It's Pure Silver That 'Gets the Picture,'" 45.

4   Eastman Kodak Company, "Silver," 17.

5   Nye, *American Technological Sublime*, 9.

6   Bratter, "A Survey of Silver: Part I," 582.

7   Hopkins, "The Gilded Gaze," 225–26.

8   Marx, *A Contribution to the Critique of Political Economy*, 60.

9   Marx, *A Contribution to the Critique of Political Economy*, 158.

10   Around 75 percent of silver is extracted as a byproduct of more commonly found materials such as lead, zinc, or copper. Nassar et al., "By-Product Metals," 1.

11   del Mar, *The History of Precious Metals*, 10.

12   Bratter, "A Survey of Silver: Part II," 704.

13   Marx reflects on the complex relationship between metals, currency, and value by showing how economists' currency theory "misunderstand[s] the function that precious metals perform." The emphasis on "money only as a crystalline product of circulation" misses the "fluid form" of money, and a contradiction is revealed as "the Catholic fact that gold and silver as the direct embodiment of social labour, and therefore as the expression of abstract wealth, confront other profane commodities, has of course violated the protestant code of honour of bourgeois economists, and from fear of the prejudices of the Monetary System, they lost for some time any sense of discrimination towards the phenomena of money circulation" (159). For more on circulation and currency, see Marx, *A Contribution to the Critique of Political Economy*, 64–98.

14   Marx, *A Contribution to the Critique of Political Economy*, 154.

15   Thank you to Jennifer Raab for this observation.

16   Both Talbot and Daguerre immediately circulated their early experiments, revealing that photography was "always conceived as mobile media." Pelizzari and Siegel, "Circulating Photographs," 1.

17   A. Smith, *The Wealth of Nations*, 225.

18   In the Inca Empire, the mit'a was a form of mandatory public service and community obligation. Between 1573 and 1812, the mit'a was adapted by the Spanish as a requisition of Andean labor, often in the silver mines in Potosí and mercury mines in Huancavelica. Under the mit'a, over two hundred Indigenous communities were required to send one-seventh of

the adult male population for forced labor. Cummins, "Silver Threads and Golden Needles," 3.

19   Bratter, "A Survey of Silver: Part I," 583.

20   del Mar, *A History of Precious Metals*, 65. Here, del Mar shows that of the $6.5 billion in silver and gold acquired by Europe before 1810, $6 billion of that was mined by enslaved labor or was seized through plunder. From 1810 to 1880, $13 billion worth were mined, of which $4.5 billion was the result of forced or coerced labor. This has implications for the value of gold and silver: the value of metals cannot be directly tied to the cost of their production, for, as del Mar concludes, how could one account for the "Crimes against Mankind" that characterized the majority of minerals extracted? Del Mar, *A History of Precious Metals*, 447–48.

21   Lane, *Potosí*, 5.

22   Davis and Todd, "On the Importance of a Date," 764. Amitav Ghosh has similarly emphasized the desire to accumulate at the root of Western capitalist-colonialism, tracing a longer history of ecological crisis that predates the Industrial Revolution or the Great Acceleration of the 1950s. Ghosh, *The Nutmeg's Curse*.

23   Whyte, "Is it Colonial Déjà Vu?"

24   Lane, *Potosí*, 9–10.

25   Farocki, *The Silver and the Cross*; Igoe, "Creative Matter," 148.

26   Mullaney, "The New World on Display," 108–9.

27   Azoulay, "Toward the Abolition of Photography's Imperial Rights," 27.

28   Azoulay, "Toward the Abolition of Photography's Imperial Rights," 28.

29   Azoulay, "The Captive Photograph."

30   Azoulay, *The Civil Contract of Photography*, 24.

31   Marx, *Capital*, 915.

32   Azoulay, "Toward the Abolition of Photography's Imperial Rights," 28. For a theorization of photography as enacting primitive accumulation, see Andermann, *The Optic of the State*.

33   As Alexander del Mar notes, having plundered the Caribbean and Central and South America, the conquistadors appeared to exhaust the supply of metals and seemed likely to abandon the countries they had devastated. However, the discovery of silver at Potosí and subsequent finds in Mexico "sealed [their] fate." del Mar, *A History of the Precious Metals*, 153.

34   Rosenblum, *A World History of Photography*, 193.

35   A. Shapiro, *Fits, Passions, and Paroxysms*, 255.

36   Kelsey, "Photography and the Ecological Imagination," 394.

37   Marx, *A Contribution to the Critique of Political Economy*, 155.

38   According to Benjamin White, the three most significant silver discoveries in history are the sixteenth-century mines in Peru and Mexico, the Comstock Lode in the nineteenth century, and Cobalt, Canada, in the early twentieth century. White, *Silver*, 7. I have written about Cobalt in Angus, "Mining the History of Photography."

39   See, for example, Trachtenberg, *Reading American Photographs*; Sandweiss, *Print the Legend*; Kelsey, *Archive Style*; Snyder, *American Frontiers*; Snyder, *One/Many*.

40   G. Smith, *The History of the Comstock Lode*, 26, 33.

41   The King and Wheeler surveys were part of a series of US War Department surveys in the American West between 1867 and 1879, alongside Ferdinand Vandiveer Hayden's US Geological and Geographical Survey of the Territories and John Wesley Powell's US Geographical and Geological Survey of the Rocky Mountain Region. Wheeler's survey was preceded by expeditions between 1869 and 1871. Photographers—including William Henry Jackson (Hayden), Timothy H. O'Sullivan (King and Wheeler), Carleton Watkins (King), E. O. Beaman (Powell), and John K. Hillers (Powell)— played an important role in these surveys. Later, these separate surveys were reorganized as the United States Geological Survey.

42   Samson, "Photographs from the High Rockies," 474. While the *Harper's* article is attributed to John Samson, Robin Kelsey argues that the narrative parallels O'Sullivan's experiences and it is likely that O'Sullivan either wrote the essay or provided information for it. Kelsey, *Archive Style*, 123.

43   Sekula, "Photography between Labor and Capital," 227.

44   Vein mining didn't begin in California until 1860 and wasn't a significant form of extraction until the 1880s. del Mar, *A History of Precious Metals*, 399.

45   Dynamite was patented in 1867 by Alfred Nobel. Nobel's invention was intended to make blasting safer for workers, but explosives manufacturing was not regulated in this period, and the strength and stability of explosives and fuses were unpredictable. Explosives remained the leading causes of mining accidents.

46   G. Smith, *The History of the Comstock Lode*, 24.

47   White, *Silver*, 52.

48   Lawrence, "Mercury in the Carson River Basin, Nevada." Cyanide leaching replaced mercury amalgamation by 1901.

49   Peck, "Manly Gambles."

50   James and James, *A Short History of Virginia City*, 52.

51 While the first strike in 1863 was met with military force ordered by Governor James Nye, the unions of Virginia City, Gold Hill, and Silver City ultimately built power by establishing community support and electing union members to legislative and law enforcement positions. The centrality of mining to the region gave workers considerable power, and they won concessions on wages, health and safety, and the right to organize.

52 Kelsey, *Archive Style*, 120.

53 The protomodernist visual forms of O'Sullivan's images have been analyzed by art historians. See, for example, Krauss, "Photography's Discursive Spaces"; Kelsey, "Viewing the Archive." More recently, Elizabeth Keto has drawn attention to the prominent picturing of water in the Wheeler survey. See Keto, "Engineering Vision."

54 Sekula, "Photography between Labor and Capital," 228.

55 Kelsey, *Archive Style*, 127.

56 Kelsey, *Archive Style*, 127.

57 Holmes, "The Doings of the Sunbeam"; Sekula, "The Traffic in Photographs"; Trachtenberg, *Reading American Photographs*, 18; Grigsby, *Enduring Truths*, 144; Roberts, *Transporting Visions*; Crary, *Techniques of the Observer*, 13; Gründig, "Ten Dollar Faces."

58 Harris, "Inventors and Manipulators."

59 Kelsey, "Viewing the Archive," 720.

60 While tracing a direct link between extraction and photography is complicated by the structure of precious-metal supply chains, there are a number of eagerly anticipated book projects in process that tackle this question. Fionn Montell-Boyd has productively explored silver supply chains and speculative networks through a focus on Talbot. See Montell-Boyd, "Speculative Projections." Monica Bravo's in-process book project *The Silver Pacific* traces the links between extraction and photography in California, https://www.acls.org/fellow-grantees/monica-bravo/.

61 These numbers were compiled by Dr. Adolf Soetbeer and R. W. Raymond, the director of the United States Mint. Raymond also totaled the commercial value of silver based on average rates at the New York market. Reproduced in White, *Silver*, 59–60.

62 Manning, Flynn, and Wang, "Silver Circulation Worldwide," 7.

63 Manning, Flynn, and Wang, "Silver Circulation Worldwide," 10.

64 Bratter, "A Survey of Silver: Part I," 582.

65 White, *Silver*, ix.

66 Bratter, "A Survey of Silver: Part I," 582; White, *Silver*, 166, 96.

67   Bratter, "A Survey of Silver: Part II," 712.

68   Marx, *A Contribution to the Critique of Political Economy*, 163.

69   Bratter, "A Survey of Silver: Part II," 712.

70   G. Smith, *The History of the Comstock Lode*, 60.

71   Twain, *Roughing It*, 220.

72   For an overview of the history of silver as currency, see Silber, *The Story of Silver*; and for an analysis of the impact of the gold standard on cultural conceptions of value, see Michaels, *The Gold Standard and the Logic of Naturalism*; and Rotman, *Signifying Nothing*.

73   For most of the nineteenth century, the ratio was closer to 15:1. The price of silver continued to fluctuate in the twentieth century, but the price remained steady around 53 percent that of gold from 1908 to 1914. China and India remained the largest purchasers of silver. See Baldwin, *Cobalt*, 52.

74   Bratter, "A Survey of Silver: Part II," 716.

75   West, *Kodak and the Lens of Nostalgia*, 2.

76   Holmes, "The Doings of the Sunbeam," 1.

77   Marx, *Capital*, 279–80.

78   Lugon, "Photography and Scale," 388.

79   Nye, *American Technological Sublime*, 9–10.

80   Weems, "Scale, a Slaughterhouse View," 111.

81   Steve Edwards writes that Marx included photography in his list of industries, arguing that photography's historical place lies within the new forms of production that emerged with industrialization. S. Edwards, *The Making of English Photography*, 2.

82   Holmes, "The Doings of the Sunbeam," 2.

83   This number includes still and moving image film. Eastman Kodak Company, *Kodak Park*, 25.

84   Butterman and Hilliard, "Mineral Commodity Profiles," 16; Eastman Kodak Company, "How Kodak Film Is Made," 7.

85   PRO-ACT, *Silver Recovery from Photographic and Imaging Wastes*.

86   Eastman Kodak Company, "Silver—Metal, Money, and Mixture," 3.

87   Eastman Kodak Company, "Silver—Metal, Money, and Mixture," 3.

88   Bratter, "A Survey of Silver: Part I," 582.

89   Bratter, "A Survey of Silver: Part I," 588.

90   "Eastman Kodak Is Increasing Prices 6.5%," *New York Times*, 51.

91   "Saving the Silver," *Photographic News*, 324. On methods for recovery, see, for example, Mahley, "On the Expenditure of Silver Used in Photographic

Operations," 245; "Recovering Silver from Residues," *Photographic News*; "Intense Negatives and Saving Silver," *Photographic News*, 558.

92   Silver has a recycling efficiency rate of 97 percent, making silver recovery viable in both photographic industries and other industrial applications. Silver has always been melted down and recycled, but on average, spanning the nineteenth to twenty-first centuries, recycling rates have ranged from 10 to 15 percent. Sibley, *Overview of Flow Studies for Recycling Metal Commodities in the United States*, 13.

93   Eastman Kodak Company, "Silver," 17.

94   Kodak Professional, "Silver Halide Photographic Paper," 11.

95   Eastman Kodak Company, "It's Pure Silver That 'Gets the Picture,'" 45.

96   Eastman Kodak Company, *The Kodak Park Works*, 5.

97   Maxwell and Miller, *Greening the Media*, 73.

98   Maxwell and Miller, *Greening the Media*, 74

99   Kelsey, "Photography and the Ecological Imagination."

100   Povinelli, "Fires, Fogs, Winds," 507.

101   Rosenblum, *A World History of Photography*, 194.

102   Pelletan, "Discovery by M. Daguerre," 1–2.

103   Sekula, "An Eternal Esthetics of Laborious Gestures," 25.

104   Coleman, James, and Sharma, "Photography and Work," 8.

105   S. Edwards, *The Making of English Photography*, 31.

106   S. Edwards, *The Making of English Photography*, 36.

107   Sekula, "The Traffic in Photographs," 23; Marx, *Capital*, 165.

108   Wilder, "Photography and the Art of Science," 166.

109   Coleman, James, and Sharma, "Photography and Work," 8.

110   Brecht, "The Threepenny Lawsuit," 469.

111   Benjamin, "A Little History of Photography."

112   Palmquist and Kailbourn, *Pioneer Photographers from the Mississippi to the Continental Divide*, 174.

113   Lebart, *Gold and Silver*.

114   Hult-Lewis, *Carleton Watkins*, 1.

115   The oldest surviving film was by Louis Le Prince called *Roundhay Garden Scene* (1888), but the Lumière brothers' film was the first to be publicly screened.

116   Farocki, "Workers Leaving the Factory," 6.

117   Farocki, "Workers Leaving the Factory," 6.

118 Farocki, "Workers Leaving the Factory," 2.

119 Cubitt, *Anecdotal Evidence*, 235.

120 Emery, "The Mirror and the Mine."

Chapter 3. Platinum and Atmosphere

1 For an analysis of South African photographers working under Apartheid, see Peffer, *Art and the End of Apartheid*; and Hayes, "Photographic Publics and Photographic Desires in 1980s South Africa."

2 Rajak, "Platinum City and the New South African Dream," 252.

3 Trangos and Bobbins, "Gold Mining Exploits and the Legacies of Johannesburg's Mining Landscapes."

4 Greenough, "Nescafé, Surlyn, and the Alchemy of Photography," 13.

5 Greenough, "Nescafé, Surlyn, and the Alchemy of Photography," 10. My thinking on photography's temporalities as they relate to materiality is indebted to Paul Messier.

6 See, for instance, J. Peters, *The Marvelous Clouds*; Horn, "Air as Medium"; Cheetham, "'Atmospheres' of Art and Art History"; Taylor, *The Sky of Our Manufacture*; Robbins, "Atmospheric Regulation in the Panorama"; Hyde, "'London Particular'"; and Arscott, "Subject and Object in Whistler."

7 Cheetham, "'Atmospheres' of Art and Art History."

8 Willis, "A New Process of Photo-Chemical Printing in Metallic Platinum," 397.

9 In photographic processes, there is some slippage in terminology between platinum and palladium. There are significant differences between the two metals, but their application in photography is similar, so they are both explored in this chapter under the broader heading of platinum.

10 Weeks, *Discovery of the Elements*, 385.

11 Scerri, *A Tale of 7 Elements*, xxvii.

12 Ross, "Facts about Platinum."

13 World Platinum Investment Council, "Platinum Facts."

14 JM Bullion, "Historic Gold, Silver, Platinum, & Palladium Price Spikes."

15 Stulik and Kaplan, "Platinotype."

16 Ware, "Light-Sensitive Chemicals," 857.

17 Ware, *Platinomicon*, 26.

18 "The New Platinum Printing Process," *British Journal of Photography*, 265.

19 "Proceedings of Societies," *Photographic News*, 322. While platinum was chemically stable, impurities in the paper were an ongoing concern, as

trace elements of other metals in paper caused yellowing. Ware, *Platinomicon*, 31.

20 E. Edwards, "Commemorating a National Past," 125.

21 Ware, "The Eighth Metal."

22 Ware, "The Eighth Metal."

23 Hinton, *Platinotype Printing*, 10.

24 Kingsley, "Art Photography and Aesthetics," 80.

25 Ware, "The Eighth Metal."

26 "The Nature of the Metals," *Photographic News*, 89.

27 "Cost of Chemicals Fifty Years Ago," *Photographic News*, 468.

28 Thomas, "The RSC Faraday Prize Lecture of 1989 on Platinum," 9192.

29 Prior to this, nitric acid was produced from imported saltpeter. The entwined histories of photography and nitrate are explored by Xavier Ribas, Louise Purbrick, and Ignacio Acosta in the ongoing project "Traces of Nitrate." Wilhelm Ostwald also developed a reproduction process that used catalysis to make reproducible prints from platinotypes. Ware, *Platinomicon*, 64–65.

30 Ware, *Platinomicon*, 66.

31 Willis developed a process that added silver to the platinum, named *satista*, from the Latin for "good enough," reflecting the reality that adding silver was perceived to weaken the prints. The process was not popular. Ware, *Platinomicon*, 68.

32 Ware, *Platinomicon*, 17.

33 Stieglitz, "Platinum Printing," 87. Stieglitz prefigures Eva Horn's suggestion that air functions as a medium. Horn, "Air as Medium."

34 Hinton, *Platinotype Printing*, 10.

35 Kelsey, *Photography and the Art of Chance*, 103.

36 Ware, *Mechanisms of Image Deterioration in Early Photographs*.

37 "Fixing and Washing Silver Prints," *Photographic News*, 150 (my emphasis).

38 Beil, *Good Pictures*, 23–24.

39 Beil, *Good Pictures*, 23–24.

40 Beil, "Photography Has Gotten Climate Change Wrong from the Start."

41 Ware, *Mechanisms of Image Deterioration in Early Photographs*.

42 Ruskin, *The Storm-Cloud of the Nineteenth Century*, 58.

43 Ruskin, *The Storm-Cloud of the Nineteenth Century*, 58.

44 "The Employment of Photography in Meteorological Science," *Photographic News*, 469.

45   Ruskin, *The Storm-Cloud of the Nineteenth Century*, 61.

46   Robbins, "Ruskin, Whistler, and the Climate of Art in 1884"; T. Barringer, "Introduction—John Ruskin."

47   J. Davis, *The Birth of the Anthropocene*, 99; Mirzoeff, "Visualizing the Anthropocene"; Gould, "The Polluted Textures of J.M.W. Turner"; Shields, *Impressionism in the Age of Industry*.

48   "Notes [July 10, 1885]," *Photographic News*, 440.

49   Dixon's lecture was part of the Cantor Lectures and was followed by a lecture by G. V. Poore, MD, on "Climate and Its Relation to Health," reflecting a concern with the impacts of burning coal. See "Notes [January 2, 1885]," *Photographic News*, 7.

50   "The Employment of Photography in Meteorological Science," *Photographic News*, 469.

51   Beil, *Good Pictures*, 87.

52   Burton, "The Whole Duty of the Photographer," 668.

53   Ware, *Platinomicon*, 10.

54   Thomas, "The RSC Faraday Prize Lecture of 1989 on Platinum," 9186.

55   Johnson Matthey, "Celebrating 200 Years of Inspiring Science," 22.

56   Thomas, "The RSC Faraday Prize Lecture of 1989 on Platinum," 9186.

57   Thomas, "The RSC Faraday Prize Lecture of 1989 on Platinum," 9192.

58   Willis, "A New Process of Photo-Chemical Printing in Metallic Platinum," 400.

59   Ware, "Platinotype Company," 1136.

60   Johnson Matthey, "Celebrating 200 Years of Inspiring Science."

61   Ware, "Platinotype Company," 1136.

62   Johnson Matthey, "Celebrating 200 Years of Inspiring Science," 6, 7.

63   Nisbet, "On Simon Starling, One Ton II (2005)," 185.

64   R. Williams, *The City and the Country*, 120.

65   DeLue, "Homer Dodge Martin's Landscapes in Reverse," 291.

66   Dixon, "The Perturbations of Drift in a Stratified World," 130–31.

67   Kimmerer, *Braiding Sweetgrass*, 341.

68   Emery, "Art of the Industrial Trace."

69   Goldblatt and Gordimer, *On the Mines*, 85.

70   My use of the term *racial capitalism* is informed by Cedric Robinson's *Black Marxism*.

71   CJPME Foundation, "Roots of Apartheid."

72   Arndt and Ganino, *Metals and Society*, 39.

73 Alexander et al., *Marikana*, 20.

74 Altman, "Upriver."

75 Nelson and Murray, "Silicosis at Autopsy in Platinum Mine Workers," 196.

76 Nelson and Murray, "Silicosis at Autopsy in Platinum Mine Workers," 197.

77 Oliver, *Dangerous Trades*, 15.

78 Tucker, "Dangerous Exposures," 130.

79 Hughes, "Medical Surveillance of Platinum Refinery Workers," 28. Palladium(II) is nonallergenic. See Heederik et al., "Exposure-Response Analyses for Platinum Salt–Exposed Workers and Sensitization."

80 Cristaudo et al., "Occupational Hypersensitivity to Metal Salts, Including Platinum, in the Secondary Industry."

81 Lesser and Weiss, "Art Hazards," 455.

82 "The Health of Photographers," *Photographic News*, 409.

83 See, for instance, Tataryn, *Dying for a Living*.

84 While the internal mutations of disease are largely invisible within the photographic frame, photographers have documented occupational health, including Eugene Smith and Aileen M. Smith's *Minamata* (1975) and Louie Palu's *A Field Guide to Asbestos* (2019), while Jo Spence's "Narratives of Dis-ease" charted her experience with cancer.

85 Environmental Justice Network, "Principles of Environmental Justice."

86 "The Pictures in This Number," *Camera Work*, 63.

87 Raygorodetsky, "Indigenous Peoples Defend Earth's Biodiversity."

88 Whyte, "Is It Colonial Déjà Vu?," 88.

89 Whyte, "Settler Colonialism, Ecology, and Environmental Justice," 129; see also Estes, *Our History Is the Future*.

90 Smithsonian National Museum of the American Indian, *Indelible*.

91 Smithsonian National Museum of the American Indian, *Indelible*.

92 Horn, "Air as Medium," 9.

93 Estes, *Our History Is the Future*, 257.

Chapter 4. Iron and Unstable Boundaries

1 Ware, "Light-Sensitive Chemicals," 857.

2 Benjamin, *The Arcades Project*, 167.

3 A. Palmer, *In the Aura of a Hole*, 86.

4 Iron fertilization is used to stimulate algal bloom. It has also been explored as a potential geoengineering solution to climate crisis. In 2012, for ex-

ample, two hundred thousand pounds of iron sulphate, a brown powder, were dumped in the ocean to stimulate the growth of algae populations to absorb carbon dioxide and produce oxygen.

5   Ware, *Cyanomicon*, 54.

6   Benjamin, *The Arcades Project*, 157.

7   Benjamin, *The Arcades Project*, 158.

8   Benjamin, *The Arcades Project*, 157.

9   Bulstrode, "Black Metallurgists and the Making of the Industrial Revolution," 18.

10   Agricola, *De Re Metallica*, 7.

11   Mumford, *Technics and Civilization*, 69, 163.

12   Cirlot, *A Dictionary of Symbols*, 5.

13   Cirlot, *A Dictionary of Symbols*, 5.

14   Benjamin, *The Arcades Project*, 170.

15   Faustini et al., "History of Organic–Inorganic Hybrid Materials," 15.

16   Niépce wrote: "This new varnish is composed of a solution of bitumen of Judea in Dippel's animal oil, which is allowed to condense at the ordinary temperature of air to the degree of consistency required. This varnish is more greasy, tougher, and more strongly colored than the other, and it can be exposed to light as soon as the plate is coated, because it seems to solidify more rapidly, owing to the great volatility of the animal oil which causes it to dry more rapidly." Niépce, "Memoire on the Heliograph," 10.

17   Ware, *Cyanomicon*, 34.

18   Ware, *Cyanomicon*, 34.

19   Harley, *Artists' Pigments*, 65–68.

20   Ware, "Light-Sensitive Chemicals," 857.

21   Potassium ferricyanide is also used to strengthen iron and steel.

22   Armstrong, "Cameraless," 94.

23   Ware, *Cyanomicon*, 7.

24   Emerson, *Naturalistic Photography for Students of the Art*, 442.

25   Bayley, *The Complete Photographer*, 392.

26   Shteir, *Cultivating Women, Cultivating Science*, 157.

27   Hume, "The Nature Print and Photography in the 1850s," 45.

28   Tucker, *Nature Exposed*, 21; Hume, "The Nature Print and Photography in the 1850s," 51.

29   Armstrong, "Cameraless," 164.

30   Atkins, *Photographs of British Algae*.

31 Ware, *Cyanomicon*, 12.

32 Hornby, "The Cameraless Optic," 94.

33 Ware, "Cyanotype," 360.

34 Hornby, "The Cameraless Optic," 89.

35 Silverman, *The Miracle of Analogy*, 167.

36 Berger, *About Looking*, 58.

37 See, for instance, Engels, *The Condition of the Working Class in England*.

38 Beckert, "Emancipation and Empire," 1408.

39 Garascia, "'Impressions of Plants Themselves,'" 272.

40 Balston, *Elder James Whatman*, xvii.

41 Balston, *Elder James Whatman*, 27.

42 Mintie, "Material Matters."

43 In 1920, a Kodak publication drew a connection between cotton, share-cropping, and film: "A darky in the cotton field today may be, for all he knows, pulling the cotton for his next season's shirt, or for a motion picture film he will later see produced when he goes to town." Eastman Kodak Company, "How Kodak Film Is Made," 7. Photography cannot be set apart from the larger structural forces that facilitate the racialized extraction of labor, materials, and likeness.

44 Mintie, "Material Matters."

45 Draper, "The City of London and Slavery," 435.

46 E. Williams, *Capitalism and Slavery*.

47 Draper, "The City of London and Slavery," 443.

48 Draper, "The City of London and Slavery," 444.

49 The biography states, "Atkins, John (c. 1760–1838). London West India merchant and significant slave-owner in Jamaica, Lord Mayor of London 1818–1819, and MP for Arundel 1802–1806 and 1826–1832 and for London 1812–1818." Legacies of British Slavery Database, "Alderman John Atkins."

50 Legacies of British Slavery Database, "Hopewell, Jamaica, Port Royal"; Legacies of British Slavery Database, "Jamaica St Andrew 136 (Dublin Castle)."

51 Taylor and Thorne, "Atkins, John."

52 Yusoff, *A Billion Black Anthropocenes or None*.

53 Yusoff, *A Billion Black Anthropocenes or None*.

54 Armstrong, "Cameraless," 112.

55 "Mid Kent and London and South-Western Junction Railways," *Railway Times*, 1347.

56 Harvey, "Between Space and Time."

57 Plunkett, "Marion and Company," 893.

58 Batchen, *Emanations*, 14.

59 Parton, "Pittsburgh."

60 Young, "A Tour to Shropshire."

61 Young, "A Tour to Shropshire."

62 Rau, "Photography as Applied to Business," 49.

63 Stilgoe, "An Opening Between the Trains," 48.

64 Panzer, "The Invisible Photographs of William H. Rau," 38.

65 See, for instance, Nye, *American Technological Sublime*; Trachtenberg, *The Incorporation of America*; M. Shapiro, *The Political Sublime*.

66 Panzer, "William H. Rau and the Archives of Photographic History," 10.

67 Cowie and Heathcott, *Beyond the Ruins*, 4.

68 Panzer, "William H. Rau and the Archives of Photographic History," 10.

69 Finkle, "William H. Rau, Philadelphia, Photography, and the Railroad," 21.

70 Gilmore, *Abolition Geography*, 303–13.

71 Clutter, "Notes on Ruin Porn."

72 Elliott, "Levi's Features a Town Trying to Recover."

73 Nakano Glenn, *Unequal Freedom*; Windham, *Knocking on Labor's Door*; Roediger and Esch, *The Production of Difference*.

74 Sugrue, *The Origins of the Urban Crisis*.

75 Frazier's series *Campaign for Braddock Hospital (Save Our Community Hospital)* criticizes the limited-access health care in the region. Following the decline of the steel industry, the largest employer became Braddock Hospital, built in 1906. In 2010, the hospital closed, shifting care to a new hospital in an affluent suburb that is not easily accessible to working-class residents of Braddock. The struggle for access to health care is particularly heightened due to the high levels of disease in the region caused by polluted soil, air, and water from industry.

76 Miranda, "Q&A: Braddock's LaToya Ruby Frazier Takes on Levi's and 'Ruin Porn' Chic."

77 C. Peters, "Neoliberal Violence," 7.

78 Gould Ford, "Coke Oven Peaches." Jenks studied mining in Pennsylvania and the living conditions in industrial communities for *The Immigration Problem: A Study of Immigration Conditions and Need* (1911), which argued for restrictions on immigration into the United States. Jenks and his coauthor concluded that immigrants drove down wages, which reduced living standards for all Americans.

79   Lanay, "From Metal to Mettle."

80   Iron construction also has a "yearning for verticality." Benjamin, *The Arcades Project*, 161.

81   Cirlot, *A Dictionary of Symbols*, 241.

82   Mavor, *Blue Mythologies*, 86.

83   Tsing et al., *Arts of Living on a Damaged Planet*, G1.

84   Tsing et al., *Arts of Living on a Damaged Planet*, G7.

85   Tuck and Ree, "A Glossary of Haunting," 643.

86   Schmelzer, Vetter, and Vansintjan, *The Future Is Degrowth*.

Chapter 5. Uranium and Photography beyond Vision

An earlier version of sections of chapter 5 appeared in "Atomic Ecology," *October*, no. 179 (Spring 2022): 110–31.

1   At Operation Crossroads at Bikini Atoll, biological specimens, including pigs, goats, rats, mice, and guinea pigs, were placed on target ships and assessed for damage. Ultimately, 35 percent of the animals used in the tests died: 10 percent from the blasts, 15 percent from radiation exposure, and 10 percent from medical experimentation.

2   Schuppli, "Radical Contact Prints," 281.

3   Kriemann, *Canopy, Canopy*.

4   Tagliabue, "A Legacy of Ashes."

5   Seiler, "The Territory of Tiredness."

6   Hagen, Gatzweiler, and Jakubick, "Status and Outlook for the WISMUT Remediation Project in the States of Thuringa and Saxony, Germany."

7   Kupsch and Strnad, "Uranium Bicentenary," 83.

8   Dahlke, "Performing GDR in Poetry?," 183, 185.

9   For further explorations of vision, see Crary, *Techniques of the Observer*; W. Mitchell, "Showing Seeing"; Foster, *Vision and Visuality*; C. Keller, *Brought to Light*.

10   Nixon, *Slow Violence*. Nixon's influential framework has become significant in photo studies. To list a few contexts where slow violence has been explored: *Devour the Land* at the Harvard Art Museum (2021) and the accompanying catalogue, documented in Best, *Devour the Land*; as well as Iheka, *African Ecomedia*; Balaschak, *The Image of Environmental Harm in American Social Documentary Photography*.

11   Nixon, *Slow Violence*, 2.

12   Benjamin used this analogy to describe Karl Blossfeldt's close-up photographs of flowers as well as Eadweard Muybridge and Étienne-Jules Marey's motion studies. Benjamin, "A Little History of Photography," 7. As Rosalind Krauss explains, Benjamin's use of the term does not reflect Freudian thought, though Shawn Michelle Smith has productively explored Benjamin's linking of camera technology and optics to explore "the edge of sight." See, for instance, Krauss, *The Optical Unconscious*; S. Smith, *At the Edge of Sight*; Smith and Sliwinski, *Photography and the Optical Unconsciousness*.

13   Hecht, "Nuclearity," 127.

14   "What Is Uranium?," World Nuclear Association.

15   Towler, *The Silver Sunbeam*, chapter XXX.

16   Kupsch and Strnad, "Uranium Bicentenary," 83.

17   Kupsch and Strnad, "Uranium Bicentenary," 84.

18   Kupsch and Strnad, "Uranium Bicentenary," 83.

19   Kupsch and Strnad, "Uranium Bicentenary," 83.

20   "Uranium Element Facts," Chemicool.

21   Wilson, *Exclusion Zones*.

22   "April 15, 1858," *Photographic Notes*, 97.

23   van Wyck, *The Highway of the Atom*, 90.

24   Sekiya and Yamasaki, "Antoine Henri Becquerel," 3.

25   Habashi, "Niépce de Saint-Victor and the Discovery of Radioactivity," 104.

26   Habashi, "Niépce de Saint-Victor and the Discovery of Radioactivity," 104.

27   d'Albe, *The Life of Sir William Crookes*, 389.

28   "April 15, 1858," *Photographic Notes*, 94.

29   "April 15, 1858," *Photographic Notes*, 97.

30   Panchbhai, "Wilhelm Conrad Röntgen and the Discovery of X-Rays," 94.

31   Panchbhai, "Wilhelm Conrad Röntgen and the Discovery of X-Rays," 94.

32   Panchbhai, "Wilhelm Conrad Röntgen and the Discovery of X-Rays," 92.

33   Lippit, *Atomic Light*, 50.

34   Barad, "Troubling Time/s," 67–68, 63.

35   Chéroux, "Photographs of Fluids," 115.

36   "A Sensational Discovery," *Die Presse*.

37   "A Sensational Discovery," *Die Presse*.

38   Henderson, "X-Rays and the Quest for Invisible Reality," 324.

39 Sekiya and Yamasaki, "Antoine Henri Becquerel," 2. These experiments were part of a broader scientific exploration of the visual. See, for instance, Tucker, *Nature Exposed*; Ramalingam, "Natural History in the Dark"; Ramalingam, "Dust Plate, Retina, Photograph"; Daston and Galison, *Objectivity*.

40 Wilder, "Visualizing Radiation," 351.

41 Sekiya and Yamasaki, "Antoine Henri Becquerel," 2.

42 Lippit, *Atomic Light*, 44.

43 Wilder, "Visualizing Radiation," 352.

44 Wilder, "Visualizing Radiation," 353.

45 Wilder, "Visualizing Radiation," 353.

46 Wilder, "Visualizing Radiation," 365.

47 Jordan Bear shows how the "non artifactuality" of Humphry Davy's photographic experiments lead them to be framed as failures, diminishing the significance of Davy's experiments to photography. See Bear, "Self-Reflections," 191.

48 "Talk in the Studio," *Photographic News*, 32.

49 See, for instance, Chéroux, *The Perfect Medium*.

50 O'Brian, *Camera Atomica*.

51 Masco, "Flashblindness," 87.

52 The Trinity nuclear test in 1945 and the detonations at Hiroshima and Nagasaki were framed as wartime actions. Operation Crossroads at Bikini Atoll is generally viewed to have ushered in the Atomic Age, which is understood to end with the collapse of the Soviet Union.

53 Schuppli, "Radical Contact Prints," 280.

54 Bryan-Wilson, "Posing by the Cloud," 114.

55 Hales, "The Atomic Sublime," 12.

56 Hales, "The Atomic Sublime," 10.

57 Maclear, *Beclouded Visions*, 9, 10.

58 O'Brian, *The Bomb in the Wilderness*, xiii.

59 Wellerstein, "Counting the Dead at Hiroshima and Nagasaki."

60 Sontag, *On Photography*; Sontag, *Regarding the Pain of Others*.

61 Berger, "Hiroshima," 315.

62 Berger, "Hiroshima," 318.

63 Hales, "The Atomic Sublime," 24.

64 Farish, "Below the Bombs," 113.

65   Maurer and Hogue, "Introduction," 32.

66   Hales, "The Atomic Sublime," 24.

67   Hales, "The Atomic Sublime," 24.

68   Bryan-Wilson, "Posing by the Cloud," 119.

69   Brown, *Plutopia*, 3.

70   M. Davis, "Los Angeles after the Storm."

71   Lippit, *Atomic Light*, 27.

72   Lippit, *Atomic Light*, 42.

73   The discovery of the X-ray coincided with Sigmund Freud and Josef Breuer's publication of *Studies on Hysteria*, laying the groundwork for psychoanalysis as a field of study, and Auguste and Louis Lumière's release of *Workers Leaving the Lumière Factory in Lyon*, widely credited as the first motion picture film (see chapter 2). Lippit, *Atomic Light*, 27. Kittler's *Gramophone, Film, Typewriter* also explores the convergence of media technology and psychoanalysis.

74   Bergstein, *Mirrors of Memory*.

75   Smith and Sliwinski, *Photography and the Optical Unconsciousness*, 3, 2.

76   Virilio, *War and Cinema*, 101.

77   van Wyck, *The Highway of the Atom*, 83.

78   Webb, "The Fogging of Photographic Film."

79   van Wyck, *The Highway of the Atom*, 83.

80   Virilio, *War and Cinema*, 101 (emphasis in original).

81   Wilson, *Exclusion Zones*, 7.

82   Chin, "Revival Field."

83   Wilson, *Exclusion Zones*, 12.

84   Casid, *Sowing Empire*, xiii.

85   Landbrecht and Schäfer, "Lamp," 51.

86   Sekula, "Photography between Labor and Capital," 214.

87   Sekula, "Reading an Archive," 445.

88   Chakrabarty, "The Climate of History," 201.

89   Demos, "The Agency of Fire."

90   Derrida, *Archive Fever*, 36.

91   Masco, "Mutant Ecologies," 533 (emphasis in original).

92   For an analysis of preservationist environmentalism, see, for example, Buell, *The Future of Environmental Criticism*; Dunaway, *Natural Visions*; Dunaway, *Seeing Green*; and Seymour, *Bad Environmentalism*. For a discussion of toxicity

and environmentalism, see Buell, "Toxic Discourse"; and Marran, *Ecology without Culture*.

93 Tsing, *The Mushroom at the End of the World*, 18.

94 Tsing et al., *Arts of Living on a Damaged Planet*, G7.

95 Haraway, *Staying with the Trouble*, 31.

Chapter 6. Rare Earth Elements and De/Materialization

1 P. Keller, "Joan Fontcuberta's Landscapes," 134–35.

2 Fontcuberta, *Pandora's Camera*, 10.

3 Moreiras, "Joan Fontcuberta."

4 The CMYK process is an early form of color photography that dates back to the three-color method invented by the Scottish physicist James Clerk Maxwell in 1855, which resulted in the first color photograph in 1861 of a tartan ribbon. The tonal range was limited to blue and violet, as silver emulsions are more sensitive to blues than other colors. In the 1880s, Hermann Vogel developed a process that made silver bromide sensitive to green, yellow, orange, and red by staining the emulsion with aniline dyes—made from coal tar—to absorb rays of each color. The first commercially successful color photographic process, the Lumière Autochrome, reached the market in 1907, and by 1935, Kodak launched Kodachrome. Today, the standard printing set still uses CMYK.

5 Wark, *Capital Is Dead*, 2.

6 Rohrig, "Smartphones."

7 For instance, Geoffrey Batchen turns to Talbot's contact print negatives of lace, where the image was imprinted with the form of the delicate fabric. Batchen describes how the images are "an abstraction of visual data; it's a fledgling form of information culture," locating new media in Talbot's early experiments, suggesting continuity rather than rupture in the shift from analog to digital. Batchen, "Electricity Made Visible," 30.

8 Cleveland and Ruth, "Indicators of Dematerialization and the Materials Intensity of Use," 16.

9 For an analysis of darkroom labor, see Cho, "Darkroom Material"; Wall, "Photography and Liquid Intelligence."

10 Drucker, *SpecLab Digital Aesthetics and Projects in Speculative Computing*, 142.

11 Klinger, *Rare Earth Frontiers*, 21.

12 The more common light rare earths (cerium, lanthanum, praseodymium, neodymium, promethium, europium, gadolinium, and samarium) make

up 85–90 percent of supply, the rest of which constitute the heavy rare earths (terbium, dysprosium, holmium, erbium, thulium, ytterbium, lutetium, and yttrium). Promethium is radioactive and not used in smartphones.

13 Ronda, "Rare Earth Phosphors."

14 Klinger shows that before the rare earths crisis of 2010, Bayan Obo, which dominated rare earth production, received very little international attention. Klinger, *Rare Earth Frontiers*, 32. Following the crisis, there was increased attention, but visual explorations of this form of mining remain incredibly rare.

15 Klinger, *Rare Earth Frontiers*, 1.

16 Klinger, *Rare Earth Frontiers*, 6.

17 Klinger, *Rare Earth Frontiers*, 55.

18 Klinger, *Rare Earth Frontiers*, 122.

19 For a description of the geopolitical machinations and implications for mining and supply chains, see Klinger, *Rare Earth Frontiers*, 4.

20 Klinger, *Rare Earth Frontiers*, 35.

21 Penn and Lipton, "The Lithium Gold Rush." The article is illustrated with photographs by Gabriella Angotti-Jones. An earlier *New York Times* article, "Green-Energy Race Draws an American Underdog to Bolivia's Lithium," by Clifford Krauss, with photographs by Meridith Kohut, is even more image-heavy, with photographs that luxuriate in the textures and rich colors of the striking desert landscape.

22 Katwala, "The Spiralling Environmental Cost of Our Lithium Battery Addiction."

23 Burtynsky, "A Good Anthropocene."

24 Burtynsky, "A Good Anthropocene."

25 Asafu-Adjaye et al., "An Ecomodernist Manifesto."

26 Demos, "To Save a World."

27 Demos, "To Save a World."

28 Smil, *Making the Modern World*, 130.

29 Efoui-Hess, *Climate Crisis*.

30 J. Peters, *The Marvelous Clouds*, 332.

31 "Submarine Cable Map," TeleGeography.

32 Paglen, "Some Sketches on Vertical Geographies."

33 Starosielski, *The Undersea Network*, 1.

34 Starosielski, *The Undersea Network*, 5.

35 Efoui-Hess, *Climate Crisis*.

36 Elegant, "The Internet Cloud Has a Dirty Secret."

37 Sverdlik, "Here's How Much Energy All US Data Centers Consume."

38 Hogan, "Big Data Ecologies," 638.

39 R. Miller, "Data Centers Forge Ahead with Shift to Renewable Energy."

40 Hogan, "Big Data Ecologies," 636.

41 Hogan, "Big Data Ecologies," 634.

42 Bogost, "Your Phone Wasn't Built for the Apocalypse."

43 The limited tonal ranges often had racial implications. See, for example, Dyer, *White*.

44 Demos, "The Agency of Fire."

45 Demos, "The Agency of Fire."

46 MacDonald, "Just 3 Percent of Broadcast TV News Segments on the California Wildfires Connected Them to Climate Change."

47 Parenti, *Tropic of Chaos*.

48 Nixon, *Slow Violence*, 1–2.

49 Demos, "The Agency of Fire."

50 Gilmore, *Abolition Geography*, 410–48.

51 Jacob Badcock reads photographs of Agbogbloshie through the lens of discard studies, arguing that the scale of Agbogbloshie is exaggerated by visual images, which directs attention away from the causes of pollution to focus on abject suffering. Badcock, "Photography after Discard Studies."

52 Ahmed, "E-Waste."

53 Sullivan, "Trash or Treasure," 98.

54 United Nations Environment Programme, *Waste Crime*.

55 Hugo, "Permanent Error," 106.

56 Milbourne, "African Photographers and the Look of (Un)Sustainability in the African Landscape," 125.

57 Iheka, *African Ecomedia*, 79, 92.

58 Pulido, "Rethinking Environmental Racism," 13.

59 Ouedraogo, "The Hell of Copper."

60 Sullivan, "Trash or Treasure," 96.

61 Ouedraogo, "The Hell of Copper."

62 Iheka, *African Ecomedia*, 81, 87.

63 Dunaway, *Seeing Green*; Sims, "Green Magic."

64 Lawsuits have been brought against Apple in Europe and the United States, arguing that programmed obsolescence driven by software upgrades forces consumers to buy new phones.

65 Blahnik and Schindelbeck, "Smartphone Imaging Technology and Its Applications."

66 See, for example, Ackerman, *Why Do We Recycle*; MacBride, *Recycling Reconsidered*; Liboiron, "Modern Waste as Strategy"; Liboiron, "Recycling as a Crisis of Meaning."

67 Crary, 24/7, 37.

68 Cakebread, "People Will Take 1.2 Trillion Digital Photos this Year—Thanks to Smartphones."

69 Cubitt, *Anecdotal Evidence*, 224.

70 See, for example, Noble, *Algorithms of Oppression*.

71 Henning, "Image Flow," 134.

72 Cubitt, *Anecdotal Evidence*, 224.

73 Cubitt, *Anecdotal Evidence*, 246.

74 Batchen, "Ectoplasm," 18.

75 Crary, 24/7, 33.

76 Tedone, "From Spectacle to Extraction."

77 For an analysis of narratives of the end of photography, see Batchen, "Ectoplasm."

78 Zylinska, *Nonhuman Photography*, 18.

79 Berardi, "Desire, Pleasure, Senility, and Evolution."

80 Marx, *Capital*, 638.

81 For example, Bastani's *Fully Automated Luxury Communism* offers a left-wing fantasy of growth with no costs, where all humanity's problems are solved by technological innovation.

82 Salgado, *Workers*, 7.

83 Michaels and Zamora, "Chris Killip and LaToya Ruby Frazier," 26.

84 Smil, *Making the Modern World*, 120.

85 Reinsel, Gantz, and Rydning, *IDC Report*.

Conclusion

1 Belcher et al., "Hidden Carbon Costs of the 'Everywhere War.'"

2 Wall, "Photography and Liquid Intelligence," 110.

3 Wall, "Photography and Liquid Intelligence," 110.

4 Batchen, "Ectoplasm," 14; Sekula, "Between the Net and the Deep Blue Sea," 4.

5 Blaschke, "The Excess of the Archive," 237.

6 Sekula, "Between the Net and the Deep Blue Sea," 10.

7 Blaschke, "The Excess of the Archive," 226.

8 Berger, "Overexposed"; Watt and Lowenthal, *Paradox of Preservation*; Spence, *Dispossessing the Wilderness*.

9 Hudson, *Fire Management in the American West*; Horton, "Fire Oppression," 67–71.

10 Moore, "The Capitalocene and Its Discontents."

11 Meadows, "The Symbol's Symbol," 272.

12 Jiménez, "Spiderweb Anthropologies," 53.

13 Cariou and Angus, "Tar Remedies."

Ackerman, Frank. *Why Do We Recycle? Markets, Values, and Public Policy*. Washington, DC: Island Press, 1996.

Agricola, Georgius. *De Re Metallica*. Translated by H. C. Hoover and L. H. Hoover. New York: Dover Publications, 1950.

Ahmed, Syed Faraz. "E-Waste: The Global Cost of Discarded Electronics." *Atlantic*, September 29, 2016.

Ahooja, Sarita. "Alberta: Oil Sands Truth." IsumaTV, January 21, 2009. http://www.isuma.tv/2010-olympic-boycott/alberta-oil-sands-truth.

Albers, Kate Palmer. *The Night Albums: Visibility and the Ephemeral Photograph*. Berkeley: University of California Press, 2021.

Alexander, Peter, Thapelo Lekgowa, Botsang Mmope, Luke Sinwell, and Bongani Xezwi. *Marikana: Voices from South Africa's Mining Massacre*. Athens: Ohio University Press, 2013.

Altman, Rebecca. "Upriver." *Orion Magazine*, June 2, 2021.

Andermann, Jens. *The Optic of the State: Visuality and Power in Argentina and Brazil*. Pittsburgh, PA: University of Pittsburgh Press, 2007.

Angus, Siobhan. "Mining the History of Photography." In *Capitalism and the Camera*, edited by Kevin Coleman and Daniel James, 55–73. New York: Verso, 2021.

"April 15, 1858." *Photographic Notes: Journal of the Birmingham Photographic Society* 3 (1858): 94–97.

Arabindan-Kesson, Anna. *Black Bodies, White Gold: Art, Cotton, and Commerce in the Atlantic World*. Durham, NC: Duke University Press, 2021.

Armstrong, Carol. "Cameraless: From Natural Illustrations and Nature Prints to Manual and Photogenic Drawings and Other Botanographies." In *Ocean Flowers: Impressions from Nature*, edited by Carol Armstrong and Catherine de Zegher, 87–180. New Haven, CT: Yale Center for British Art, 2004.

Arndt, N. T., and Clément Ganino. *Metals and Society: An Introduction to Economic Geology*. Berlin: Springer, 2012.

Arscott, Caroline. "Subject and Object in Whistler: The Context of Physiological Aesthetics." In *Palaces of Art: Whistler and the Art Worlds of Aestheticism*, edited by Lee Glazer and Linda Merrill, 55–66. Washington, DC: Smithsonian Institution Scholarly Press, 2013.

Asafu-Adjaye, John, et al. "An Ecomodernist Manifesto." Ecomodernism. April 2015. http://www.ecomodernism.org/.

Atkins, Anna. *Photographs of British Algae: Cyanotype Impressions*. Halstead Place, UK: Self-published, 1843–53.

Azoulay, Ariella. "The Captive Photograph." *Boston Review*, September 23, 2021.

Azoulay, Ariella. *The Civil Contract of Photography*. New York: Zone Books, 2008.

Azoulay, Ariella. "Toward the Abolition of Photography's Imperial Rights." In *Capitalism and the Camera*, edited by Kevin Coleman and Daniel James, 27–54. New York: Verso, 2021.

Badcock, Jacob. "Photography after Discard Studies: The Case of Agbogbloshie." *Burlington Magazine*, November 2022.

Balaschak, Chris. *The Image of Environmental Harm in American Social Documentary Photography*. New York: Routledge, 2021.

Baldwin, Douglas. *Cobalt: Canada's Forgotten Silver Boom Town*. Charlottetown, PE: Indigo Press, 2016.

Balston, J. N. *Elder James Whatman*. Kent, UK: J. N. Balston, 1992.

Barad, Karen. "Troubling Time/s and Ecologies of Nothingness: Re-turning, Remembering, and Facing the Incalculable." *New Formations* 92 (2018): 56–86.

Barnes, Julian. *Keeping an Eye Open*. London: Jonathan Cape, 2015.

Barrett, Ross, and Daniel Worden. "Introduction." In *Oil Culture*, edited by Ross Barrett and Daniel Worden. Minneapolis: University of Minnesota Press, 2014.

Barringer, Mark Daniel. *Selling Yellowstone: Capitalism and the Construction of Nature*. Lawrence: University Press of Kansas, 2002.

Barringer, Tim. "Introduction—John Ruskin: Seer of the Storm-Cloud." In *Unto This Last: Two Hundred Years of John Ruskin*, edited by Tim Barringer, Tara Contractor, Victoria Hepburn, Judith Stapleton, and Courtney Skipton Long, 13–29. New Haven, CT: Yale Center for British Art, 2019.

Barthes, Roland. *Camera Lucida*. New York: Hill & Wang, 1980.

Barthes, Roland. *The Neutral.* Translated by Rosalind Krauss and Denis Hollier. New York: Columbia University Press, 2007.

Barthes, Roland. "The Rhetoric of the Image." In *Image, Music, Text,* translated by Stephen Heath, 32–51. New York: Hill & Wang, 1977.

Bastani, Aaron. *Fully Automated Luxury Communism: A Manifesto.* London: Verso, 2019.

Batchen, Geoffrey. *Burning with Desire: The Conception of Photography.* Cambridge, MA: MIT Press, 1997.

Batchen, Geoffrey. "Ectoplasm: Photography in the Digital Age." In *Over Exposed: Essays on Contemporary Photography,* edited by Carol Squiers, 9–23. New York: New Press, 1999.

Batchen, Geoffrey. "Electricity Made Visible." In *New Media, Old Media,* edited by Wendy Hui Kyong Chun and Thomas Keenan, 27–44. London: Routledge, 2006.

Batchen, Geoffrey. *Emanations: The Art of the Cameraless Photograph.* Kassel, Germany: Prestel, 2016.

Batchen, Geoffrey. "The Naming of Photography: 'A Mass of Metaphor.'" *History of Photography* 17, no. 1 (1993): 22–32.

Bayley, R. Child. *The Complete Photographer.* London: Methuen, 1932.

Bear, Jordan. "Self-Reflections: The Nature of Sir Humphry Davy's Photographic 'Failures.'" In *Photography and Its Origins,* edited by Tanya Sheehan and Andrés Mario Zervigón, 185–94. London: Routledge, 2015.

Beckert, Sven. "Emancipation and Empire: Reconstructing the Worldwide Web of Cotton Production in the Age of the American Civil War." *American Historical Review* 109, no. 5 (2004): 1405–38.

Beil, Kim. *Good Pictures: A History of Popular Photography.* Stanford, CA: Stanford University Press, 2020.

Beil, Kim. "Photography Has Gotten Climate Change Wrong from the Start." *Atlantic,* November 27, 2020.

Belcher, Oliver, Patrick Bigger, Ben Neimark, and Cara Kennelly. "Hidden Carbon Costs of the 'Everywhere War': Logistics, Geopolitical Ecology, and the Carbon Boot-Print of the US Military." *Transactions of the Institute of British Geographers* 45, no. 1 (2019): 65–80.

Belting, Hans. *The Invisible Masterpiece.* Chicago: University of Chicago Press, 2001.

Benjamin, Walter. *The Arcades Project.* Translated by Howard Eiland and Kevin McLaughlin. Cambridge, MA: Belknap Press of Harvard University Press, 2002.

Benjamin, Walter. "The Author as Producer." *New Left Review* 62 (July–August 1970): 83–96.

Benjamin, Walter. "A Little History of Photography." In *Walter Benjamin, Selected Writings, Volume 2: 1927–1934*, edited by Michael W. Jennings, Howard Eiland, Gary Smith, and Rodney Livingstone. Cambridge, MA: Belknap Press of Harvard University Press, 1996.

Benjamin, Walter. "Theses on the Philosophy of History." In *Illuminations*, edited by Hannah Arendt, translated by Harry Zohn, 253–64. New York: Schocken Books, 2007.

Benjamin, Walter. "The Work of Art in the Age of Mechanical Reproduction." In *Illuminations*, edited by Hannah Arendt, translated by Harry Zohn, 217–51. New York: Schocken Books, 2007.

Berardi, Franco. "Desire, Pleasure, Senility, and Evolution." *e-flux journal* 106 (February 2020). https://www.e-flux.com/journal/106/312516/desire -pleasure-senility-and-evolution/.

Berger, John. *About Looking*. London: Bloomsbury, 2009.

Berger, John. "Hiroshima." In *The Sense of Sight: Writings*, 287–95. New York: Pantheon, 1985.

Berger, Martin A. "Overexposed: Whiteness and the Landscape Photography of Carleton Watkins." *Oxford Art Journal* 26, no. 1 (2003): 3–23.

Bergstein, Mary. *Mirrors of Memory: Freud, Photography, and the History of Art.* Ithaca, NY: Cornell University Press, 2010.

Best, Makeda, ed. *Devour the Land: War and American Landscape Photography since 1970.* Cambridge, MA: Harvard Art Museums, 2021.

Blahnik, Vladan, and Oliver Schindelbeck. "Smartphone Imaging Technology and Its Applications." *Advanced Optical Technologies* 10, no. 3 (2021): 145–232.

Blake, William. *Milton: A Poem in Two Books*. London: Self-published, 1810.

Blaschke, Estelle. "The Excess of the Archive." In *Documenting the World: Film, Photography, and the Scientific Record*, edited by Gregg Mitman and Kelley Wilder. Chicago: University of Chicago Press, 2017.

Bogost, Ian. "Your Phone Wasn't Built for the Apocalypse." *Atlantic*, September 11, 2020.

Braddock, Alan C. "Poaching Pictures: Yellowstone, Buffalo, and the Art of Wildlife Conservation." *American Art* 23, no. 3 (2009): 36–59.

Braddock, Alan C., and Karl Kusserow. "Introduction." In *Nature's Nation: American Art and Environment*, edited by Alan C. Braddock and Karl Kusserow, 12–30. Princeton, NJ: Princeton University Art Museum, 2018.

Bratter, Herbert M. "A Survey of Silver: Part I." *Pacific Affairs* 5, no. 7 (1932): 581–99.

Bratter, Herbert M. "A Survey of Silver: Part II." *Pacific Affairs* 5, no. 8 (1932): 704–19.

Brecht, Bertolt. "On Form and Subject-Matter." In *Brecht on Theatre: The Development of an Aesthetic*, edited by John Willett, 29–30. New York: Hill & Wang, 1964.

Brecht, Bertolt. "A Short Organum for the Theatre." In *Brecht on Theatre: The Development of an Aesthetic*, edited by John Willett, 179–205. New York: Hill & Wang, 1964.

Brecht, Bertolt. "The Threepenny Lawsuit." In *Brecht on Film and Radio*. London: Methuen, 2000.

Brower, Matthew. "Trophy Shots: Early North American Photographs of Nonhuman Animals and the Display of Masculine Prowess." *Society & Animals* 13, no. 1 (2005): 13–31.

Brown, Kate. *Plutopia: Nuclear Families, Atomic Cities, and the Great Soviet and American Plutonium Disasters*. Oxford: Oxford University Press, 2013.

Bryan-Wilson, Julia. "Posing by the Cloud: US Nuclear Test Site Photography in Process." In *Camera Atomica*, edited by John O'Brian, 107–23. London: Black Dog Press and Art Gallery of Ontario, 2015.

Buell, Lawrence. *The Future of Environmental Criticism*. Malden, MA: Wiley-Blackwell, 2005.

Buell, Lawrence. "Toxic Discourse." *Critical Inquiry* 24, no. 3 (1998): 639–65.

Bulstrode, Jenny. "Black Metallurgists and the Making of the Industrial Revolution." *History and Technology* (2023): 1–41.

Burton, Cosmo. "The Whole Duty of the Photographer." *British Journal of Photography* 36 (October 11, 1889): 668.

Burtynsky, Edward. "A Good Anthropocene." Edwardburtynsky.com. Accessed September 5, 2018. https://www.edwardburtynsky.com/news-hub/2018/9/5/a-good-anthropocene.

Burtynsky, Edward. "Life in the Anthropocene." In *Anthropocene*, edited by Sophie Hackett, Andrea Kunard, and Urs Stahel, 189–95. Göttingen: Steidl, 2018.

Burtynsky, Edward. "Oil: Artist's Statement." Edwardburtynsky.com. Accessed December 9, 2020. https://www.edwardburtynsky.com/projects/photographs/oil.

Butterman, W. C., and H. E. Hilliard. "Mineral Commodity Profiles: Silver." Reston, VA: US Department of the Interior and US Geological Survey, 2005.

Cadava, Eduardo. *Words of Light: Theses on the Photography of History*. Princeton, NJ: Princeton University Press, 1996.

Cakebread, Caroline. "People Will Take 1.2 Trillion Digital Photos This Year—Thanks to Smartphones." *Business Insider*, August 31, 2017.

Cameron, D. S. "Recycling: Noble Metal Recycling." In *Encyclopedia of Electrochemical Power Sources*, edited by J. Garche, C. Dyer, P. Moseley, Z. Ogumi, D. Rand, and B. Scrosati, 209–14. Amsterdam: Elsevier, 2009.

Campbell, Claire, ed. *A Century of Parks Canada, 1911–2011*. Calgary: University of Calgary Press, 2011.

Campt, Tina. *Image Matters: Archive, Photography, and the African Diaspora in Europe*. Durham, NC: Duke University Press, 2012.

Cariou, Warren. "Petrography." Warrencariou.com. Accessed November 15, 2019. http://www.warrencariou.com/petrography.

Cariou, Warren. "Portfolio: Petrographs." *Photography and Culture* 13, no. 2 (2020): 253–56.

Cariou, Warren, and Siobhan Angus. "Tar Remedies: Methods of Return and Re-vision on Colonized/Contaminated Land." *Environmental Humanities* (forthcoming).

Cariou, Warren, and Jon Gordon. "Petrography, the Tar Sands Paradise, and the Medium of Modernity." *Goose* 14, no. 2 (2016): 1–29.

Casid, Jill. *Sowing Empire: Landscape and Colonization*. Minneapolis: University of Minnesota Press, 2004.

Chakrabarty, Dipesh. "The Climate of History: Four Theses." *Critical Inquiry* 35, no. 2 (2009): 197–222.

Cheetham, Mark. "'Atmospheres' of Art and Art History." *Venti* 1, no. 1 (2020): 50–61.

Chéroux, Clément. *The Perfect Medium: Photography and the Occult*. New Haven, CT: Yale University Press, 2005.

Chéroux, Clément. "Photographs of Fluids." In *The Perfect Medium: Photography and the Occult*. New Haven, CT: Yale University Press, 2005.

Chin, Mel. "Revival Field." ART21. Accessed January 21, 2020. https://art21.org/read/mel-chin-revival-field/.

Cho, Lily. "Darkroom Material: Race and the Chromogenic Print Process." *Postmodern Culture* 28, no. 2 (2018). https://doi.org/10.1353/pmc.2018.0010.

Cirlot, J. E. *A Dictionary of Symbols*. Translated by Jack Sage. London: Routledge, 2001.

CJPME Foundation. "Roots of Apartheid: South Africa's Mining Industry." CJPME Foundation, May 2014. https://www.cjpmefoundation.org/roots _of_apartheid_south_africa_s_mining_industry.

Clark, K. A., S. Ikram, and R. P. Evershed. "The Significance of Petroleum Bitumen in Ancient Egyptian Mummies." *Philosophical Transactions: Mathematical, Physical and Engineering Sciences* 374, no. 2079 (2016): 1–15.

Cleveland, Cutler J., and Matthias Ruth. "Indicators of Dematerialization and the Materials Intensity of Use." *Journal of Industrial Ecology* 2, no. 3 (1998): 15–50.

Clutter, McLain. "Notes on Ruin Porn." *Avery Review* 18 (October 2016): 1–9.

Coleman, Kevin, and Daniel James, eds. *Capitalism and the Camera*. New York: Verso, 2021.

Coleman, Kevin, Daniel James, and Jayeeta Sharma. "Photography and Work." *Radical History Review* 132 (October 2018): 1–22.

"Cost of Chemicals Fifty Years Ago." *Photographic News* 10 (September 28, 1866): 468.

Coulthard, Glen Sean. *Red Skin, White Masks: Rejecting the Colonial Politics of Recognition*. Minneapolis: University of Minnesota Press, 2014.

Cowen, Deborah. *The Deadly Life of Logistics: Mapping the Violence of Global Trade*. Minneapolis: University of Minnesota Press, 2014.

Cowie, Jefferson R., and Joseph Heathcott. *Beyond the Ruins: The Meanings of Deindustrialization*. Ithaca, NY: ILR Press, 2003.

Crary, Jonathan. *Techniques of the Observer: On Vision and Modernity in the Nineteenth Century*. Cambridge, MA: MIT Press, 1990.

Crary, Jonathan. *24/7: Late Capitalism and the Ends of Sleep*. New York: Verso, 2014.

Cristaudo, A., F. Sera, V. Severino, M. De Rocco, E. Di Lella, and M. Picardo. "Occupational Hypersensitivity to Metal Salts, Including Platinum, in the Secondary Industry." *Allergy (Copenhagen)* 60, no. 2 (2005): 159–64.

Cronin, J. Keri. *Manufacturing National Park Nature: Photography, Ecology, and the Wilderness Industry of Jasper*. Vancouver: University of British Columbia Press, 2011.

Cubitt, Sean. *Anecdotal Evidence: Ecocritique from Hollywood to the Mass Image*. New York: Oxford University Press, 2020.

Cummins, Thomas. "Silver Threads and Golden Needles: The Inca, the Spanish, and the Sacred World." In *The Colonial Andes: Tapestry and Silverwork, 1530–1830*, edited by Elena Phipps, Johanna Hecht, and Cristina Esteras Martin, 3–15. New York: Metropolitan Museum of Art, 2004.

Dahlke, Birgit. "Performing GDR in Poetry? The Literary Significance of 'East German' Poetry in Unified Germany." In *German Literature in a New Century: Trends, Traditions, Transitions, Transformations*, edited by Katharina Gerstenberger and Patricia Herminghouse, 178–95. New York: Berghahn Books, 2008.

d'Albe, Edmund Edward Fournier. *The Life of Sir William Crookes, OM, FRS*. Cambridge: Cambridge University Press, 1923.

Daschuk, James. *Clearing the Plains: Disease, Politics of Starvation, and the Loss of Indigenous Life*. Regina, SK: University of Regina Press, 2014.

Daston, Lorraine, and Peter Galison. *Objectivity*. New York: Zone Books, 2007.

Davis, Heather. *Plastic Matter*. Durham, NC: Duke University Press, 2022.

Davis, Heather, and Zoe Todd. "On the Importance of a Date." *ACME: An International Journal for Critical Geographies* 16, no. 4 (2017): 761–80.

Davis, Jeremy. *The Birth of the Anthropocene*. Berkeley: University of California Press, 2016.

Davis, Mike. "Los Angeles after the Storm: The Dialectic of Ordinary Disasters." *Antipode* 27, no. 3 (1995): 221–41.

de la Cadena, Marisol. "Uncommoning Nature." *e-flux journal* 65 (August 22, 2015). https://www.e-flux.com/journal/65/336365/uncommoning-nature/.

del Mar, Alexander. *A History of Precious Metals.* New York: Cambridge Encyclopedia, 1902.

DeLue, Rachael Z. "Homer Dodge Martin's Landscapes in Reverse." In *Nature's Nation: American Art and Environment,* edited by Karl Kusserow and Alan Braddock, 290–93. Princeton, NJ: Princeton University Art Museum, 2018.

Demos, T. J. *Against the Anthropocene: Visual Culture and Environment Today.* Berlin: Sternberg Press, 2017.

Demos, T. J. "The Agency of Fire: Burning Aesthetics." *e-flux journal* 98 (February 2019). https://www.e-flux.com/journal/98/256882/the-agency-of-fire-burning-aesthetics/.

Demos, T. J. "To Save a World: Geoengineering, Conflictual Futurisms, and the Unthinkable." *e-flux journal* 94 (October 2018). https://www.e-flux.com/journal/94/221148/to-save-a-world-geoengineering-conflictual-futurisms-and-the-unthinkable/.

Derrida, Jacques. *Archive Fever: A Freudian Impression.* Translated by Eric Prenowitz. Chicago: University of Chicago Press, 1996.

Dinkar, Niharika. "'Our Best Machines Are Made of Sunlight': Photography and Technologies of Light." In *Ubiquity: Photography's Multitudes,* edited by Jacob W. Lewis and Kyle Parry, 93–113. Leuven, Belgium: Leuven University Press, 2021.

Dixon, Deborah. "The Perturbations of Drift in a Stratified World." *Performance Research* 23, no. 7 (2018): 130–35.

Dootson, Kirsty Sinclair. *The Rainbow's Gravity: Colour, Materiality and British Modernity.* London: Paul Mellon Centre/Yale University Press, 2023.

Draper, Nicholas. "The City of London and Slavery: Evidence from the First Dock Companies, 1795–1800." *Economic History Review* 61, no. 2 (2008): 432–66.

Drucker, Johanna. *SpecLab Digital Aesthetics and Projects in Speculative Computing.* Chicago: University of Chicago Press, 2009.

Dunaway, Finnis. *Natural Visions: The Power of Images in American Environmental Reform.* Chicago: University of Chicago Press, 2005.

Dunaway, Finnis. *Seeing Green: The Use and Abuse of American Environmental Images.* Chicago: University of Chicago Press, 2015.

Dyer, Richard. *White.* London: Routledge, 1997.

Eastman Kodak Company, "Conflict Minerals Report." United States Securities and Exchange Commission Report, May 31, 2022, 1–19. https://www.kodak.com/content/products-brochures/EKC-Conflict-Minerals-Report-2021.pdf.

Eastman Kodak Company. *The Home of Kodak: Facts about the World's Largest Photographic Organization*. Rochester, NY: Eastman Kodak Company, 1929.

Eastman Kodak Company. "How Kodak Film Is Made." *Kodak Magazine*, October 1920. https://archive.org/details/kodakmagazine01eastuoft/page/n167/mode/2up.

Eastman Kodak Company. "It's Pure Silver That 'Gets the Picture' on Verichrome and Other Kodak Films." *LIFE*, January 15, 1945, 45.

Eastman Kodak Company. *Kodak Park: Its Products and People*. Rochester, NY: Eastman Kodak Company, 1951, 1953, 1954.

Eastman Kodak Company. *The Kodak Park Works: Where Kodak Film, Papers, and Chemicals Are Made*. Rochester, NY: Eastman Kodak Company, 1964.

Eastman Kodak Company. "Silver." *Kodak Photo Magazine*, February and March 1947.

Eastman Kodak Company. "Silver—Metal, Money, and Mixture." *Kodak: A Magazine for Eastman Employees*, April 1938, 1–3.

"Eastman Kodak Is Increasing Prices 6.5%." *New York Times*, April 16, 1974, 51.

Eder, Josef Maria. *History of Photography*. New York: Dover Publications, 1978.

Edwards, Elizabeth. "Commemorating a National Past: The National Photographic Record Association, 1897–1910." *Journal of Victorian Culture: JVC* 10, no. 1 (2005): 123–29.

Edwards, Elizabeth. "Photographs as Objects of Memory." In *Material Memories: Design and Evocation*, edited by Marius Kwint, Christopher Breward, and Jeremy Aynsley, 221–36. Oxford: Berg Publishers, 1999.

Edwards, Elizabeth, and Janice Hart. "Photographs as Objects." In *Photographs Objects Histories: On the Materiality of Images*, edited by Elizabeth Edwards and Janice Hart, 1–15. London: Routledge, 2004.

Edwards, Elizabeth, and Janice Hart, eds. *Photographs Objects Histories: On the Materiality of Images*. London: Routledge, 2004.

Edwards, Steve. *The Making of English Photography: Allegories*. University Park: Pennsylvania State University Press, 2006.

Edwards, Steve. *Photography: A Very Short Introduction*. Oxford: Oxford University Press, 2006.

Efoui-Hess, Maxime. *Climate Crisis: The Unsustainable Use of Online Video*. Shift Project, July 11, 2019. https://theshiftproject.org/en/article/unsustainable-use-online-video/.

Elegant, Naomi Xu. "The Internet Cloud Has a Dirty Secret." *Fortune*, September 18, 2019.

Elliott, Stuart. "Levi's Features a Town Trying to Recover." *New York Times*, June 23, 2010.

Emerson, Peter Henry. *Naturalistic Photography for Students of the Art*. London: Sampson Low, Marston, Searle, and Rivington, 1889.

Emery, Jacob. "Art of the Industrial Trace." *New Left Review* 71 (September–October 2011): 117–33.

Emery, Jacob. "The Mirror and the Mine." In *Capitalism and the Camera*, edited by Kevin Coleman and Daniel James, 226–53. New York: Verso, 2021.

"The Employment of Photography in Meteorological Science." *Photographic News* 22 (October 4, 1878): 469.

Energy Mix. "Canada Boosts Fossil Subsidies to $14.3B Per Year, Joins Saudi Arabia as G20's Top Two Oil and Gas Funders." Energy Mix, November 16, 2020. https://www.theenergymix.com/2020/11/16/canada-boosts-fossil-subsidies-to-14-3b-per-year-joins-saudi-arabia-as-g20s-top-two-oil-and-gas-funders/.

Engels, Friedrich. *The Condition of the Working Class in England*. London: Electric Book Co., 2001.

Environmental Justice Network. "Principles of Environmental Justice." First National People of Color Environmental Leadership Summit, Washington, DC, October 24–27, 1991. https://www.ejnet.org/ej/principles.html.

Estes, Nick. *Our History Is the Future: Standing Rock Versus the Dakota Access Pipeline, and the Long Tradition of Indigenous Resistance*. London: Verso, 2019.

Falconer, John, ed. *The Waterhouse Albums: Central Indian Provinces*. Ahmedabad, India: Mapin Publishing, 2009.

Farish, Matthew. "Below the Bombs." In *Through Post-Atomic Eyes*, edited by Claudette Lauzon and John O'Brian, 113–32. Montreal: McGill-Queens University Press, 2020.

Farocki, Harun. *The Silver and the Cross*. 2010. Art installation. Thaddeus Ropac Collection. https://www.harunfarocki.de/installations/2010s/2010/the-silver-and-the-cross.html.

Farocki, Harun. "Workers Leaving the Factory." Translated by Laurent Faasch-Ibrahim. *Senses of Cinema* 21 (2001): 1–6.

Farrier, David. "How the Concept of Deep Time Is Changing." *Atlantic*, October 31, 2016.

Faustini, Marco, Lionel Nicole, Eduardo Ruiz-Hitzky, and Clement Sanchez. "History of Organic–Inorganic Hybrid Materials: Prehistory, Art, Science, and Advanced Applications." *Advanced Functional Materials* 28, no. 27 (April 2018): 1–78.

Finkle, Kenneth. "William H. Rau, Philadelphia, Photography, and the Railroad." In *Traveling the Pennsylvania Railroad: Photographs of William H. Rau*, edited by John C. Van Horne, 15–26. Philadelphia: University of Pennsylvania Press, 2002.

Finlay, Victoria. *Color: A Natural History of the Palette*. London: Random House, 2004.

Fisher, Mark. *Capitalist Realism: Is There No Alternative?* New Alresford, UK: Zero Books, 2012.

"Fixing and Washing Silver Prints." *Photographic News* 24 (March 25, 1880): 150.

Flint, Kate. *Flash! Photography, Writing, and Surprising Illumination.* Oxford: Oxford University Press, 2017.

Fontcuberta, Joan. *Pandora's Camera: Photogr@phy after Photography.* London: Mack, 2014.

Foster, Hal, ed. *Vision and Visuality.* Seattle: Bay Press, 1988.

Gabara, Esther. *Errant Modernism: The Ethos of Photography in Mexico and Brazil.* Durham, NC: Duke University Press, 2008.

Galison, Peter, and Caroline Jones. "Representations of Oil Spills." *Artforum,* November 2010.

Garascia, Ann. "'Impressions of Plants Themselves': Materializing Eco-Archival Practices with Anna Atkins's Photographs of British Algae." *Victorian Literature and Culture* 47, no. 2 (2019): 267–303.

Garnier, Christine. "Framing Silver's Void in Timothy O'Sullivan's Photographs of the Gould & Curry Mine." *Panorama: Journal of the Associations of Historians of American Art* 9, no. 1 (Spring 2023): 1–25.

Gaudet, Manon. "Patterns of Possession: H. R. Mallinson & Co.'s American National Park Silks." Unpublished manuscript, 2022.

Gaudreault, André. "One and Many: Cinema as a Series of Series." *History of Photography* 31, no. 1 (2007): 31–39.

Ghosh, Amitav. *The Nutmeg's Curse: Parables for a Planet in Crisis.* Chicago: University of Chicago Press, 2021.

Ghosh, Amitav. "Petrofiction." *New Republic* 2 (March 1992): 29–33.

Gilmore, Ruth Wilson. *Abolition Geography: Essays towards Liberation.* London: Verso, 2022.

Goffe, Tao Leigh. "Human Resources: Art's History and the Ecology of Black Extraction." *British Art Studies* 18 (2020).

Goldblatt, David, and Nadine Gordimer. *On The Mines.* Göttingen: Steidl, 2012.

Gómez-Barris, Macarena. *The Extractive Zone: Social Ecologies and Decolonial Perspectives.* Durham, NC: Duke University Press, 2017.

Gordon, Paul. *Book of Film Care.* Rochester, NY: Eastman Kodak Company, 1983.

Gould, Sarah. "Painting Vile Air in the Age of Turner and Ruskin." Paper presented at College Art Association Annual Conference, Online, February 13, 2021.

Gould, Sarah. "The Polluted Textures of J.M.W. Turner." *Victorian Network* 10 (2021): 77–105.

Gould Ford, Sandra. "Coke Oven Peaches." Steel Genesis, November 21, 2019. https://steelgenesis.com/coke-oven-peaches/.

Greenough, Sarah. "Nescafé, Surlyn, and the Alchemy of Photography: Irving Penn's Platinum Prints." In *Irving Penn: Platinum Prints*, 1–19. New Haven, CT: Yale University Press, 2005.

Grieveson, Lee. *Cinema and the Wealth of Nations: Media, Capital, and the Liberal World System*. Berkeley: University of California Press, 2017.

Grigsby, Darcy Grimaldo. *Enduring Truths: Sojourner's Shadows and Substance*. Chicago: University of Chicago Press, 2015.

Gründig, Matthias. "Ten Dollar Faces: On Photographic Portraiture and Paper Money in the 1860s." *History of Photography* 45, no. 1 (2021): 5–19.

"Gun Cotton and Collodion." *Scientific American*, December 12, 1857, 109.

Habashi, Fathi. "Niépce de Saint-Victor and the Discovery of Radioactivity." *Bulletin for the History of Chemistry* 26, no. 2 (2001): 104–5.

Hackett, Sophie, Andrea Kunard, and Urs Stahel. *Anthropocene: Burtynsky, Baichwal, de Pencier*. Toronto: Art Gallery of Ontario, 2018.

Hagen, M., R. Gatzweiler, and A. T. Jakubick. "Status and Outlook for the WISMUT Remediation Project in the States of Thuringia and Saxony, Germany." In *Proceedings of NEA/IAEA Workshop RADLEG-142-O, Moscow*, 233–43. Vienna: International Atomic Energy Agency, 2002.

Hales, Peter B. "The Atomic Sublime." *American Studies* 32, no. 1 (1991): 5–31.

Hall, D. J. *Clifford Sifton, Volume I: The Young Napoleon, 1861–1900*. Vancouver: University of British Columbia Press, 1981.

Haraway, Donna. "Anthropocene, Capitalocene, Plantationocene, Chthulucene: Making Kin." *Environmental Humanities* 6 (2015): 159–65.

Haraway, Donna. *Staying with the Trouble: Making Kin in the Chthulucene*. Durham, NC: Duke University Press, 2016.

Harley, Rosamond D. *Artists' Pigments c. 1600–1835*. London: Butterworths, 1970.

Harris, Mazie M. "Inventors and Manipulators: Photography as Intellectual Property in Nineteenth-Century New York." PhD diss., Brown University, 2014.

Harvey, David. "Between Space and Time: Reflections on the Geographical Imagination." *Annals of the Association of American Geographers* 80, no. 3 (1990): 418–34.

Hayes, Patricia. "Photographic Publics and Photographic Desires in 1980s South Africa." *photographies* 10, no. 3 (2017): 303–27.

"The Health of Photographers." *Photographic News* 8, no. 312 (August 26, 1864): 409.

Hecht, Gabrielle. "Nuclearity." In *The Nuclear Culture Source Book*, edited by Elle Carpenter. London: Black Dog Publishing, 2016.

Heederik, Dick, Jose Jacobs, Sadegh Samadi, Frits van Rooy, Lützen Portengen, and Remko Houba. "Exposure-Response Analyses for Platinum Salt-Exposed Workers and Sensitization: A Retrospective Cohort Study among

Newly Exposed Workers Using Routinely Collected Surveillance Data." *Journal of Allergy and Clinical Immunology* 137, no. 3 (2015): 922–29.

Heidegger, Martin. "The Age of the World Picture." In *Heidegger: Off the Beaten Track*, 57–84. Cambridge: Cambridge University Press, 2002.

Heise, Ursula K. *Sense of Place and Sense of Planet: The Environmental Imagination of the Global*. Oxford: Oxford University Press, 2008.

Henderson, Linda Dalrymple. "X-Rays and the Quest for Invisible Reality in the Art of Kupka, Duchamp, and the Cubists." *Art Journal* 47, no. 4 (1988): 323–40.

Henning, Michelle. "Image Flow." *photographies* 11, no. 2-3 (2018): 133–44.

Hinton, A. Horsley. *Platinotype Printing*. London: Hazell, Watson & Viney, 1899.

Hodgins, Peter, and Peter Thompson. "Taking the Romance out of Extraction: Contemporary Canadian Artists and the Subversion of the Romantic/Extractive Gaze." *Environmental Communication* 5, no. 4 (2011): 393–410.

Hogan, Mel. "Big Data Ecologies." *Ephemera: Theory & Politics in Organisation* 18, no. 3 (2018): 631–57.

Holmes, Oliver Wendell. "The Doings of the Sunbeam." *Atlantic Monthly*, July 1863, 1–15.

Holmes, Oliver Wendell. "The Stereoscope and the Stereograph." *Atlantic Monthly*, June 1859.

Hopkins, Candice. "The Gilded Gaze: Wealth and Economies on the Colonial Frontier." In *The documenta 14 Reader*, edited by Quinn Latimer and Adam Szymczyk, 219–50. Kassel, Germany: Prestel, 2017.

Horn, Eva. "Air as Medium." *Grey Room* 73, no. 96 (2018): 6–25.

Hornby, Louise. "The Cameraless Optic: Anna Atkins and Virginia Woolf." In "Photography and Literature," edited by Karen Jacobs, special issue, *English Language Notes* 44, no. 2 (Fall/Winter 2006): 87–100.

Horton, Jessica L. "Fire Oppression: Burning and Weaving in Indigenous California." In *Humans*, edited by Laura Bieger, Joshua Shannon, and Jason Weems, 64–85. Chicago: Terra Foundation for American Art, 2021.

Hoskins, Gareth. "People Like Us: Historical Geographies of Industrial-Environmental Crisis at Malakoff Diggins State Historic Park." *Journal of Historical Geography* 50 (2015): 14–24.

Huber, Matthew. *Lifeblood: Oil, Freedom, and the Forces of Capital*. Minneapolis: University of Minnesota Press, 2013.

Hudson, Mark. *Fire Management in the American West: Forest Politics and the Rise of Megafires*. Boulder: University Press of Colorado, 2011.

Hughes, Glyn. "Medical Surveillance of Platinum Refinery Workers." *Occupational Medicine* 30, no. 1 (1980): 27–30.

Hugo, Pieter. "Permanent Error." *Etnofoor* 24 (2012): 105–16.

Hult-Lewis, Christine. *Carleton Watkins: The Complete Mammoth Photographs*. Los Angeles: J. Paul Getty Museum, 2011.

Hult-Lewis, Christine. "The Mining Photographs of Carleton Watkins, 1858–1891, and the Origins of Corporate Photography." PhD diss., Boston University, 2011.

Hume, Naomi. "The Nature Print and Photography in the 1850s." *History of Photography* 35, no. 1 (2011): 44–58.

Hunter, Matthew C. *Painting with Fire: Sir Joshua Reynolds, Photography, and the Temporally Evolving Chemical Object*. Chicago: University of Chicago Press, 2020.

Hutton, James. *Theory of the Earth*. Vol. 1. New York: Hafner Publishing, 1959.

Hyde, Timothy. "'London Particular': The City, Its Atmosphere and the Visibility of Its Objects." *Journal of Architecture* 21, no. 8 (2016): 1274–98.

Igoe, Laura Turner. "Creative Matter: Tracing the Environmental Context of Materials in American Art." In *Nature's Nation: American Art and Environment*, edited by Karl Kusserow and Alan Braddock, 140–69. Princeton, NJ: Princeton University Art Museum, 2018.

Iheka, Cajetan. *African Ecomedia: Network Forms, Planetary Politics*. Durham, NC: Duke University Press, 2021.

"Intense Negatives and Saving Silver." *Photographic News* 9, no. 377 (November 24, 1865): 558.

International Resource Panel. *Global Resources Outlook 2019: Natural Resources for the Future We Want*. New York: United Nations Environment Programme, 2019. https://wedocs.unep.org/handle/20.500.11822/27517;jsessionid=9C579B3492DA17C6D2DE78FC992303DF.

James, Ronald M., and Susan A. James. *A Short History of Virginia City*. Reno: University of Nevada Press, 2014.

Jiménez, Alberto Corsín. "Spiderweb Anthropologies." In *A World of Many Worlds*, edited by Marisol de la Cadena and Mario Blaser, 53–82. Durham, NC: Duke University Press, 2020.

JM Bullion. "Historic Gold, Silver, Platinum, & Palladium Price Spikes." JM Bullion. Accessed May 25, 2021. https://www.jmbullion.com/investing-guide/pricing-payments/historic-gold-silver-platinum-palladium-price-spikes/.

Johnson Matthey. "Celebrating 200 Years of Inspiring Science." Johnson Matthey, 2017. https://matthey.com/documents/161599/162233/JM-200-History-Book.pdf/f726b817-c053-565f-0fbf-f9057aba89f9?t=1650968141116.

Katwala, Amit. "The Spiralling Environmental Cost of Our Lithium Battery Addiction." *Wired*, August 5, 2018.

Keller, Corey, ed. *Brought to Light: Photography and the Invisible, 1840–1900*. New Haven, CT: Yale University Press, 2008.

Keller, Patricia. "Joan Fontcuberta's Landscapes—Remapping Photography and the Technological Image." *Journal of Spanish Cultural Studies* 12, no. 2 (2011): 129–53.

Kelsey, Robin. *Archive Style: Photographs and Illustrations for US Surveys, 1850–1890.* Berkeley: University of California Press, 2007.

Kelsey, Robin. *Photography and the Art of Chance.* Cambridge, MA: Harvard University Press, 2015.

Kelsey, Robin. "Photography and the Ecological Imagination." In *Nature's Nation: American Art and Environment,* edited by Karl Kusserow and Alan Braddock, 394–405. Princeton, NJ: Princeton University Art Museum, 2018.

Kelsey, Robin. "Sierra Club Photography and the Exclusive Property of Vision." *RCC Perspectives* 1 (2013): 11–26.

Kelsey, Robin. "Viewing the Archive: Timothy O'Sullivan's Photographs for the Wheeler Survey, 1871–74." *Art Bulletin* 85 (December 2003): 702–23.

Keto, Elizabeth. "Engineering Vision: Seeing Settler-Colonial Capitalism in a Survey Stereograph." Unpublished manuscript, 2022.

Kimmerer, Robin Wall. *Braiding Sweetgrass: Indigenous Wisdom, Scientific Knowledge and the Teachings of Plants.* Minneapolis, MN: Milkweed Editions, 2013.

King, Tiffany Lethabo. *The Black Shoals: Offshore Formations of Black and Native Studies.* Durham, NC: Duke University Press, 2019.

Kingsley, Hope. "Art Photography and Aesthetics." In *Encyclopedia of Nineteenth-Century Photography,* edited by John Hannavy, 76–81. New York: Taylor & Francis, 2008.

Kittler, Friedrich A. *Gramophone, Film, Typewriter.* Stanford, CA: Stanford University Press, 1999.

Klein, Naomi. *This Changes Everything: Capitalism vs. the Climate.* New York: Simon & Schuster, 2015.

Klinger, Julie. *Rare Earth Frontiers: From Terrestrial Subsoils to Lunar Landscapes.* Ithaca, NY: Cornell University Press, 2017.

Knight, Jacqui. "A Relational Ecology of Photographic Practices." *AVANT* 8 (2017): 285–99.

Kodak Professional. "Silver Halide Photographic Paper: It Still Makes Sense in the Digital Age." Rochester, NY: Eastman Kodak Company, 2015.

Krauss, Clifford. "Green-Energy Race Draws an American Underdog to Bolivia's Lithium." *New York Times,* December 16, 2021.

Krauss, Rosalind. *The Optical Unconscious.* Cambridge, MA: MIT Press, 1993.

Krauss, Rosalind. "Photography's Discursive Spaces: Landscape/View." *Art Journal* 42, no. 4 (Winter 1982): 311–19.

Kriemann, Susanne. *Canopy, Canopy.* San Francisco: CCA Wattis Institute, 2018. https://wattis.org/media/01770.pdf.

Kupsch, Walter O., and Jiří Strnad. "Uranium Bicentenary." *Earth Sciences History* 8, no. 1 (January 1, 1989): 83–86.

Lanay, Jessica. "From Metal to Mettle: An Interview with Sandra Gould Ford." *Bomb Magazine*, December 1, 2017.

Lanay, Jessica. "Intimate Debris: Nature, Industry, and the Body." *ArtSlant*, November 30, 2017.

Landbrecht, Christina, and Friederike Schäfer. "Lamp." In *Susanne Kriemann— P(ech) B(lende): Library for Radioactive Afterlife*, edited by Heike Catherina Mertens, 49–63. Leipzig: Spector Books, 2016.

Lane, Kris. *Potosí: The Silver City That Changed the World*. Berkeley: University of California Press, 2019.

Langley, S. P. "The Sun's Energy." *Photographic News* 29 (January 9, 1886): 28–29.

Lawrence, Stephen J. "Mercury in the Carson River Basin, Nevada." In *Geologic Studies of Mercury by the U.S. Geological Survey, U.S. Geological Survey Circular 1248*, 29–33. Denver, CO: US Department of the Interior, US Geological Survey, 2003.

Leahy, Stephen. "This Is the World's Most Destructive Oil Operation—and It's Growing." *National Geographic*, April 11, 2019.

Lebart, Luce. *Gold and Silver: Daguerreotypes, Ambrotypes and Tintypes from the Gold Rush*. Paris: RVB Books and Canadian Photography Institute, 2017.

Lee, Pamela M. *Forgetting the Art World*. Cambridge, MA: MIT Press, 2012.

Legacies of British Slavery Database. "Alderman John Atkins." University College London. Accessed June 22, 2021. http://wwwdepts-live.ucl.ac.uk/lbs /person/view/22420.

Legacies of British Slavery Database. "Hopewell, Jamaica, Port Royal." University College London. Accessed June 22, 2021. http://wwwdepts-live.ucl.ac .uk/lbs/estate/view/2963.

Legacies of British Slavery Database. "Jamaica St Andrew 136 (Dublin Castle)." University College London. Accessed June 22, 2021. http://wwwdepts-live .ucl.ac.uk/lbs/claim/view/16547.

Le Guin, Ursula. "The Carrier Bag Theory of Fiction." In *Women of Vision: Essays by Women Writing Science Fiction*, edited by Denise Dupont, 1–12. New York: St. Martin's Press, 1988.

LeMenager, Stephanie. *Living Oil: Petroleum Culture in the American Century*. Oxford: Oxford University Press, 2014.

LeMenager, Stephanie. "Petro-Melancholia: The BP Blowout and the Arts of Grief." *Qui Parle* 19, no. 2 (2011): 25–56.

Lesser, Steven H., and Steven J. Weiss. "Art Hazards." *American Journal of Emergency Medicine* 13, no. 4 (1995): 451–58.

Liboiron, Max. "Modern Waste as Strategy." *Lo Squaderno: Explorations in Space and Society* 29 (2013): 9–12.

Liboiron, Max. *Pollution Is Colonialism*. Durham, NC: Duke University Press, 2021.

Liboiron, Max. "Recycling as a Crisis of Meaning." *E-Topia: Canadian Journal of Cultural Studies* (2010): 1–9.

Lippit, Akira Mizuta. *Atomic Light (Shadow Optics)*. Minneapolis: University of Minnesota Press, 2005.

Lorke, Melanie. *Liminal Semiotics: Boundary Phenomena in Romanticism*. Berlin: De Gruyter, 2013.

Lovejoy, Alice. "Celluloid Geopolitics: Film Stock and the War Economy, 1939–47." *Screen* 60, no. 2 (Summer 2019): 224–41.

Lowe, Lisa. *The Intimacy of Four Continents*. Durham, NC: Duke University Press, 2015.

Lugon, Olivier. "Photography and Scale: Projection, Exhibition, Collection." *Art History* 38, no. 2 (2015): 386–403.

Mabley, William Tudor. "On the Expenditure of Silver Used in Photographic Operations, and on the Savings That May Be Effected." *Photographic News* (May 23, 1862): 245.

MacBride, Samantha. *Recycling Reconsidered: The Present Failure and Future Promise of Environmental Action in the United States*. Cambridge, MA: MIT Press, 2011.

MacDonald, Ted. "Just 3 Percent of Broadcast TV News Segments on the California Wildfires Connected Them to Climate Change." Media Matters for America, November 11, 2019. https://www.mediamatters.org/broadcast-networks/just-3-broadcast-tv-news-segments-california-wildfires-connected-them-climate.

Mackert, Gabriele. "Sisyphus's Prestige: On Allan Sekula's Marea Negra." In *Allan Sekula: OKEANOS*, edited by Daniela Zyman and Cory Scozzari. Berlin: Sternberg Press, 2017.

Maclear, Kyo. *Beclouded Visions: Hiroshima-Nagasaki and the Art of Witness*. Albany: State University of New York Press, 1999.

Malm, Andreas. *Fossil Capital: The Rise of Steam Power and the Roots of Global Warming*. New York: Verso, 2016.

Manning, Patrick, Dennis O. Flynn, and Qiyao Wang. "Silver Circulation Worldwide: Initial Steps in Comprehensive Research." *Journal of World-Historical Information* 3–4, no. 1 (2016–17): 1–18.

Marien, Mary Warner. "Imaging the Corporate Sublime." In *Carleton Watkins: Selected Texts and Bibliography*, edited by Amy Rule. Boston: G. K. Hall, 1993.

Marran, Christine L. *Ecology without Culture: Aesthetics for a Toxic World*. Minneapolis: University of Minnesota Press, 2017.

Marx, Karl. *Capital*. Vol. 1. London: Penguin, 1990.

Marx, Karl. *A Contribution to the Critique of Political Economy*. London: Lawrence & Wishart, 1971.

Marx, Karl. *Economic and Philosophic Manuscripts of 1844.* Translated by Martin Milligan. Buffalo, NY: Prometheus Books, 2009.

Marx, Karl. *The German Ideology.* Buffalo, NY: Prometheus Books, 2011. Kindle ed.

Marx, Karl, and Friedrich Engels. *The Communist Manifesto.* London: Penguin Books, 2002.

Masco, Joseph. "Flashblindness." In *Through Post-Atomic Eyes*, edited by Claudette Lauzon and John O'Brian, 86–106. Montreal and Kingston: McGill-Queens University Press, 2020.

Masco, Joseph. "Mutant Ecologies: Radioactive Life in Post–Cold War New Mexico." *Cultural Anthropology* 19, no. 4 (2004): 517–50.

Mathiot, George. "The Capability of Photography." *Philadelphia Photographer*, February 1868, 44–47.

Maurer, Anaïs, and Rebecca H. Hogue. "Introduction: Transnational Nuclear Imperialisms." *Journal of Transnational American Studies* 11, no. 2 (2020): 25–43.

Mavor, Carol. *Blue Mythologies: Reflections on a Colour.* London: Reaktion, 2013.

Maxwell, Richard, and Toby Miller. *Greening the Media.* Oxford: Oxford University Press, 2012.

McLaren, Ian, ed. *Culturing Wilderness in Jasper National Park: Two Centuries of Human History in the Upper Athabasca River Watershed.* Edmonton: University of Alberta Press, 2007.

McLaurin, John J. *Sketches in Crude Oil.* Harrisburg, PA: Self-published, 1896. https://www.gutenberg.org/files/53672/53672-h/53672-h.htm.

Meadows, Patrick Alan. "The Symbol's Symbol: Spider Webs in French Literature." *Symposium* 44, no. 4 (Winter 1990): 272–90.

Mees, C. E. Kenneth. "The History of Sensitisers." *Australasian Photo-Review* 53, no. 3 (1946).

Merchant, Carolyn. *The Death of Nature: Women, Ecology, and the Scientific Revolution.* San Francisco: HarperOne, 1983.

Messier, Paul. "Image Isn't Everything: Revealing Affinities across Collections through the Language of the Photographic Print." In *Object:Photo. Modern Photographs: The Thomas Walther Collection 1909–1949*, edited by Mitra Abbaspour, Lee Ann Daffner, and Maria Morris Hambourg, 332–39. New York: Museum of Modern Art, 2014.

Michaels, Walter Benn. *The Gold Standard and the Logic of Naturalism.* Berkeley: University of California Press, 1988.

Michaels, Walter Benn. "Photographs and Fossils." In *Photography Theory*, edited by James Elkins, 431–50. London: Routledge, 2007.

Michaels, Walter Benn, and Daniel Zamora. "Chris Killip and LaToya Ruby Frazier: The Promise of a Class Aesthetic." *Radical History Review* 132 (October 2018): 23–45.

"Mid Kent and London and South-Western Junction Railways." *Railway Times*, December 18, 1852.

Milbourne, Karen. "African Photographers and the Look of (Un)Sustainability in the African Landscape." *Africa Today* 61, no. 1 (Fall 2014): 115–40.

Miles, Melissa. "The Burning Mirror: Photography in an Ambivalent Light." *Journal of Visual Culture* 4, no. 3 (2005): 329–49.

Miller, Elizabeth. *Extraction Ecologies and the Literature of the Long Exhaustion*. Princeton, NJ: Princeton University Press, 2021.

Miller, Rich. "Data Centers Forge Ahead with Shift to Renewable Energy." Data Center Frontier, June 2, 2017. https://www.datacenterfrontier.com/energy/article/11430735/data-centers-forge-ahead-with-shift-to-renewable-energy.

Mintie, Katherine. "Material Matters: The Transatlantic Trade in Photographic Materials during the Nineteenth Century." *Panorama* 6, no. 2 (Fall 2020): 1–12.

Miranda, Carolina A. "Q&A: Braddock's LaToya Ruby Frazier Takes on Levi's and 'Ruin Porn' Chic." *Los Angeles Times*, December 5, 2014.

Mirzoeff, Nicholas. *How to See the World*. London: Pelican, 2015.

Mirzoeff, Nicholas. "It's Not the Anthropocene, It's the White Supremacy Scene; or, The Geological Color Line." In *After Extinction*, edited by Richard A. Grusin, 123–50. Minneapolis: University of Minnesota Press, 2018.

Mirzoeff, Nicholas. "Visualizing the Anthropocene." *Public Culture* 26, no. 2 (2014): 213–32.

Mitchell, Timothy. *Carbon Democracy*. New York: Verso, 2011.

Mitchell, W. J .T. *The Last Dinosaur Book: The Life and Times of a Cultural Icon*. Chicago: University of Chicago Press, 1998.

Mitchell, W. J. T. "Showing Seeing: A Critique of Visual Culture." In *What Do Pictures Want? The Lives and Loves of Images*, 336–56. Chicago: University of Chicago Press, 2005.

Montell-Boyd, Fionn. "Speculative Projections: The Political Economy of Silver and the Emergence of Photography in Britain, 1780–1841." PhD diss., University of Oxford, forthcoming.

Moore, Jason W. *Capitalism in the Web of Life*. New York: Verso, 2015.

Moore, Jason W. "The Capitalocene and Its Discontents: Towards a Revolutionary Ecology." Paper presented at Extraction: Decolonial Visual Cultures in the Age of the Capitalocene, University of California, Santa Cruz, May 12, 2017.

Moreiras, Camila. "Joan Fontcuberta: Post-Photography and the Spectral Image of Saturation." *Journal of Spanish Cultural Studies* 18, no. 1 (2017): 1–21.

Mullaney, Steven. "The New World on Display." In *New World of Wonders: European Images of the Americas, 1492–1700*, edited by Rachel Doggett, 105–13. Washington, DC: Folger Shakespeare Library, 1992.

Mumford, Lewis. *Technics and Civilization*. New York: Harcourt, Brace, 1934.

Nakano Glenn, Evelyn. *Unequal Freedom: How Race and Gender Shaped American Citizenship and Labor*. Cambridge, MA: Harvard University Press, 2004.

Nassar, N. T., T. E. Graedel, and E. M. Harper. "By-Product Metals Are Technologically Essential but Have Problematic Supply." *Scientific Advances* 1, no. 3 (April 2015): 1–10.

"The Nature of the Metals." *Photographic News* 1, no. 8 (October 29, 1858): 89.

Nelson, G., and J. Murray. "Silicosis at Autopsy in Platinum Mine Workers." *Occupational Medicine* 63 (2013): 196–202.

"The New Platinum Printing Process." *British Journal of Photography* 22, no. 787 (June 4, 1875): 264–67.

Nickel, Douglas. "Talbot's Natural Magic." *History of Photography* 26, no. 2 (Summer 2002): 132–40.

Niépce, Joseph Nicéphore. "Heliography: Designs and Engravings." Obelisk Art History, February 24, 2015. http://arthistoryproject.com/artists/nicephore-niepce/heliography-designs-and-engravings/.

Niépce, Joseph Nicéphore. "Memoire on the Heliograph." In *Classic Essays on Photography*, edited by Alan Trachtenberg, 5–10. New Haven, CT: Leete's Island Books, 1980.

Nisbet, James. "On Simon Starling, One Ton II (2005)." In *Critical Landscapes Art, Space, Politics*, edited by Emily Eliza Scott and Kirsten Swenson, 185–87. Berkeley: University of California Press, 2015.

Nixon, Rob. *Slow Violence and the Environmentalism of the Poor*. Cambridge, MA: Harvard University Press, 2011.

Noble, Safiya Umoja. *Algorithms of Oppression: How Search Engines Reinforce Racism*. New York: New York University Press, 2018.

"Notes." *Photographic News* 29 (January 2, 1885): 7.

"Notes." *Photographic News* 29 (July 10, 1885): 440.

Nye, David. *American Technological Sublime*. Cambridge, MA: MIT Press, 1994.

O'Brian, John. *The Bomb in the Wilderness: Photography and the Nuclear Era in Canada*. Vancouver: University of British Columbia Press, 2020.

O'Brian, John, ed. *Camera Atomica*. London: Black Dog Press and Art Gallery of Ontario, 2015.

Olin, Margaret. *Touching Photographs*. Chicago: University of Chicago Press, 2012.

Oliver, Thomas. *Dangerous Trades*. London: John Murray, 1902.

Ouedraogo, Nyaba. "The Hell of Copper." Afikaris. Accessed July 1, 2021. https://afikaris.com/pages/the-hell-of-copper.

Paglen, Trevor. "Some Sketches on Vertical Geographies." *e-flux Architecture: Superhumanity*, October 2016.

Palmer, A. Laurie. *In the Aura of a Hole*. London: Black Dog Publishing, 2014.

Palmer, Daniel. *Photography and Collaboration*. London: Routledge, 2017.

Palmquist, Peter E., and Thomas R. Kailbourn. *Pioneer Photographers from the Mississippi to the Continental Divide*. Stanford, CA: Stanford University Press, 2005.

Panchbhai, Arati S. "Wilhelm Conrad Röntgen and the Discovery of X-Rays: Revisited after Centennial." *Journal of Indian Academy of Oral Medicine and Radiology* 27, no. 1 (January 1, 2015): 90–95.

Panzer, Mary. "The Invisible Photographs of William H. Rau." In *Traveling the Pennsylvania Railroad: Photographs of William H. Rau*, edited by John C. Van Horne, 27–42. Philadelphia: University of Pennsylvania Press, 2002.

Panzer, Mary. "William H. Rau and the Archives of Photographic History." *Afterimage* 15, no. 9 (April 1988): 9–11.

Parenti, Christian. *Tropic of Chaos: Climate Change and the New Geography of Violence*. New York: Bold Type Books, 2011.

Parikka, Jussi. *A Geology of Media*. Minneapolis: University of Minnesota Press, 2015.

Parton, James. "Pittsburgh." *Atlantic Monthly*, January 1868.

Peck, Gunther. "Manly Gambles: The Politics of Risk on the Comstock Lode, 1860–1880." *Journal of Social History* 26, no. 4 (1993): 701–23.

Peffer, John. *Art and the End of Apartheid*. Minneapolis: University of Minnesota Press, 2009.

Peirce, Charles Sanders. *The Collected Papers of Charles Sanders Peirce*, vol. 7: *Science and Philosophy*. Edited by Arthur W. Burks. Cambridge, MA: Harvard University Press, 1958.

Pelizzari, Maria Antonella, and Steffen Siegel. "Circulating Photographs: A Special Issue of *History of Photography*." *History of Photography* 45, no. 1 (2021): 1–4.

Pelletan, Jules. "Discovery by M. Daguerre [excerpt], *La Presse*, January 24, 1839, 1–2." Reproduced in Steffen Siegel, *First Exposures: Writings from the Beginning of Photography*. Los Angeles: J. Paul Getty Museum, 2017.

Penn, Ivan, and Eric Lipton. "The Lithium Gold Rush: Inside the Race to Power Electric Vehicles." *New York Times*, May 7, 2021.

Peters, Clorinde. "Neoliberal Violence, Online Protest Images, and Constructing the Photography of the Commons." *Afterimage* 45, no. 5 (2018): 16–21.

Peters, John Durham. *The Marvelous Clouds: Toward a Philosophy of Elemental Media*. Chicago: University of Chicago Press, 2015.

Peters, John Durham. "Space, Time, and Communication Theory." *Canadian Journal of Communication* 28, no. 4 (2003): 397–411.

"The Pictures in This Number." *Camera Work: A Photographic Quarterly*, no. 1 (January 1903): 63. https://doi.org/10.11588/diglit.29887#0001.

Pinson, Stephen C. *Speculating Daguerre: Art and Enterprise in the Work of L. J. M. Daguerre*. Chicago: University of Chicago Press, 2012.

Plunkett, John. "Marion and Company." In *Encyclopedia of Nineteenth-Century Photography*, edited by John Hannavy, 892–94. New York: Taylor & Francis, 2008.

Povinelli, Elizabeth. "Fires, Fogs, Winds." *Cultural Anthropology* 32, no. 4 (2017): 504–13.

PRO-ACT. *Silver Recovery from Photographic and Imaging Wastes*. Washington, DC: United States Air Force, 1998. https://p2infohouse.org/ref/20/19926/Fact_Sheets/DSDATA/AirForce/Silver%20Recovery.pdf.

"Proceedings of Societies." *Photographic News* 21 (July 6, 1877): 322.

Pulido, Laura. "Rethinking Environmental Racism: White Privilege and Urban Development in Southern California." *Annals of the Association of American Geographers* 90, no. 1 (2000): 12–40.

Raab, Jennifer. "Landscape and the Risk of Metaphor." *American Art* 31, no. 2 (2017): 56–58.

Rajak, Dinah. "Platinum City and the New South African Dream." *Africa* 82, no. 2 (2012): 252–71.

Ramalingam, Chitra. "Dust Plate, Retina, Photograph: Imaging on Experimental Surfaces in Early Nineteenth-Century Physics." *Science in Context* 28, no. 3 (September 2015): 317–55.

Ramalingam, Chitra. "Natural History in the Dark: Seriality and the Electric Discharge in Victorian Physics." *History of Science* 48 (September 2010): 371–98.

Rau, William H. "Photography as Applied to Business." *Photographic Times* 1, no. 4 (April 1871): 49.

Raygorodetsky, Gleb. "Indigenous Peoples Defend Earth's Biodiversity—But They're in Danger." *National Geographic*, November 16, 2018.

Ray 2015. "Interview with Simon Starling." Museum Angewandte Kunst, 2015. https://www.museumangewandtekunst.de/en/events/interview-with-simon-starling/.

"Recovering Silver from Residues." *Photographic News* 9, no. 332 (July 13, 1865): 13.

Reinsel, David, John Gantz, and John Rydning. *IDC Report: The Digitization of the World From Edge to Core*. Needham, MA: International Data Corporation, November 2018.

Rich, Nathaniel. "Everything You Thought You Knew, and Why You're Wrong." *New York Times*, May 11, 2022.

Riofrancos, Thea. *Resource Radicals: From Petro-Nationalism to Post-Extractivism in Ecuador*. Durham, NC: Duke University Press, 2020.

Robbins, Nicholas. "Atmospheric Regulation in the Panorama." *Grey Room* 83 (Spring 2021): 56–81.

Robbins, Nicholas. "Ruskin, Whistler, and the Climate of Art in 1884." In *Ruskin's Ecologies*, edited by Kelly Freeman and Thomas Hughes, 203–23. London: Courtauld Books Online, 2021.

Roberts, Jennifer L. *Scale*. Chicago: Terra Foundation for American Art, 2016.

Roberts, Jennifer L. *Transporting Visions: The Movement of Images in Early America*. Berkeley: University of California Press, 2014.

Robinson, Cedric J. *Black Marxism: The Making of the Black Radical Tradition*. Chapel Hill: University of North Carolina Press, 2021.

Robinson, Michael A. "The Techniques and Material Aesthetics of the Daguerreotype." PhD diss., De Montfort University, 2017.

Roediger, David, and Elizabeth Esch. *The Production of Difference: Race and the Management of Labor in US History*. Oxford: Oxford University Press, 2014.

Rogers, David N. *Chemistry of Photography from Classical to Digital Technologies*. Cambridge, UK: Royal Society of Chemistry, 2006.

Rohrig, Brian. "Smartphones: Smart Chemistry." *American Chemical Society*, 2015.

Ronan, Annie. "Capturing Cruelty: Camera Hunting, Water Killing, and Winslow Homer's Adirondack Deer." *American Art* 31, no. 3 (Fall 2017): 52–79.

Ronda, C. "Rare Earth Phosphors: Fundamentals and Applications." In *Encyclopedia of Materials: Science and Technology*, edited by K. H. Jürgen Buschow, Robert W. Cahn, Merton C. Flemings, Bernhard Ilschner, Edward J. Kramer, Subhash Mahajan, and Patrick Veyssière, 8026–33. Amsterdam: Elsevier, 2001.

Rosenblum, Naomi. *A World History of Photography*. New York: Abbeville Press, 2007.

Ross, Rachel. "Facts about Platinum." *Live Science*, August 1, 2016.

Rotman, Brian. *Signifying Nothing: The Semiotics of Zero*. Basingstoke, UK: Macmillan, 1987.

Rudwick, Martin. "The Emergence of a Visual Language for Geological Science 1760–1840." *History of Science* 14, no. 3 (1976): 149–95.

Ruffles, Tom. "Waterhouse, James (1842–1922)." In *Encyclopedia of Nineteenth-Century Photography*, edited by John Hannavy, 1474–76. New York: Taylor & Francis, 2008.

Ruskin, John. *The Storm-Cloud of the Nineteenth Century*. Kent, UK: George Allen, 1884.

Salgado, Sebastião. *Workers: An Archaeology of the Industrial Age*. New York: Aperture, 1993.

Samson, John. "Photographs from the High Rockies." *Harper's New Monthly Magazine*, September 1869, 465–75.

Sandweiss, Martha A. *Print the Legend: Photography and the American West*. New Haven, CT: Yale University Press, 2002.

"Saving the Silver." *Photographic News*, July 4, 1862, 324.

Sawyer, J. "Woodruff v. North Bloomfield Gravel Mining Co., 16 F. 25." United States Circuit Court for the District of California, 1883. https://cite.case.law/f/16/25/.

Scerri, Eric. *A Tale of 7 Elements*. Oxford: Oxford University Press, 2013.

Schaefer, William. "Photographic Ecologies." *October* 161 (2017): 42–68.

Schmelzer, Matthias, Andrea Vetter, and Aaron Vansintjan. *The Future Is Degrowth: A Guide to a World beyond Capitalism*. London: Verso, 2022.

Schuppli, Susan. "Radical Contact Prints." In *Camera Atomica*, edited by John O'Brian, 277–91. London: Black Dog Press and Art Gallery of Ontario, 2015.

Schuster, Joshua. "Between Manufacturing and Landscapes: Edward Burtynsky and the Photography of Ecology." *Photography and Culture* 6, no. 2 (2013): 193–212.

Scott, Conohar. *Photography and Environmental Activism: Visualising the Struggle against Industrial Pollution*. New York: Routledge, 2022.

Seiler, Lutz. "The Territory of Tiredness." In *Susanne Kriemann—P(ech)B(lende): Library for Radioactive Afterlife*, edited by Heike Catherina Mertens, 193–207. Leipzig: Spector Books, 2016.

Sekiya, Masaru, and Michio Yamasaki. "Antoine Henri Becquerel (1852–1908): A Scientist Who Endeavored to Discover Natural Radioactivity." *Radiological Physics and Technology* 8, no. 1 (January 2015): 1–3.

Sekula, Allan. "Between the Net and the Deep Blue Sea (Rethinking the Traffic of Photographs)." *October* 102 (Fall 2002): 3–34.

Sekula, Allan. "The Body and the Archive," *October* 39 (1986): 3–64.

Sekula, Allan. "An Eternal Esthetics of Laborious Gestures." *Grey Room* 55 (2014): 16–27.

Sekula, Allan. *Fish Story*. Düsseldorf: Richter Verlag, 2003.

Sekula, Allan. "Photography between Labor and Capital: The Emerging Picture Language of Industrial Capitalism." In *Mining Photographs and Other Pictures, 1948–1968*, edited by Benjamin H. D. Buchloh and Robert Wilkie, 193–268. Halifax: Press of the Nova Scotia College of Art and Design, 1983.

Sekula, Allan. "Reading an Archive: Photography between Labor and Capital." In *The Photography Reader*, edited by Liz Wells, 443–52. London: Routledge, 2002.

Sekula, Allan. "The Traffic in Photographs." *Art Journal* 41, no. 1 (1981): 15–25.

"A Sensational Discovery." *Die Presse*, January 5, 1896.

Seymour, Nicole. *Bad Environmentalism: Irony and Irreverence in the Ecological Age*. Minneapolis: University of Minnesota Press, 2018.

Shapiro, Alan E. *Fits, Passions, and Paroxysms: Physics, Method, and Chemistry and Newton's Theories of Colored Bodies and Fits of Easy Reflection*. Cambridge: Cambridge University Press, 1993.

Shapiro, Michael J. *The Political Sublime*. Durham, NC: Duke University Press, 2018.

Sheppard, S. E. *Gelatin in Photography*. New York: D. Van Nostrand, 1923.

Shields, Caroline. *Impressionism in the Age of Industry*. Toronto: Art Gallery of Ontario, 2020.

Shteir, Ann B. *Cultivating Women, Cultivating Science: Flora's Daughters and Botany in England, 1760-1860*. Baltimore, MD: Johns Hopkins University Press, 1996.

Shukin, Nicole. *Animal Capital: Rendering Life in Biopolitical Times*. Minneapolis: University of Minnesota Press, 2009.

Sibley, Scott F. *Overview of Flow Studies for Recycling Metal Commodities in the United States*. Reston, VA: US Geological Survey, 2011. https://pubs.usgs.gov/circ /circ1196-AA/pdf/circ1196-AA.pdf.

Sierra Club Canada Foundation. "Tar Sands." Sierra Club Canada Foundation. Accessed December 21, 2021. https://www.sierraclub.ca/en/tar-sands.

Silber, William L. *The Story of Silver: How the White Metal Shaped America and the Modern World*. Princeton, NJ: Princeton University Press, 2019.

Silverman, Kaja. *The Miracle of Analogy, or, The History of Photography*. Stanford, CA: Stanford University Press, 2015.

Sims, Christo. "Green Magic: On Technologies of Enchantment at Apple's Corporate Headquarters." *Public Culture* 34, no. 2 (2022): 291-317.

Smil, Vaclav. *How the World Really Works*. New York: Viking, 2022.

Smil, Vaclav. *Making the Modern World: Materials and Dematerialization*. Hoboken, NJ: John Wiley & Sons, 2013.

Smith, Adam. *The Wealth of Nations*. Indianapolis: Liberty Classics, 1981.

Smith, Grant H. *The History of the Comstock Lode 1850-1920*. Reno: Nevada State Bureau of Mines and the Mackay School of Mines, 1943.

Smith, Shawn Michelle. *At the Edge of Sight: Photography and the Unseen*. Durham, NC: Duke University Press, 2013.

Smith, Shawn Michelle, and Sharon Sliwinski, eds. *Photography and the Optical Unconsciousness*. Durham, NC: Duke University Press, 2017.

Smithsonian National Museum of the American Indian. *Indelible: The Platinum Photographs of Larry McNeil and Will Wilson*. Washington, DC: Smithsonian Institution, 2014.

Snyder, Joel. *American Frontiers: The Photographs of Timothy H. O'Sullivan, 1867-1874*. Millertown, NY: Aperture Foundation and Philadelphia Museum of Art, 1981.

Snyder, Joel. *One/Many: Western American Survey Photographs, 1860-1876*. Chicago: David and Alfred Smart Museum of Art, 2006.

Solomon-Godeau, Abigail. "Who Is Speaking Thus? Some Questions on Documentary Photography." In *Photography at the Dock: Essays on Photographic History, Institutions, and Practices*, 169-83. Minneapolis: University of Minnesota Press, 1991.

Sontag, Susan. *On Photography*. New York: Delta Books, 1977.

Sontag, Susan. *Regarding the Pain of Others*. New York: Farrar, Straus & Giroux, 2003.

Spady, Samantha, and Siobhan Angus. "Histories of the Present: Tar Sands Photography and Colonial Cultural Production." In *Energy Humanities and Energy Transition: Current State and Future Directions*, edited by Matúš Mišík and Nada Kujundžić, 121–48. New York: Springer, 2020.

Spence, Mark David. *Dispossessing the Wilderness: Indian Removal and the Making of National Parks*. New York: Oxford University Press, 2000.

Starosielski, Nicole. *The Undersea Network*. Durham, NC: Duke University Press, 2015.

Stieglitz, Alfred. "Pictorial Photography." In *Classic Essays on Photography*, edited by Alan Tratchenberg, 115–23. New Haven, CT: Leete's Island Books, 1980.

Stieglitz, Alfred. "Platinum Printing." In *Stieglitz on Photography: His Selected Essays and Notes*, edited by Richard Whelan, 87–96. New York: Aperture, 2000.

Stilgoe, John R. "An Opening Between the Trains." In *Traveling the Pennsylvania Railroad: Photographs of William H. Rau*, edited by John C. Van Horne, 43–54. Philadelphia: University of Pennsylvania Press, 2002.

Stulik, Dusan C., and Art Kaplan. "Platinotype." In *The Atlas of Analytical Signatures of Photographic Processes*. Los Angeles: Getty Conservation Institute, 2013.

"Submarine Cable Map." TeleGeography. Accessed April 21, 2021. https://www.submarinecablemap.com/.

Sugimoto, Hiroshi. "Pre-Photography Time-Recording Device." Photography series. sugimotohiroshi.com, 2008. https://www.sugimotohiroshi.com/pptrd.

Sugrue, Thomas. *The Origins of the Urban Crisis*. Princeton, NJ: Princeton University Press, 1996.

Sullivan, Jack. "Trash or Treasure: Global Trade and the Accumulation of E-Waste in Lagos, Nigeria." *Africa Today* 61, no. 1 (2014): 89–112.

Sverdlik, Yevgeniy. "Here's How Much Energy All US Data Centers Consume." Data Center Knowledge, June 27, 2016. https://www.datacenterknowledge.com/archives/2016/06/27/heres-how-much-energy-all-us-data-centers-consume.

Szeman, Imre, and Jennifer Wenzel, "What Do We Talk about When We Talk about Extractivism?" *Textual Practice* 35, no. 3 (2021): 505–23.

Tagliabue, John. "A Legacy of Ashes: The Uranium Mines of Eastern Germany." *New York Times*, March 19, 1991.

Talbot, William Henry Fox. *The Pencil of Nature*. London: Longman, Brown, Green & Longmans, 1844.

Talbot, William Henry Fox. "Some Account of the Art of Photogenic Drawing, or the Process by Which Natural Objects May Be Made to Delineate

Themselves without the Aid of the Artist's Pencil." *Proceedings of the Royal Society of London* 4 (1843): 120–21.

"Talk in the Studio: Professor Tyndall on Light and Mining Accidents." *Photographic News* 27 (January 12, 1883): 31–32.

Tataryn, Lloyd. *Dying for a Living*. Ottawa: Deneau & Greenberg, 1979.

Taylor, Jesse Oak. "Auras and Ice Cores: Atmospheric Archives and the Anthropocene." *Minnesota Review* 83 (2014): 73–82.

Taylor, Jesse Oak. *The Sky of Our Manufacture*. Charlottesville: University of Virginia Press, 2016.

Taylor, Lawrence, and R. G. Thorne. "Atkins, John." In *The History of Parliament: The House of Commons 1820–1832*, edited by David R. Fisher. Cambridge: Cambridge University Press, 2009. https://www.historyofparliamentonline .org/volume/1820–1832/member/atkins-john-1754–1838.

Tedone, Gaia. "From Spectacle to Extraction. And All Over Again." *Unthinking Photography*, November 2019. https://unthinking.photography/articles /from-spectacle-to-extraction-and-all-over-again.

Temple, Robert. *The Genius of China: 3000 Years of Science, Discovery, and Invention*. Rochester, VT: Inner Traditions, 2007.

Thomas, John Meurig. "The RSC Faraday Prize Lecture of 1989 on Platinum." *Chemical Communications* 53, no. 66 (2017): 9185–92.

Todd, Zoe. "Fish, Kin and Hope: Tending to Water Violations in *amiskwaciwâskahikan* and Treaty Six Territory." *Afterall* 43 (2017): 102–7.

Todhunter, Isaac. *William Whewell: An Account of His Writings with Selections from His Literary and Scientific Correspondence, Volume I*. London: MacMillan, 1876.

Towler, John. *The Silver Sunbeam*. New York: Joseph H. Ladd, 1864.

Trachtenberg, Alan. *The Incorporation of America: Culture and Society in the Gilded Age*. New York: Hill & Wang, 1982.

Trachtenberg, Alan. *Reading American Photographs: Images as History, Mathew Brady to Walker Evans*. New York: Hill & Wang, 1989.

Trangos, Guy, and Kerry Bobbins. "Gold Mining Exploits and the Legacies of Johannesburg's Mining Landscapes." *Scenario* 5 (Fall 2015).

Tsēmā. "Artist Statement." Tsema.ca. Accessed June 21, 2021. https://www.tsema .ca/about.

Tsēmā. "Black Gold." Tsema.ca. Accessed June 21, 2021. https://www.tsema.ca /798817030644.

Tsing, Anna. *The Mushroom at the End of the World: On the Possibility of Life in Capitalist Ruins*. Princeton, NJ: Princeton University Press, 2015.

Tsing, Anna, Heather Swanson, Elaine Gan, and Nils Bubandt, eds. *Arts of Living on a Damaged Planet*. Minneapolis: University of Minnesota Press, 2017

Tuck, Eve, and C. Ree. "A Glossary of Haunting." In *Handbook of Autoethnography*, edited by Stacey Holman Jones, Tony E. Adams, and Carolyn Ellis, 639–58. Walnut Creek, CA: Left Coast Press, 2013.

Tucker, Jennifer. "Dangerous Exposures: Visualizing Work and Waste in the Victorian Chemical Trades." *International Labor and Working-Class History* 95 (2019): 130–65.

Tucker, Jennifer. *Nature Exposed: Photography as Eyewitness in Victorian Science.* Baltimore, MD: Johns Hopkins University Press, 2005.

Turkey Mill Business Park. "History." Turkey Mill. Accessed July 29, 2020. https://www.turkeymill.co.uk/useful-information/history/.

Twain, Mark. *Roughing It.* Hartford, CT: American Publishing, 1872.

United Nations Environment Programme. *Waste Crime—Waste Risks Gaps in Meeting the Global Waste Challenge: A Rapid Response Assessment.* Nairobi: United Nations Environmental Programme and GRID-Arendal, 2015.

"Uranium Element Facts." Chemicool. Accessed November 1, 2022. https://www.chemicool.com/elements/uranium.html.

van Wyck, Peter C. *The Highway of the Atom.* Montreal: McGill-Queen's University Press, 2010.

Virilio, Paul. *War and Cinema: The Logistics of Perception.* London: Verso, 1989.

Wall, Jeff. "Photography and Liquid Intelligence," In *Jeff Wall: Selected Essays and Interviews*, edited by Peter Galassi, 109–10. New York: Museum of Modern Art, 2007.

Ware, Mike. *Cyanomicon: History, Science and Art of Cyanotype: Photographic Printing in Prussian Blue.* n.p.: Self-published, 2016. https://www.mikeware.co.uk/downloads/Cyanomicon_II.pdf.

Ware, Mike. "Cyanotype." In *Encyclopedia of Nineteenth-Century Photography*, edited by John Hannavy, 360–61. New York: Taylor & Francis, 2008.

Ware, Mike. "The Eighth Metal: The Rise of the Platinotype Process." In *Photography 1900: The Edinburgh Symposium.* Edinburgh: National Museums of Scotland and National Galleries of Scotland, 1993. https://www.mikeware.co.uk/mikeware/Eighth_Metal.html.

Ware, Mike. "Light-Sensitive Chemicals." In *Encyclopedia of Nineteenth-Century Photography*, edited by John Hannavy, 857. New York: Taylor & Francis, 2008.

Ware, Mike. *Mechanisms of Image Deterioration in Early Photographs: The Sensitivity to Light of W. H. F. Talbot's Halide-Fixed Images 1834–1844.* London: Science Museum and National Museum of Photography, Film, and Television, 1994.

Ware, Mike. *Platinomicon: A Technical Account of Photographic Printing in Platinum and Palladium.* n.p.: Self-published, 2017.

Ware, Mike. "Platinotype Company." In *Encyclopedia of Nineteenth-Century Photography*, edited by John Hannavy, 1134–36. New York: Taylor & Francis, 2008.

Wark, McKenzie. *Capital Is Dead. Is This Something Worse?* New York: Verso, 2021.

Waterhouse, James. "Nicéphore Niépce's Early Photographic Work with Bitumen." In *A Dictionary of Arts, Manufactures, and Mines,* edited by E. F. W. Rudler. London: Longmans, Green, 1867.

Waterhouse, James. "Photo-Lithography and Photo-Zincography: Chap. XII. Direct Collo-Chromate Methods." *Photographic Times,* 1883.

Watt, Laura Alice, and David Lowenthal. *Paradox of Preservation: Wilderness and Working Landscapes at Point Reyes National Seashore.* Berkeley: University of California Press, 2016.

Watts, Jonathan. "Resource Extraction Responsible for Half World's Carbon Emissions." *Guardian,* March 12, 2019.

Webb, J. H. "The Fogging of Photographic Film by Radioactive Contaminants in Cardboard Packaging Materials." *Physical Review* 76, no. 3 (1949): 375–80.

Weeks, M. E. *Discovery of the Elements.* Easton, PA: Journal of Chemical Education, 1968.

Weems, Jason. "Scale, a Slaughterhouse View: Industry, Corporeality, and Being in Turn-of-the-Century Chicago." In *Seeing Scale,* edited by Jennifer L. Roberts, 106–43. Chicago: Terra Foundation for American Art, 2016.

Wellerstein, Alex. "Counting the Dead at Hiroshima and Nagasaki." *Bulletin of the Atomic Scientists,* August 4, 2020. https://thebulletin.org/2020/08/counting-the-dead-at-hiroshima-and-nagasaki/.

West, Nancy Martha. *Kodak and the Lens of Nostalgia.* Charlottesville: University of Virginia Press, 2000.

"What Is Uranium? How Does It Work?" World Nuclear Association, August 2022. https://www.world-nuclear.org/information-library/nuclear-fuel-cycle/introduction/what-is-uranium-how-does-it-work.aspx.

White, Benjamin. *Silver: Its History and Romance.* London: Waterlow & Sons, 1920.

Whyte, Kyle Powys. "Is It Colonial Déjà Vu? Indigenous Peoples and Climate Injustice." In *Humanities for the Environment: Integrating Knowledges, Forging New Constellations of Practice,* edited by Joni Adamson, Michael Davis, and Hsinya Huang, 88–104. Oxford: Earthscan Publications, 2016.

Whyte, Kyle Powys. "Settler Colonialism, Ecology, and Environmental Justice." *Environment and Society* 9, no. 1 (2018): 125–44.

Wilder, Kelley. "Photography and the Art of Science." *Visual Studies* 24, no. 2 (2009): 163–68.

Wilder, Kelley. "Visualizing Radiation: The Photographs of Henri Becquerel." In *Histories of Scientific Observation,* edited by Lorraine Daston and Elizabeth Lunbeck, 349–68. Chicago: University of Chicago Press, 2010.

Wilder, Kelley, and Martin Kemp. "Proof Positive in Sir John Herschel's Concept of Photography." *History of Photography* 26, no. 4 (December 1, 2002): 358–66.

Williams, Eric. *Capitalism and Slavery*. Chapel Hill: University of North Carolina Press, 1944.

Williams, Raymond. *The City and the Country*. London: Chatto & Windus, 1973.

Willis, William. "A New Process of Photo-Chemical Printing in Metallic Platinum." *Photographic News*, August 23, 1878, 397–98.

Wilson, Eva. *Exclusion Zones*. Vancouver: Fillip, 2017.

Wilson, Sheena, Adam Carlson, and Imre Szeman, eds. *Petrocultures: Oil, Politics, Culture*. Montreal: McGill-Queen's University Press, 2017.

Windham, Lane. *Knocking on Labor's Door: Union Organizing in the 1970s and the Roots of a New Economic Divide*. Chapel Hill: University of North Carolina Press, 2017.

World Platinum Investment Council. "Platinum Facts." World Platinum Investment Council. Accessed May 25, 2021. https://platinuminvestment.com /about/platinum-facts.

Wrigley, E. A. "The Supply of Raw Materials in the Industrial Revolution." *Economic History Review* 15 (1962): 1–16.

Young, Arthur. "A Tour to Shropshire." In *Tours in England and Wales, Selected from the Annals of Agriculture*. London: London School of Economics, 1932.

Yusoff, Kathryn. *A Billion Black Anthropocenes or None*. Minneapolis: University of Minnesota Press, 2019.

Ziolkowski, Theodore. *German Romanticism and Its Institutions*. Princeton, NJ: Princeton University Press, 1990.

Zuromskis, Catherine. "Petroaesthetics and Landscape Photography." In *Oil Culture*, edited by Ross Barrett and Daniel Worden, 289–308. Minneapolis: University of Minnesota Press, 2014.

Zylinska, Joanna. *Nonhuman Photography*. Cambridge, MA: MIT Press, 2017.